POPULATION ECOLOGY

A Unified Study of Animals and Plants

POPULATION ECOLOGY
A Unified Study of Animals and Plants

MICHAEL BEGON BSc, PhD
Department of Zoology
University of Liverpool

MARTIN MORTIMER BSc, PhD
Department of Botany
University of Liverpool

SECOND EDITION

BLACKWELL SCIENTIFIC PUBLICATIONS
OXFORD LONDON EDINBURGH
BOSTON PALO ALTO MELBOURNE

Editorial offices:
Osney Mead, Oxford, OX2 0EL
8 John Street, London, WC1N 2ES
23 Ainslie Place, Edinburgh, EH3 6AJ
52 Beacon Street, Boston
Massachusetts 02108, USA
667 Lytton Avenue, Palo Alto
California 94301, USA
107 Barry Street, Carlton
Victoria 3053, Australia

First published 1981
Reprinted 1982, 1983, 1985
Second edition 1986
Reprinted 1987

Photoset by Enset (Photosetting)
Midsomer Norton, Bath, Avon
Printed and bound by
Butler & Tanner Ltd,
Frome, Somerset

DISTRIBUTORS

USA and Canada
Blackwell Scientific Publications Inc
P.O. Box 50009, Palo Alto
California 94303

Australia
Blackwell Scientific Publications
(Australia) Pty Ltd
107 Barry Street,
Carlton, Victoria 3053

British Library Cataloguing in Publication Data

Begon, Michael
Population ecology: a unified study of animals and plants.—2nd ed.
1. Population biology
I. Title II. Mortimer, Martin
574.5′248 QH352

ISBN 0-632-01443-1

Contents

Preface

This book is intended primarily for students. It is designed to describe the present state of Population Ecology in terms which can be readily understood by undergraduates with little or no prior knowledge of the subject. We have, however, presented *our* view, rather than some definitive view of the subject, and, consequently, we have tried to provide sufficient information for everybody (student and expert alike) to disagree with us wherever they think fit.

Population Ecology is, to us, the study of the sizes (and to a lesser extent the distributions) of plant and animal populations, and of the processes—particularly the biological processes—which determine these sizes. As such, it must inevitably be a numerical and quantitative subject. Nevertheless, we have avoided complex mathematics, and we have, wherever possible, relegated the mathematical aspects of a topic to the final parts of the section in which that topic is examined. This will, we hope, make Population Ecology more generally accessible, and more palatable. But this is not to say that the mathematics have been played down. Rather, we have tried to play *up* the importance of real data from the real world: it is these, and not some mathematical abstraction, which must always be the major and ultimate concern of the population ecologist.

Developing the subject in this way, however, emphasizes that mathematical models *do* have an essential role to play. Time and again they crystallize our understanding of a topic, or actually tell us more about the real world than we can learn directly from the real world itself. Nature may be the ultimate concern of Population Ecology, but mathematical models, laboratory experiments and field experiments and observations can all help to further our understanding.

We have also tried to establish the point implied by the subtitle: that Population Ecology is a unified study of animals *and* plants. We are, of course, aware of the differences between the two. We feel, however, that plant and animal populations have had their own, independent ecologists for too long, and that, since the same fundamental principles apply to both, there is most to be gained at present from a concentration on similarities rather than differences.

In this second edition, we have retained the basic structure of the first edition; but we have sought to evolve the text in areas where we feel particular progress has been made and consolidated. Specifically we have addressed 'modularity' in plants and colonial animals and the importance of this in their population ecology and the definition of 'the individual'. We have also tried to cement some of the links between animal and plant populations by paying attention to descriptive equations common to both.

The book is set out in three parts. The first starts from the simplest first principles and examines the dynamics and interactions occurring within single-species populations. The second part, occupying approximately half of the book, is concerned with interspecific interactions: interspecific competition and predation. 'Predation', however, is defined very broadly, and includes the plant–herbivore, host–parasite, host–parasitoid and prey–predator interactions. The third part of the book synthesizes and expands upon the topics from the preceding chapters, and does so at three levels: individual life-history 'strategies', the regulation and determination of population size, and the importance of intra- and interpopulation interactions in determining community structure.

A number of people read all or most of this book in manuscript, and made generous and helpful suggestions, many of which we have now incorporated. We are deeply grateful to Professor Tony Bradshaw, Professor J. L. Harper, Professor Michael Hassell, Dr Richard Law, Professor Geoffrey Sagar, Dr Bryan Shorrocks, Dr David Thompson and, most especially, Dr John Lawton.

We also thank Mrs Barbara Cotgreave for drawing the figures, Mr Brian Lewis for his photography and Anita Callaghan, Susan Scott and Miss D. S. Paterson for typing the manuscript.

Population Ecology has come a long way since its inception, and the rate of progress has never been faster than at present. Nevertheless, there are few, if any, populations for which we can claim to fully comprehend the underlying causes of abundance. Much remains to be understood, and a great deal more remains to be done.

Michael Begon
Martin Mortimer

PART 1 SINGLE-SPECIES POPULATIONS

Chapter 1 Describing Populations

1.1 Introduction

Few would disagree with the proposition that nature is immensely complex. However, if we wish to understand this complexity, we will be well advised to abstract relatively simple facets from nature and examine these first. Species–habitat interrelations could be considered, or interactions between species within communities; but before they are, we must examine the basic components: single-species populations.

Even here there is a wealth of general questions: 'What causes species to be common or rare?', 'What underlies the fluctuations in their numbers?', 'Why do populations of the same species vary in their size and age-structure?', and so on. The dynamics of single-species populations, therefore, must be described in a way that allows such questions to be approached; for it is only when actual populations are encapsulated by the appropriate description that we can go on to consider underlying causes. This first chapter, then, is concerned with description, and with abstracting from populations the common properties that link them together.

1.2 Population processes

Although studies of animal and plant populations have developed quite separately, these two life forms have much in common when examined from a demographic viewpoint. At the simplest level, plants are born from seeds just as birds are born from eggs; and old animals exhibit signs of senility just as old oak trees bear dead branches. Moreover, if we were to catalogue the ages of every dandelion plant and every vole living in a field, we would probably find a range of ages in each; and, as time passed, individuals would either die, or survive to reach the next age-group; and in some age-groups, at certain times, individuals would produce offspring of their own. From the outset, therefore, it would seem sensible to suggest that, even though life forms and stages of development may differ substantially amongst species, certain basic *population processes* are common to all of them.

We can start considering these population processes by imagining a study of the numbers of voles inhabiting a meadow. Let us suppose that the vole numbers increase. We know that there has *either* been an influx of voles from adjoining meadows, *or* young voles have been born, *or* both of these events have occurred. We have, therefore, pin-pointed two very basic processes which affect the size of a population: *immigration* and *birth*. If, on the other hand, vole numbers decline, then our explanation would be that voles must have either died, or simply left the meadow, or both. These processes, which reduce population numbers, are *death* and *emigration*.

Of course, there is no reason to suggest that all four processes are not occurring simultaneously in the population. If the population declines, then the reason is simply that death and emigration together have *outweighed* birth and immigration, and vice versa if the population increases. We can certainly say that birth, death, immigration and emigration are the four fundamental demographic parameters in any study of population dynamics. Moreover, they can be combined in a simple algebraic equation describing the change in numbers in a population between two points in time.

$$N_{t+1} = N_t + B - D + I - E \qquad (1.1)$$

where N_t is the population size (number of individuals) at time t, N_{t+1} is the population size one time-period later, at time $t+1$, B is the number of new individuals born between t and $t+1$, D is the number of individuals which die between t and $t+1$, and I and E, respectively, are the numbers of immigrants and emigrants during the same period of time.

If the population is so large that our study cannot encompass the whole of it, then this equation must be constructed in terms of densities rather than absolute numbers. Thus, samples are taken, and N_t, for instance, becomes 'the number of plants per square metre at time

t' or 'the number of insects per leaf'. Nevertheless, equation 1.1 indicates that, at its simplest, the task of the demographer is to measure these four parameters and account for their values—yet the translation of this into practice is rarely straightforward. Almost all species pass through a number of stages in their life cycle. Insects metamorphose from eggs to larvae to adults, and some have a pupal stage as well; plants pass from seeds to seedlings and then to photosynthesizing adult plants, and so on. In all such cases the different stages must be studied individually. Also, in reality, the four 'basic' parameters are themselves often compounded from several other component processes. Equation 1.1, therefore, cannot be considered as anything more than a basis upon which more realistic descriptions can be built.

1.3 The diagrammatic life-table

1.3.1 General form

The description we require is one which retains the generality of equation 1.1, but can also reflect the complexities of most actual populations. One such description is the *diagrammatic life-table* (Sagar & Mortimer 1976), which is applied to an idealized higher plant in Fig. 1.1. The numbers at the *start* of each of the stages—seeds, seedlings and adults—are given in the square boxes. Thus, the N_{t+1} adults alive at time $t+1$ are derived from two sources. Some are the survivors of the N_t adults alive at time t. Their probability of survival (or, equivalently, the proportion of them that survive) is placed inside a triangle (or arrow) in Fig. 1.1, and denoted by p. So, for instance, if N_t is 100 and p, the survival-rate, is 0.9, then there are 100×0.9 or 90 survivors contributing to N_{t+1} at time $t+1$ (10 individuals have died; the mortality-rate $(1-p)$ between t and $t+1$ is clearly 0.1).

The other source of the N_{t+1} adults is 'birth', which in the present case can be viewed as a multi-stage process involving seed production, seed germination and the growth and survival of seedlings. The average number of seeds produced per adult—the average *fecundity* of the plant population—is noted by F in Fig. 1.1 and placed in a diamond. The total number of seeds produced is, therefore $N_t \times F$. The proportion of these seeds that

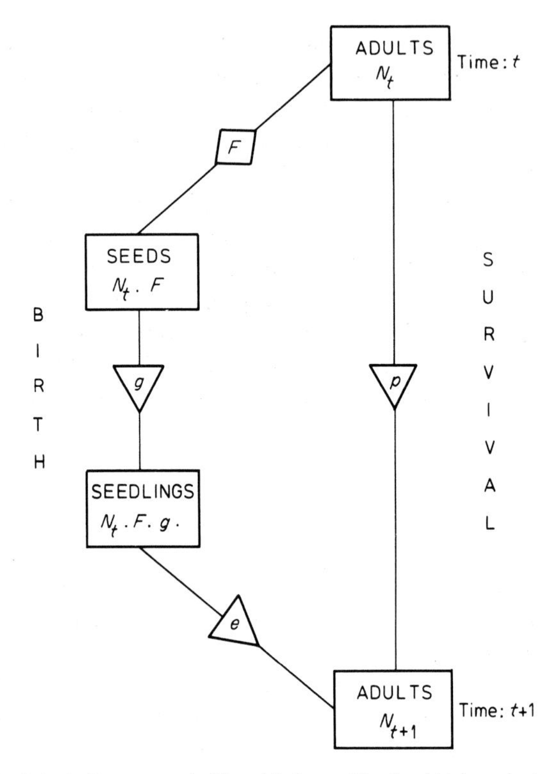

Fig. 1.1. A diagrammatic life-table for an idealized higher plant. F: number of seeds per plant; g: chance of a seed germinating $(0 \leqq g \leqq 1)$; e: chance of a seedling establishing itself as an adult $(0 \leqq e \leqq 1)$; p: chance of an adult surviving $(0 \leqq p \leqq 1)$.

actually germinate on average is denoted by g, which, being essentially a survival-rate, is placed in an arrow in Fig. 1.1. Multiplying $N_t \times F$ by g gives us the number of seedlings which germinate successfully. The final part of the process is the physiological establishment of seedlings as independently photosynthesizing adults. The probability of surviving this very risky phase of plant growth is denoted by e (once again in an arrow), and the total number of 'births' is, therefore, $N_t \times F \times g \times e$. The number in the population at time $t+1$ is then the sum of this and the number of surviving adults, $N_t \times p$.

We can now substitute the terms from the life-table into our basic equation of population growth (equation 1.1) as follows:

$$N_{t+1} = \overbrace{N_t - \underbrace{N_t(1-p)}_{\text{Death}}}^{\text{Surviving}} + \underbrace{N_t \times F \times g \times e.}_{\text{Birth}} \qquad (1.2)$$

There are several points to note about this equation. The first is that both here and in Fig. 1.1 immigration and emigration have, for simplicity, been ignored, and our description of how a plant population may change in size is essentially incomplete. The second is that 'death' has been calculated as the product of N_t and the mortality-rate $(1-p)$—survival and mortality are opposite sides of the same coin. The third point is that birth is quite clearly a complex product of 'birth-proper' and subsequent survival. This is frequently the case: even human 'birth'-rates are the product of the rate at which fertilized eggs implant in the womb and the rate of pre-natal survival.

1.3.2 *The common field grasshopper, an annual species*

In practice, careful and meticulous field-work is necessary to build a diagrammatic life-table of the type illustrated in Fig. 1.1. Reliable estimates of the transition probabilities (p, g and e in Fig. 1.1) are required, as well as measurements of the fecundity of adults. Such data for the common field grasshopper, *Chorthippus brunneus*, are illustrated in Fig. 1.2. These were obtained by a combination of field samples and back-up laboratory observations on a population near Ascot in Berkshire (Richards & Waloff 1954). The population was isolated so that immigration and emigration *could* be ignored.

The first point to note about Fig. 1.2 is that no adults survive from one year to the next ($p = 0$). *Ch. brunneus* is, therefore, an 'annual' species; each generation lasts for just one year, and generations are discrete, i.e. they do not overlap. It is also clear that the 'birth' of adults is a complex process involving at least six stages. The first stage is the laying of egg-pods in the soil by adult females. On average, each female laid 7.3 pods, each containing 11 eggs. F is, therefore, 80.3. These eggs remain dormant over winter, and by early summer only 0.079 of them had survived to hatch into first-instar nymphs. Subsequently the transition probabilities between instars were fairly constant, taking a remorseless toll on the surviving population; less than a third of the first-instar nymphs survived to be 'born' into the adult population. Despite their apparently high fecundity, therefore, the adults of 1947 did little more than replace themselves with newly born adults in the following year.

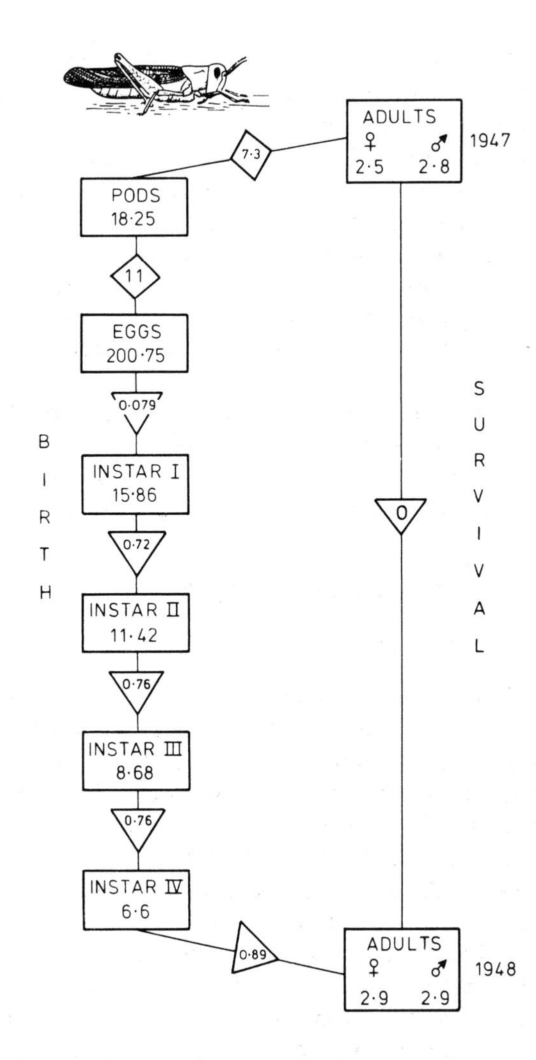

Fig. 1.2. Diagrammatic life-table of the field grasshopper, *Chorthippus brunneus*. (Population sizes are per 10 m^2; data from Richards & Waloff 1954.)

Ch. brunneus' diagrammatic life-table is illustrated in a simplified form in Fig. 1.3a. This life-table is appropriate for all species which breed at a discrete period in their life cycle, and whose generations do not overlap. If the time between t_0 and t_1 is one year, the life history is referred to as annual.

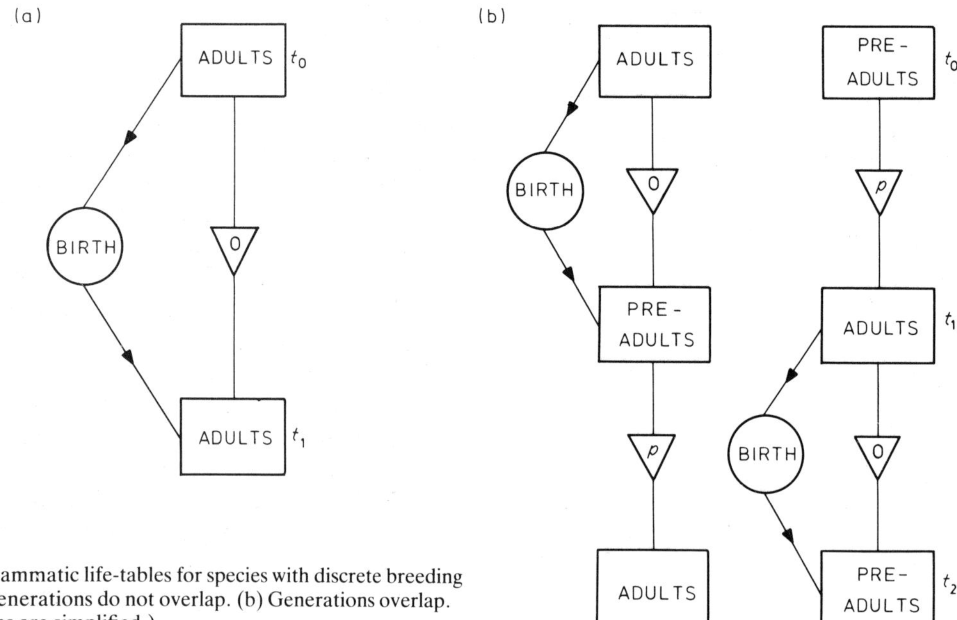

Fig. 1.3. Diagrammatic life-tables for species with discrete breeding seasons. (a) Generations do not overlap. (b) Generations overlap. (Birth processes are simplified.)

1.3.3 Ragwort, a biennial

An annual life history is only one of a number of possible patterns. If we consider species that live for two years rather than one, reproducing only in the final year, then we have a life history that involves breeding at one discrete time in the life cycle, but in which generations of adults may well overlap; this is illustrated in Fig. 1.3b. If the time periods are years, then this life cycle is referred to as 'biennial'. During any one summer, the population contains both young adults which will not reproduce until the following year, and mature, reproducing adults.

Ragwort, *Senecio jacobaea,* is a biennial plant with a life cycle in which seeds germinate principally in the autumn. Then, during the next year, young plants form a rosette of leaves. In the second year a flowering stem is formed. A diagrammatic life-table for *S. jacobaea* is shown in Fig. 1.4, in which the birth-process has been expanded to include some extra stages which are specific to plants. The data come from measurements made on a population living in sand dune environments in the Netherlands (van der Meijden 1971). Of the 5040 seeds that are produced, 62% fall on to the ground; the other 38% are dispersed by the wind to other areas. By the same token there is quite a high chance that immigrants enter this population. This necessitates a further modification of our life-table, indicated in Fig. 1.4 by the inclusion of invading seeds, which may contribute either to the seed banks or to the incoming seed 'rain'.

Having arrived on the ground, various potential fates await ragwort seeds. They lie on the surface of the sand in the 'surface seed bank', where they may germinate, be eaten or just die. Alternatively, wind or insects, acting as migratory agents, may transport them to neighbouring areas; or they may become buried. The detailed fates of ragwort seeds in sand dune environments are not fully known, but only 11.4% stay in the surface seed bank; and of the 3124 seeds that rain on to the soil only 40 actually germinate successfully. However, seedlings can arise from an additional source: the buried seed bank. We do not know how many seeds are buried in the sand profile, but for many plant species, especially weeds, the numbers of buried seeds can be very high (up to *c.* 50 000

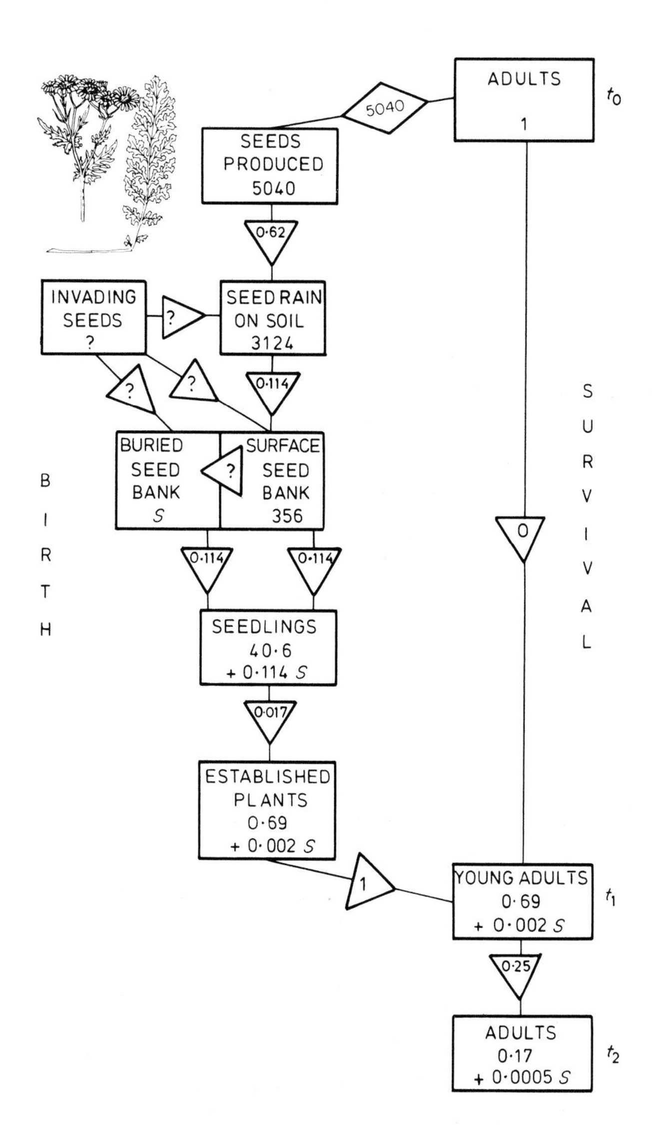

Fig. 1.4. Diagrammatic life-table of the biennial ragwort, *Senecio jacobaea*. (Population sizes are per 4 m^2; data from van der Meijden 1971.)

m^{-2}), and a proportion of each season's seed crop does become buried. To indicate that this is a birth route we have added to the 'seedling' and 'established plant' components a fraction of the buried seed bank, denoted by S. Finally, the transition from seedlings to young, established, photosynthetically independent adults in sand dune environments is also an exceedingly risky phase for ragwort: only 1.7% of the seedlings actually become established.

The life-table in Fig. 1.4, therefore, illustrates the importance of additional seed sources to the 4 m^2 area, since from t_0 to t_1 the original ragwort density of 1 becomes reduced to 0.69. Thus, to keep the number of young adults at t_1 up to exactly 1 we might argue that there are 155 seeds in the buried bank which germinate (since if $0.69+0.002S = 1$, then $S = 155$). Alternatively, some of these 155 might enter the 'birth process' as immigrants; and if we recall that 38% of the 5040 seeds were dispersed, we can see that there are ample numbers to rely upon. To complete this life-table, however, we should note that the chance of a young adult surviving to become a mature one *producing seed* is only 0.25. To ensure that the population size at t_2 is still 1, therefore, we will have to imagine a further input of seeds into the birth process.

Such data as these emphasize the extreme severity of the sand dune habitat to plant life, and the considerable mobility of seeds in the life-cycle of ragwort: individual seeds may travel at least 15 m. Since sand dunes, by their nature, offer shifting and temporarily suitable habitats for ragwort, we can infer that seed movement by dispersal on or above the sand is a very necessary feature in the life of this plant.

1.3.4 More complex life cycles

Overlapping generations are not confined to biennials. Consider the population of great tits (*Parus major*) near Oxford studied by Perrins (1965) and illustrated in Fig. 1.5. Adult birds build their nests and lay eggs in the early summer, but of these eggs only a proportion (0.84 in this case) survive to hatch as nestlings. These nestlings are themselves subject to many dangers, and by the late summer only 71% of them survive to fledge—leaving the nest and fending for themselves. Of these fledglings, only a small proportion live through the winter to become breeding adults. However, a rather larger proportion of the previous generation's adults have also survived. The population of breeding adults, therefore, consists of individuals of various ages, from one to five or more years old. As Fig. 1.5 shows, this situation is

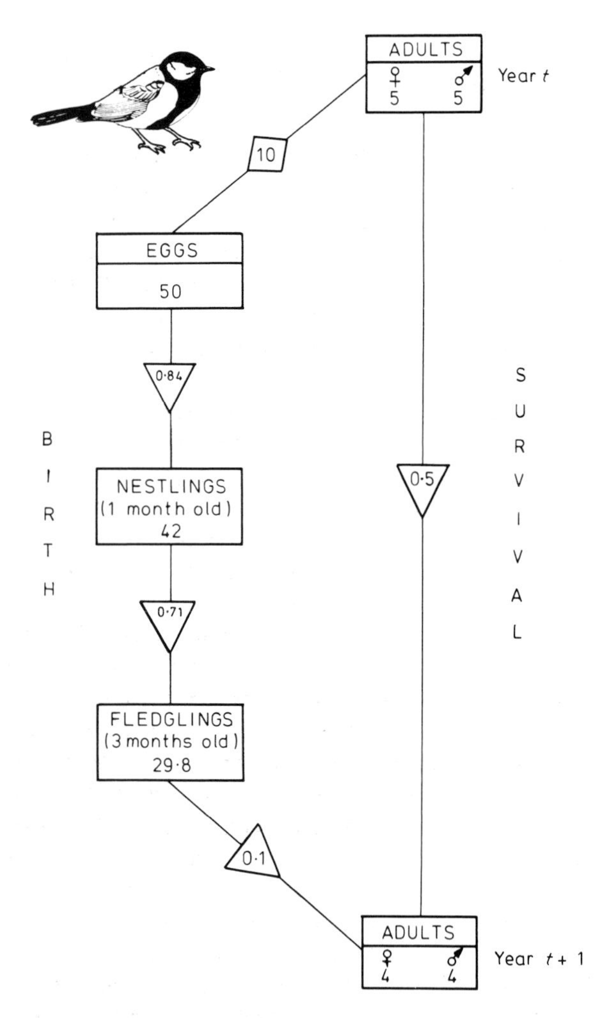

Fig. 1.5. Diagrammatic life-table of the perennial great tit, *Parus major*. (Population sizes are per hectare; data from Perrins 1965.)

readily described by a variant of our diagrammatic life-table very similar to Fig. 1.1. We are dealing with a population in which breeding occurs at discrete time periods, but in which the individuals are potentially long-lived so that *many* generations overlap.

We have assumed with our great tits, however, that adults of different ages are equivalent and may be treated as equal members of a common pool. Yet there will be many instances in which their demographic characteristics will be 'age-dependent' or 'age-specific'. In such cases, a diagrammatic life-table of the type shown in Fig. 1.6 may be more appropriate. In Fig. 1.6, the population at any one time consists of individuals in a range of age-classes: a_0 individuals are in the youngest age-class, a_1 individuals in the next oldest, and so on. With the passage of one unit of time a proportion of the individuals in one age group survive to become individuals in the next oldest age group. Thus, p_{01} is the proportion of the a_0 individuals surviving to become a_1 individuals one time-unit later, p_{12} is the proportion of the a_1 individuals surviving to become a_2 individuals, and so on (though in practice these p-values will, of course, vary with the changing circumstances of the population). Fig. 1.6 also shows that each age group has the potential to contribute to the youngest age-class via

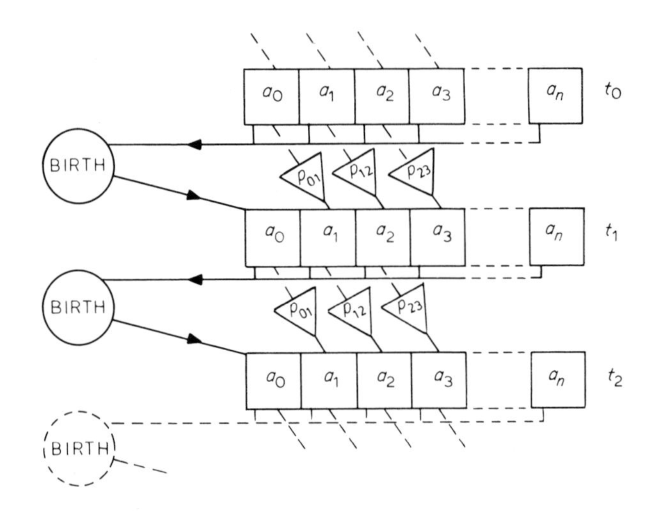

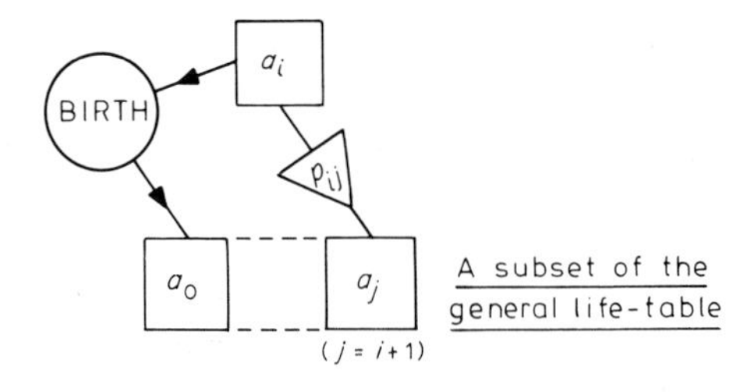

Fig. 1.6. A diagrammatic life-table for any species that reproduces continuously with overlapping generations. a_0, a_1–a_n represent age-groups of individuals, a_n being the oldest group. p_{ij} is an age-specific probability of survival, where, for example, p_{01} is the probability of individuals in a_0 at one time surviving to reach a_1 by the next time period ($0 \leqq p_{ij} \leqq 1$). The inset shows a subset of the general life-table.

the birth process. For simplicity, birth from all age groups has been fused together; in reality, fecundity, like survival, would vary from age-class to age-class. Nevertheless, despite this increased sophistication, inspection of the life-table in Fig. 1.6 reveals that it is built up of units which are little more than the diagrammatic life-table with which we are already familiar. One such unit is illustrated in the inset in Fig. 1.6.

The implication in Fig. 1.6 is that breeding occurs at discrete periods, even though generations overlap and there are many age-classes each with their own birth- and survival-rate. In many species, however, birth (and death) occur continuously within a population. Fig. 1.6 is still appropriate in such cases, but time must be split arbitrarily into intervals, and the various terms take on slightly different meanings. Suppose, for instance, that we consider the numbers in a population at monthly intervals. At t_0, a_2 is the total number of individuals between two and three months old. One month later (at t_1), p_{23} of these will survive to become the a_3 individuals that are between three and four months old. Thus, even though birth and death are occurring continuously, they are considered 'one month at a time'.

1.3.5 *Age and stage: the problems of describing some plant and animal populations*

Fig. 1.6 illustrates the age-dependent transitions that may occur in populations with overlapping generations. As a means of describing flux in populations, this approach is only justifiable if individuals can be classified meaningfully by *age* alone. For many plant species, however, especially those that are perennial, the fate of an individual is not so much dependent on its absolute age as on its size or stage of growth. In grasses, for instance, the chance of being grazed may be crucially size-dependent, and an individual with a few tillers may escape the notice of a herbivore whilst a conspicuous one bearing many tillers may not. Equally important is the fact that the grazed individual, though reduced in size, may not necessarily die, but may generate from basal growing points (buds) to achieve its former size. Age of the plant under such circumstances may have little relevance. Size of the plant is the more important determinant of its fate. Considerations of this sort have led a number of plant ecologists, particularly in the Soviet school (Gatsuk *et al.* 1980) to reject age *per se* in favour of 'age states' as a useful criterion for describing individuals (Uranov 1975). Thus, individuals may be classified on an ontogenetic or developmental basis and categories might include seed, seedling, juvenile, immature, virginile, reproductive, subsenile and senile states. Such a classification recognizes that there are broad morphological changes that occur during the growth and development of a plant species but that the duration of time spent in each may differ widely. Mertz & Boyce (1956), for instance, have shown that almost 75% of the oak tree 'seedlings' developing after a forest felling were in fact sprouts attached to roots up to forty years old. Presumably this 'seedling' population had remained suppressed at an early stage or 'age-state' due to factors such as grazing, trampling and poor conditions for growth.

An alternative classification of individuals is by size alone and indeed this can be superimposed upon an age-state classification. Terrestrial plants that can 'reiterate' ther growth form (see Section 1.6) can not only regress or advance from one age-state to another, but within an age-state they may change in size. The choice between a strictly size-based classification and an age-state one, or indeed a hybrid between the two, is very much dependent upon the organism under study.

In studying the population dynamics of reef coral *Agaricia agaricites*, off the coast of Jamaica, Hughes (1984) classified individual coral colonies on the basis of age-state—'larva' or 'coral colony'—and on the size of colony. By repeated photography of the reef he was able to record size changes as well as larval recruitment. A diagrammatic representation of the population structure is shown in Fig. 1.7. Coral colonies may remain static in size, they can grow, they can shrink, and they can reproduce sexually by free-swimming larvae as well as asexually by fission. The chances of their doing so for each size-class are shown in this figure, for a year when storms were absent and conditions ideal for coral growth. This coral species has a mean annual lateral extension rate of less than 2 cm. Thus, a large proportion of the population remained in the same size class from one season to the next. Moreover, whilst only a small fraction of the colonies increased in size, a more likely

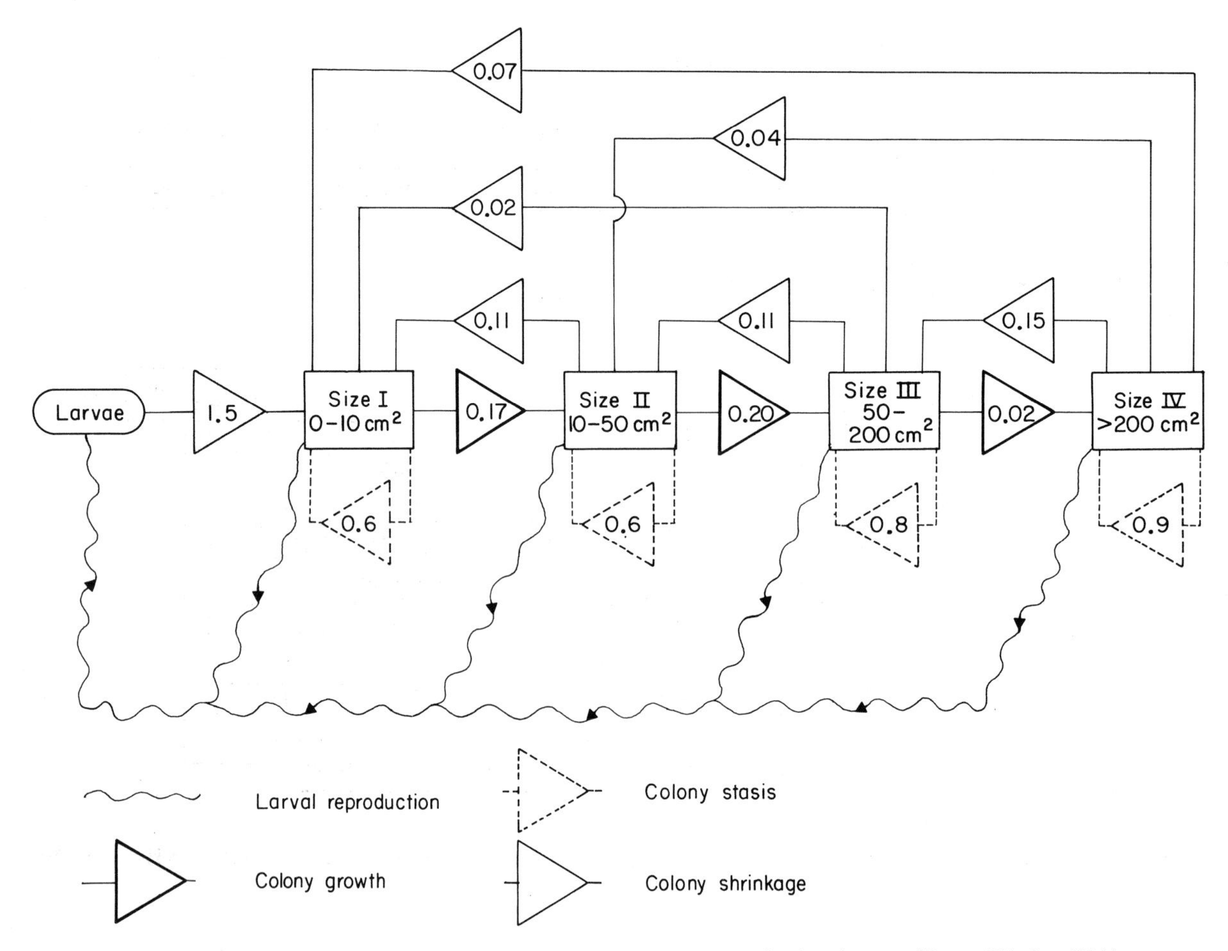

Fig. 1.7. A size-based diagrammatic life-table for the reef coral *Agaricia agaricites* growing in calm seas. (Data of Hughes 1984.)

occurrence was that they should shrink to smaller size-classes—a fate moreover that was size-dependent. This shrinkage may represent a series of remorseless steps towards mortality of the individual, but equally it may also reflect fission of the colony and hence *asexual reproduction*. Indeed, in the case of size-class 4, it must do so since the sum of the fractional transitions exceeds unity.

This sort of description, based primarily on size alone, has limitations in so far as it does not readily distinguish between survival and fecundity, especially when asexual reproduction is involved. Moreover, in the majority of classifications of this type, at least one additional category, as we have already mentioned, must be included to complete the life cycle: larvae. Hughes recorded 1.5 larvae m^{-2} settling as new recruits during a year in his study. In *Agaricia* the probability of reproducing sexually by larvae is unknown (Connell 1973), but even if size-specific fecundities were measurable, they might have little value in describing the local populations of this coral since larvae tend to be very widely dispersed. Using this figure for larval recruitment, Hughes was able to calculate by matrix methods (Chapter 3) that the population was almost static in size.

A comparable approach describing a plant species is the one used by Sarukhan & Gadgil (1974) for the 'creeping buttercup', *Ranunculus repens* in Britain. This species reproduces sexually by seed and by asexual

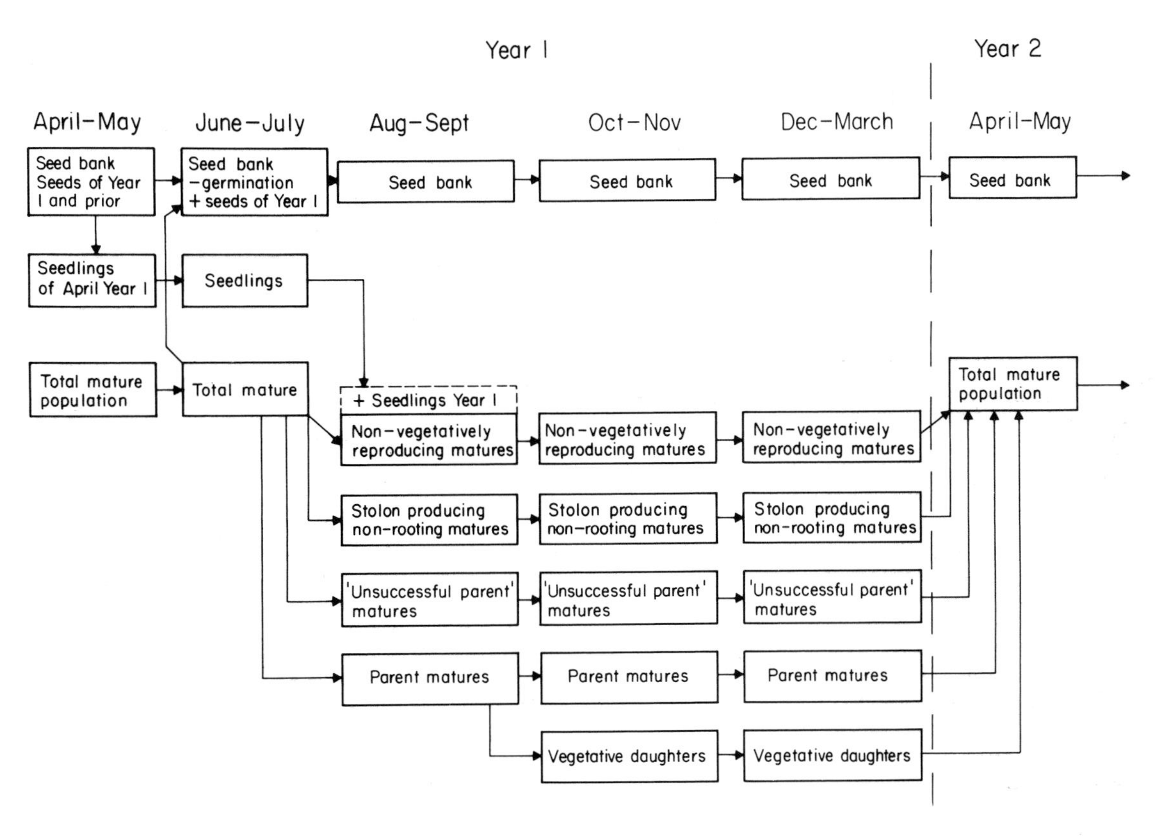

Fig. 1.8. The transitions occurring throughout the year in a buttercup population (*Ranunculus repens*) as envisaged by Sarukhan & Gadgil (1974).

propagation, though recruitment by these means occurs at different times during the season. Seeds germinate in late spring and early summer, whilst in late summer new 'daughter' plants become established as separate adult plants from shoots borne at nodes along creeping stolons. If we require a description of these events within a season, and hence a more faithful representation of the biological events that occur, we must recognize that the generalized life-table diagram is inadequate and accept a more complex flow diagram (Fig. 1.8). In essence this is an age-state classification in which the fluxes are precisely defined chronologically. This approach makes an additional important distinction, in that asexually produced 'vegetative' daughters are classified separately from sexually produced seedlings, at least during the first year of life. For the purposes of generality, Sarukhan and Gadgil lumped these recruits together once they attained one year of age, but there is no reason why this distinction could not be maintained if continued resolution was required.

In describing populations by diagrammatic means, then, we have considered a range of forms of description. Some have been based on generation-to-generation changes classifying by age, while others have been based on size taking for convenience season-to-season or month-to-month time-steps. In this latter instance, it is easy to preclude particular life-cycle transitions (if the biology of the species demands this) by setting them to zero.

Additionally we can impose a genetic subclassification on our life-table. Individuals in a population may be known to be genetically distinct. On the other hand,

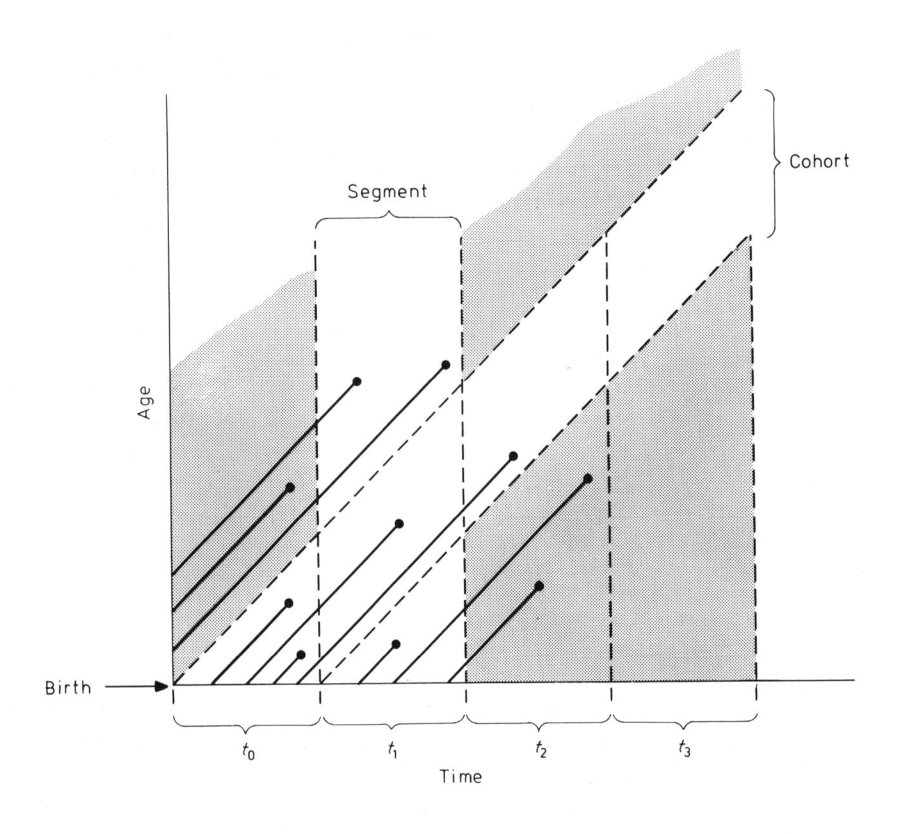

Fig. 1.9. A population portrayed as a set of diagonal lines, each line representing the life 'track' of an individual. As time progresses, each individual ages and eventually dies. Three individuals are born prior to t_0, four during t_0, and three during t_1. To construct a 'fixed cohort' life-table, a 'searchlight' is directed onto the cohort of individuals born during t_0 and the subsequent development of the cohort is monitored. Two of the four individuals have survived to the beginning of t_1; only one of these is alive at the beginning of t_2; and none survives to the start of t_3. To construct a 'static' life-table, the searchlight is directed onto the whole population during a single segment of time (t_1). The ages of the seven individuals alive at some time during t_1 may be taken as an indication of the age-specific survival-rates if we assume that the rates of birth and survival are constant. (After Skellam 1972.)

Table 1.1 A cohort life-table for *Poa annua*. (Adapted from Law 1975.)

Age (in 3-month periods) x	Number observed alive at each quarter year a_x	Standardized number surviving at the start of age interval x l_x	Standardized number dying between x and $x+1$ d_x	Mortality-rate q_x	$\log_{10} a_x$	$\log_{10} l_x$	$\log_{10} a_x - \log_{10} a_{x+1}$ k_x	Average number of seeds per individual aged x B_x
0	843	1000	143	0.143	2.926	3.000	0.067	0
1	722	857	232	0.271	2.859	2.933	0.137	300
2	527	625	250	0.400	2.722	2.796	0.222	620
3	316	375	204	0.544	2.500	2.574	0.342	430
4	144	171	107	0.626	2.158	2.232	0.426	210
5	54	64	46.2	0.722	1.732	1.806	0.556	60
6	15	17.8	14.24	0.800	1.176	1.250	0.699	30
7	3	3.56	3.56	1.000	0.477	0.551		10
8	0	0	—					—

'individuals' may actually be asexually produced ramifications of the same genotype. Kays & Harper (1974) recognized this dichotomy and introduced the terms *genet* and *ramet* to avoid confusion. Thus, individuals that are produced asexually (e.g. 'daughter' plants of buttercups, bulbs, tillers, polyps of *Obelia,* corals) and have the potential for growth independent of the parent are called ramets; and a population of ramets with the same ('maternal') parentage constitutes a *clone*. A genet on the other hand, is an organism, *however much ramified* which has arisen from a single zygote—all parts having the same genotype.

1.4 Conventional life-tables

1.4.1 The cohort life-table

The most reliable method of determining age-specific mortality and fecundity for a continuously breeding population, or simply one in which generations are overlapping, is to follow the fate of a group of individuals, all born during the same time interval. Such a group is called a *cohort*. The process is essentially a journey from the top left-hand corner of Fig. 1.6 to its bottom right-hand corner, and, in many respects, it is similar to following the fate of an annual species throughout its yearly cycle. The difference in this case is that each individual has to be recognized and distinguished from those individuals belonging to other cohorts which are in the population at the same time. The situation is described diagrammatically in Fig. 1.9 in which individuals are represented by solid lines, ageing with time, and eventually dying (a 'spot' in Fig. 1.9). The cohort of four individuals (born at t_0) is observed again at t_1 (when there are two survivors), at t_2 (one survivor), and at t_3 (no survivors).

Plants are ideal subjects for such study, since they are generally sessile and can be tagged or mapped, enabling the fates of individuals to be precisely recorded and their reproductive output measured. Law (1975), for instance, followed the fate of a cohort of the annual meadow grass, *Poa annua,* from initial establishment to the ultimate death of the last individual. Recording the number alive at successive time-periods and the number of offspring (seeds) produced per plant, he was able to compile a table of data showing survivorship and fecundity (Table 1.1). The first (left-hand) column gives the age at the beginning of each time interval. Thereafter, only the second and last columns (a_x and B_x) actually contain field data. All other columns are derived from the a_x column. We can see that this (conventional) life-table contains essentially the same information as the diagrammatic life-tables previously described.

The a_x column summarizes the raw data collected in the field by mapping the positions of 843 *Poa annua* plants that arose from naturally sown seeds in a number of metre-square quadrats. From this raw data 'l_x' values are calculated, by converting the numbers observed at the start of each time interval to the equivalent number that would have occurred had the starting density of the cohort been 1000; e.g. $l_3 = 316 \times 1000/843 = 375$. The value of this procedure is that l_x-values can be compared between populations, or between species, since they do not depend on the actual number of individuals considered in each study. In other words an a_0 value of 843 is peculiar to *this* set of observations, whereas *all* studies have an l_0 value of 1000.

To consider mortality more explicitly, the standardized numbers dying in each time interval (d_x) must be computed, being simply the difference between l_x and l_{x+1}; e.g. $d_1 = 857-625 = 232$. q_x—the age-specific mortality-*rate*—has also been calculated. This relates d_x to l_x in proportional terms, so that, for instance, q_2—the proportion of the six-month-old individuals that die in the subsequent three-month period—is 250/625 or 0.4. q_x can also be thought of as the 'chance of death', and is equivalent to $(1-p_x)$ where 'p' refers to the survival-probability considered previously.

The advantage of the d_x-values is that they can be summed over a period of time: the number dying in the first nine months is $d_0+d_1+d_2$ (= 625). The disadvantage is that the individual values give no real idea of the intensity or importance of mortality at a particular time. This is because the d_x-values are larger, the more individuals there are to die. q_x-values, on the other hand, are a good measure of the intensity of mortality. Thus, in the present example, it is clear from the q_x column that the mortality-rate rose consistently with increasing age; this is not clear from the d_x column. The q_x-values, however, have the disadvantage of not being liable to summation: $q_0+q_1+q_2$ does *not* give us the overall mortality-rate for the first nine months. These advantages are combined, however, in the penultimate column of Table 1.1 in which 'k'-values (Haldane 1949; Varley & Gradwell 1970) are listed. k_x is defined, simply, as $\log_{10} a_x - \log_{10} a_{x+1}$ (or, equivalently, $\log_{10} a_x/a_{x+1}$), and is sometimes referred to as 'killing-power'. Like q, k-values reflect the intensity or rate of mortality, and, in the present case, they increase consistently with age. But, unlike q, summing the k-values is a meaningful procedure. Thus the killing-power or k-value of the first nine months is $0.067+0.137+0.222 = 0.426$, which is also the value of $\log_{10} a_0 - \log_{10} a_3$. Note, furthermore, that the k_x-values can be computed from the l_x-values as well as the a_x-values; and that, like l_x, k_x is standardized and is, therefore, appropriate for comparing quite separate studies. k-values will be of considerable use to us in later chapters.

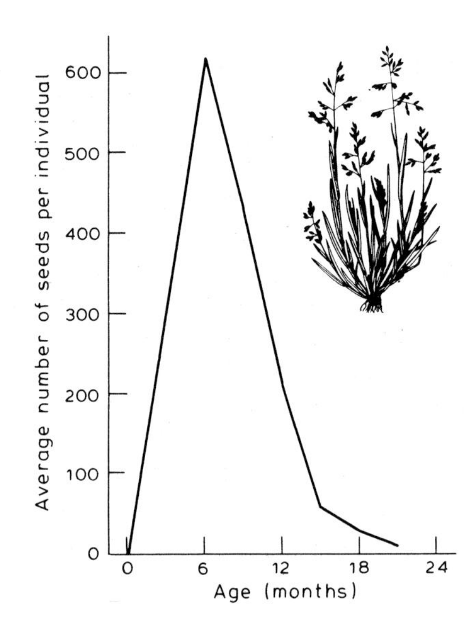

Fig. 1.10. Age-specific fecundity (B_x) for the annual meadow grass, *Poa annua*. (Data from Law 1975.)

The age-specific patterns of fecundity and mortality have been plotted in Figs 1.10 and 1.11. Fig. 1.10 indicates quite clearly an initial sharp rise in fecundity reaching a peak at six months, followed by a gradual decline until the death of the last individual after two years. Fig. 1.11 illustrates a single pattern in three different ways. Fig. 1.11a is a 'survivorship curve'—$\log_{10} l_x$ plotted against age—while Fig. 1.11b contains two mortality curves, q_x and k_x, plotted against age. All show a consistent rise in the rate of mortality, leading to an increasingly rapid decline in survivorship. Of the three, Fig. 1.11a—the survivorship curve—probably shows this most clearly.

The use of logarithms in the survivorship curve deserves further comment. Consider, for instance, the halving of a population over 1 unit of time, in one case

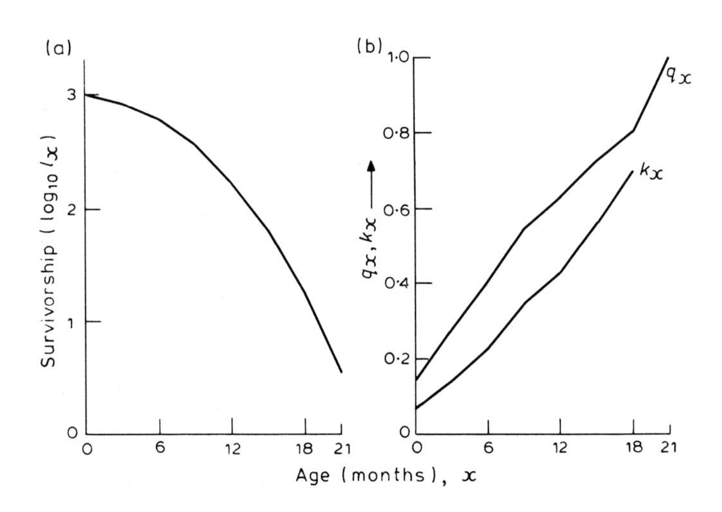

Fig. 1.11. (a) Age-specific survivorship ($\log_{10} l_x$), and (b) age-specific mortality-rates (q_x) and killing-powers (k_x) for the annual meadow grass, *Poa annua*. (Data from Law 1975.)

from 100 to 50 individuals, and in another case from 10 to 5 individuals. In both cases there has been a reduction by half, i.e. the rate or probability of death *per individual* (usually referred to as the 'per capita rate') is the same. Nevertheless, the slope of an *arithmetic* survivorship curve would be −50 in the first case but −5 in the second. With logarithmic survivorship curves, however, the slopes in these two, equivalent cases are identical. In fact, equivalent advantages are gained by the use of k_x-values: being based on logarithms, they, too, allow recognition of cases in which per capita rates of mortality are the same. Moreover, logarithms also indicate when per capita rates of *increase* are identical. 'Log numbers' should therefore be used in preference to 'numbers' when numerical change is being plotted.

1.4.2 The static life-table

Unfortunately, it is not always possible to monitor the dynamics of a population by constructing a 'fixed cohort' life-table. It is, in fact, rarely possible with natural populations of animals, since the individuals are often highly mobile, highly cryptic or both. There is, however, a rather imperfect alternative, which is also illustrated diagrammatically in Fig. 1.7. It involves examining the age structure of the whole population at one particular time, or, since these things cannot be done instantaneously, during one short 'segment' of time.

As an example, we can consider the results, reported by Lowe (1969), of an extensive study of the red deer (*Cervus elaphus*) on the small island of Rhum, Scotland. Each year from 1957 onwards, Lowe and his co-workers examined every one of the deer that was shot under the rigorously controlled conditions of this Nature Conservancy Council reserve. They also made extensive searches for the carcasses of deer that had died from natural causes. Thus, they had access to a large proportion of the deer that died from 1957 onwards. Deer can be reliably aged by the examination of tooth replacement, eruption and wear, and Lowe and his co-workers carried out such examinations on all of the dead deer. If, for instance, they examined a six-year-old deer in 1961, they were able to conclude that, in 1957, this deer was alive and two years old. Thus, by examining carcasses, they were able to reconstruct the age structure of the 1957 population. (Their results did not represent the total numbers alive, because some carcasses must have decomposed before they could be discovered and examined.) Of course, the age structure of the 1957 population could have been ascertained by shooting and examining large numbers of deer in 1957; but, since the ultimate aim of the project was enlightened conser-

Table 1.2 A static life-table for red deer. (From Lowe 1969.)

						smoothed				
x (years)	x_x	l_x	d_x	q_x	B_x	l_x	d_x	q_x	$\log_{10} l_x$	k_x
1	129	1000	116	0.116	0	1000	137	0.137	3.000	0.064
2	114	884	8	0.009	0	863	85	0.097	2.936	0.045
3	113	876	48	0.055	0.311	778	84	0.108	2.891	0.050
4	81	625	23	0.037	0.278	694	84	0.121	2.841	0.056
5	78	605	148	0.245	0.302	610	84	0.137	2.785	0.064
6	59	457	−47	—	0.400	526	84	0.159	2.721	0.076
7	65	504	78	0.155	0.476	442	85	0.190	2.645	0.092
8	55	426	232	0.545	0.358	357	176	0.502	2.553	0.295
9	25	194	124	0.639	0.447	181	122	0.672	2.258	0.487
10	9	70	8	0.114	0.289	59	8	0.141	1.771	0.063
11	8	62	8	0.129	0.283	51	9	0.165	1.708	0.085
12	7	54	38	0.704	0.285	42	8	0.198	1.623	0.092
13	2	16	8	0.500	0.283	34	9	0.247	1.531	0.133
14	1	8	−23	—	0.282	25	8	0.329	1.398	0.168
15	4	31	15	0.484	0.285	17	8	0.492	1.230	0.276
16	2	16	—	—	0.284	9	9	1.000	0.954	

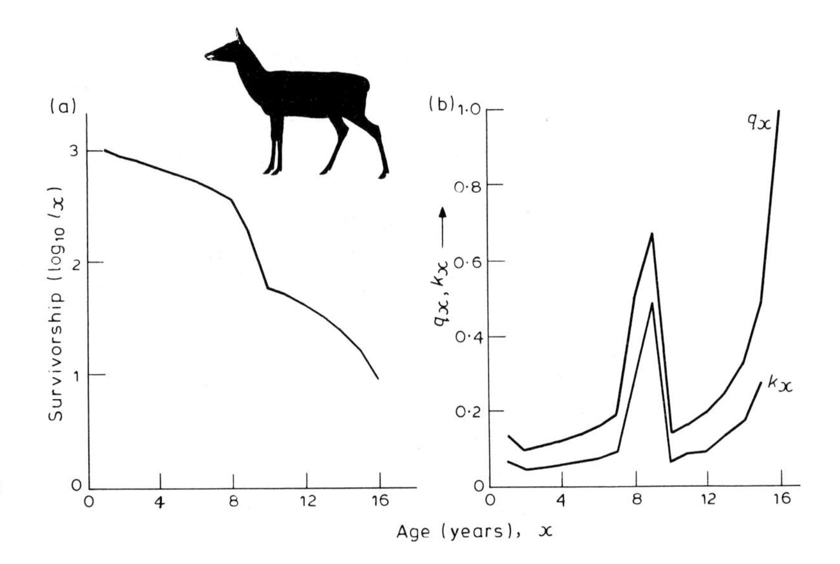

Fig. 1.12. (a) Age-specific survivorship ($\log_{10} l_x$), and (b) age-specific mortality rates (q_x) and killing powers (k_x) for the red deer, *Cervus elaphus*. (Data from Lowe 1969.)

vation of the deer, this method would have been somewhat inappropriate.

Lowe's raw data for red deer hinds are presented in column two of Table 1.2. As expected, there were many young deer and rather fewer old deer, but we can treat these raw data as the basis for a life-table *only* if we make a certain set of assumptions. We must assume that the 59 six-year-old deer alive in 1957 were the survivors of 78 five-year-old deer alive in 1956, which were themselves the survivors of 81 four-year-olds in 1955, and so on. In other words, we must assume that the numbers of births and age-specific survival-rates had remained the same from year to year, or, equivalently, that the a_x column of Table 1.2 is essentially the same as *would* have been obtained if we *had* followed a single cohort. Having made this assumption, l_x, d_x and q_x columns have been constructed. It is clear from Table 1.2, however, that our assumption is false. The 'cohort' actually increases in size from years 6 to 7 and 14 to 15, leading to 'negative' deaths and meaningless mortality-rates. The pitfalls of constructing such 'static' life-tables are, therefore, amply illustrated.

Nevertheless, such data are by no means valueless. Lowe's aim was to provide a *general* idea of the population's age-specific survival-rate (and birth-rate) prior to 1957 (when culling of the population began), and then to compare this with the situation after 1957. He was more concerned with general trends than with the particular changes occurring from one year to the next. He therefore 'smoothed out' the variations in population size between ages 2–8 and 10–16, and *created* a steady decline in both of these periods. The results of this process are shown in the final five columns of Table 1.2, and the mortality schedules are plotted in Fig. 1.12. They do, indeed, provide the general picture Lowe required: there is a fairly gentle but increasing decline in survivorship up to year 8, followed by two years of very

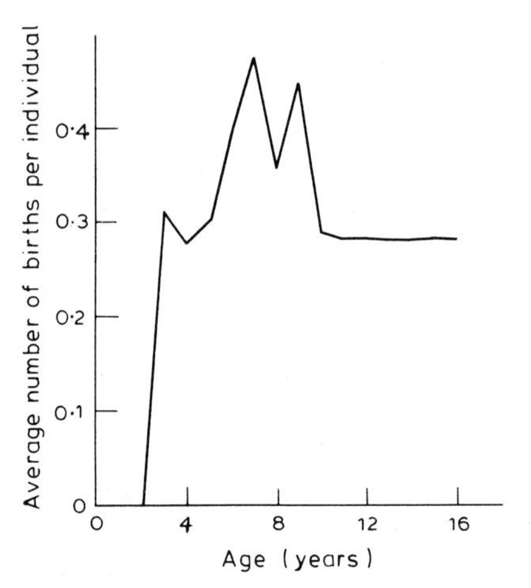

Fig. 1.13. Age-specific fecundity for the red deer, *Cervus elaphus*. (Data from Lowe 1969.)

heavy mortality, and then a return to a gentler, though again increasing, decline.

Moreover, by examining the internal reproductive organs of the hinds, Lowe was also able to derive a sequence of age-specific birth-rates. This is shown in the sixth column of Table 1.2, and illustrated in Fig. 1.13. There is, clearly, an initial pre-reproductive period of two years, followed by a sudden increase in birth-rate which is maintained for three years. There is then a period of four years during which the birth-rate is higher still, followed by a return to the previous level. It is interesting to note that the period of high birth-rates is immediately followed by a period of high mortality-rates; such apparent 'costs of reproduction' will be considered further in Chapter 6.

1.4.3 Resumé

Conventional (as opposed to diagrammatic) life-tables are the medium through which age-specific schedules of death (and birth) can be constructed, and it is obvious that the compilation of such information is vital if the dynamics of populations are to be understood. These life-tables can be of two quite separate types (Fig. 1.7).

The *fixed cohort* (or 'dynamic', or 'horizontal') life-table is derived by actually following a cohort of individuals from birth to extinction. It provides reliable information on *that* cohort; but its construction may be beset by practical difficulties, which in certain cases will be insuperable.

The *static* (or 'time-specific', or 'vertical') life-table, on the other hand, is derived by estimating the age structure of a population at one point in time. It is equivalent to a fixed cohort life-table only when the survival rates in the population are constant. Otherwise the static life-table compounds and confuses two quite separate things: the age-specific changes in birth- and mortality-rate, and the year-to-year variations in these rates in the past. Nevertheless, it can provide a general idea of age-specific birth- and mortality-rates in a population, which is particularly valuable when a fixed cohort life-table *cannot* be derived.

It should also be stressed that in either case it is often necessary to collect life-table data over a period of time for a number of generations. This allows the natural variability in the rates of birth and survival to be monitored and assessed.

1.5 Some generalizations

One of the reasons for using life-tables to monitor these age-specific rates is that this allows us to discover patterns of birth and mortality which are repeated in a variety of species in a variety of circumstances. In turn, this allows us, hopefully, to uncover the common properties shared by these various populations, leading ultimately to a deeper understanding of population dynamics *in general*. Age-specific mortality-rates have been classified by Pearl (1928), and his classification is illustrated in Fig. 1.14 in the form of survivorship curves. It is very difficult to generalize about the shape of survivorship curves, partly due to the continuing paucity of data, but Pearl argued that we can recognize three broad types. The first—epitomized, perhaps, by humans in the developed world or cosseted animals in a zoo—describes the situation in which mortality is concentrated at the end of the maximum life-span. In the second, the probability of death remains constant with age, leading to a linear decline in survivorship; this may well apply to the buried seed bank of many plant populations. In the

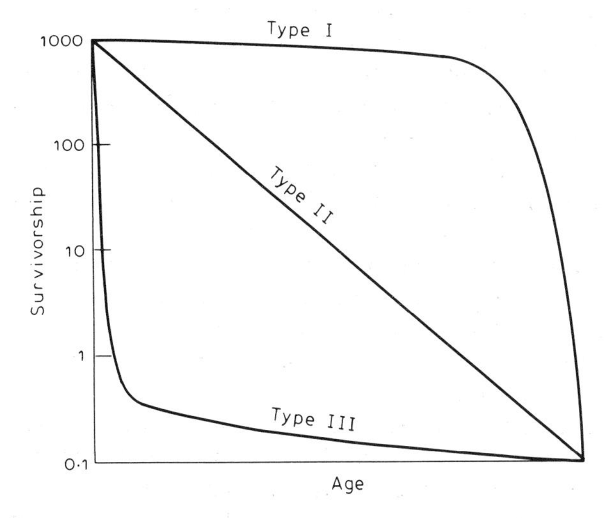

Fig. 1.14. Hypothetical standard survivorship curves. (After Pearl 1928.) For further discussion, see text.

third type there is extensive early mortality, but those that remain have a high rate of survival subsequently. This is true, for instance, of many marine fish which produce millions of eggs of which very few survive to become adults.

The difficulty with Pearl's generalizations is that as a cohort ages it may well follow, successively, more than one type of survivorship curve. It is now known, for instance, that for many grassland plants the survivorship curve of seedlings establishing into adults is type III, whereas that of the adults themselves is type II. Nevertheless, these patterns of survivorship will be of value to us when we consider 'life-history tactics' in Chapter 6.

Generalizations regarding age-specific birth-rates are, in many ways, more straightforward. The most basic distinction, perhaps, is between species which are *semelparous*, reproducing only once, and those which are *iteroparous*, reproducing many times. In either case there is likely to be a *pre-reproductive period*, which can, of course, vary in length (cf. Figs 1.13 and 1.15). Age-specific fecundity may then rise, either to a peak (*P. annua*, Fig. 1.10), or to a plateau (the deer in Fig. 1.15). As Fig. 1.13 shows, however, many species combine elements of the two in a more complex pattern. Further consideration of these patterns is reserved for Chapter 6.

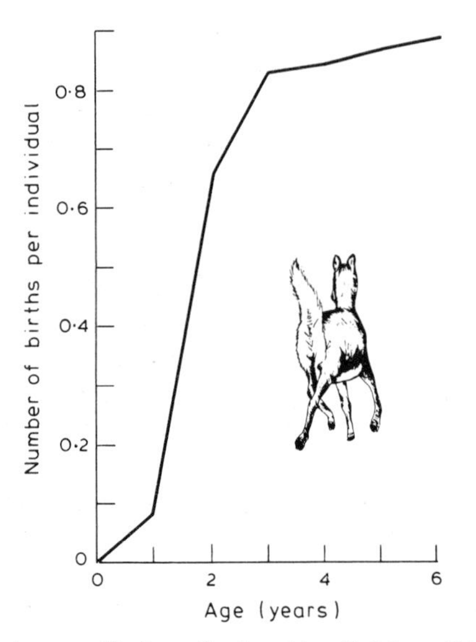

Fig. 1.15. Age-specific fecundity for whitetailed deer, *Odocoileus hemionus* in Michigan. (Data from Eberhardt 1960.)

Finally, certain restricted generalizations can be made regarding population size itself. On the one hand, it is indisputable that all populations vary in size: temporal fluctuations are the universal rule. But at the same time it is equally true that these fluctuations are usually limited in amplitude. Populations rarely increase in size so much that they utterly overrun their environment; and even localized extinctions, though by no means unknown, are also comparatively rare. Thus, population size, and the processes affecting it, are variable—but of limited variability. One of the major aims in the study of population dynamics is to understand these limitations, and this is the topic to which we shall turn in the next chapter. Before doing so we must first consider the implications of the ways in which some plants and animals grow, to our means of describing populations.

1.6 The modular growth of organisms

A major distinction amongst species of both plant and animal kingdoms lies in the organization and differentiation of tissues. This fundamental distinction divides organisms according to their growth form into those that are *unitary* and those that are *modular* (Harper & Bell 1979; Harper 1981). Most animals are unitary organisms. Development from the zygote (the fertilized egg) through to the adult involves an irreversible process of growth and tissue differentiation leading to organ development according to a highly regulated schedule. Most plants on the other hand are modular. Growth and differentiation are normally initiated in 'meristems' at the apices of shoots and roots (Esau 1953). Cell divisions occur in these meristems, and they result in root and shoot elongation and the laying down of further meristems. Growth from meristems in this way unaccompanied by any further differentiation leads to a repetitive or reiterative modular structure in the plant body (see Gottleib (1984) for a succinct discussion). Botanically, a 'module' is an axis (essentially a length of tissue) with an 'apical meristem' at its distal end. The axis is subdivided by nodes at which leaves, axillary meristems and vegetative outgrowths (e.g. tendrils) may occur. If and when the apical meristem differentiates into a terminal flower, extension growth of the axis ceases.

Modular organisms increase in size by a programme of growth and development that is structurally and functionally repetitive; unitary organisms by contrast do not. The distinction, however, is not simply one between animals and plants. In colonial animals such as corals, hydroids, bryozoans and colonial ascidians similar modular identities can be seen (Rosen 1979). We saw in Section 1.3.5 that both corals and buttercups increased in size through the addition of successive segments or 'modular units'. Thus, put simply, the buttercup becomes larger as additional stolons and ramets are produced while the coral becomes larger as polyps grow and bud. Each increment in growth can be measured by the number of modules produced in a period of time.

Three important demographic consequences arise from recognizing the modular construction of higher terrestrial plants and colonial animals. The first is that the addition of modules tends to lead to a branched structural form. Generally this is because of the placement of meristems in plants at acute angles to the main axis, which continue growth when extension of the parental axis ceases. The exact architecture of the organism will depend on: (i) whether modules vary in form, as in the short and long branches of trees or the vegetative and generative (i.e. reproductive) polyps of hydroids, (ii) their rate of production, and (iii) their position relative to one another. Nevertheless, the overall form of the organism is a *colony* of repeating modules. The form is important demographically since size and shape will influence the nature of interactions amongst static organisms (Horn 1971).

Secondly, removal of modules (through damage, or herbivory or predation for instance) may harm an organism but not kill it. A modular architectural arrangement of relatively autonomous meristems allows lost parts to be reiterated. The potential for this very much depends upon the degree of permanent physiological and morphological differentiation that has already occurred. Removal of vegetative tillers from a grass plant will often lead to the reiteration of the tiller module; removal of an inflorescence (a model that has differentiated into a sexually generative structure) may not—often because of the absence of further growth points. In unitary organisms, on the other hand, although tissue regeneration does occur, removal of a whole organ will precipitate death.

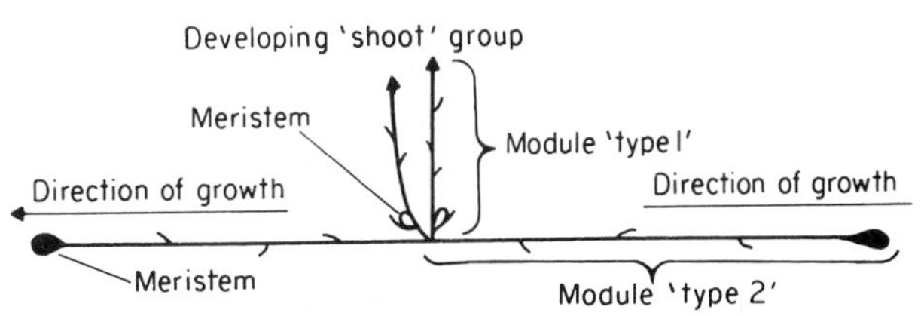

Fig. 1.16. A schematic view of modular growth in a plant that displays two types of module in its constructional organization. Natural cloning results from the fragmentation of either type of module. Module 'type 1' is a 'shoot' bearing foliage leaves and one meristem capable of generating either 'type 1' or 'type 2' modules. Module 'type 2' is a horizontal stem (rhizome or stolon) bearing a meristem with the potential to give rise to a 'type 1' module, axillary meristems and roots. (after Harper & Bell 1979).

A third consequence arises from the fact that modularity affords the opportunity of natural cloning (Harper 1984). Natural cloning arises when a genet fragments and the ramets establish into physiologically independent parts (Fig. 1.16). This is only possible when meristems at nodes retain totipotency: the ability to produce both shoots and roots. Fragmentation may arise through physical agencies (e.g. sand movement in the sand sedge *Carex arenaria*), the trampling and grazing of herbivores (in rhizomatous grasses), or it may be genetically determined (as in *Ranunculus repens* (Bell & Tomlinson 1980)). The important demographic point is that whatever the agency, it may lead to a colony of physiologically independent plants of the same genotype which are potential competitors.

These observations on modularity have prompted the suggestion that the fundamental equation of population biology (1.1) applies not only at the level of the genet (expressed through the growth of the clone as a whole) but also at the (lower) modular level (Harper 1977). Harper & Bell (1979) argue then that the study of the dynamics of modules themselves is an essential component in describing the population ecology of modular organisms. Demographic approaches to modular dynamics have employed the same techniques that we have examined earlier for populations of unitary organisms. Thus, Fig. 1.17 shows a diagrammatic life-table for a population of meristems on a *Fuchsia* plant. As the plant grows, some meristems develop into shoots which may be vegetative (branches bearing new meristems) or generative (branches bearing inflorescences), some meristems remain dormant, whilst others abort. The transitions given in Fig. 1.17 are those occurring during

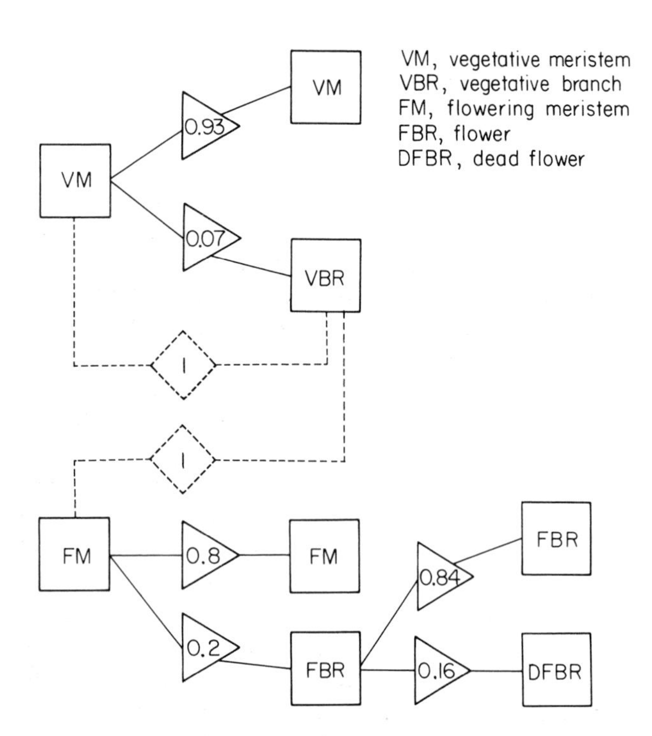

Fig. 1.17. Transitions occurring over a time period amongst meristems in Royal Velvet, a cultivar of *Fuchsia*. Measurements were made 75 days after planting on the main shoot when plants were growing exponentially. At this stage of growth a constant fraction of vegetative meristems become vegetative branches which in turn produce one vegetative and one flower meristem. (Data from Porter 1983a.)

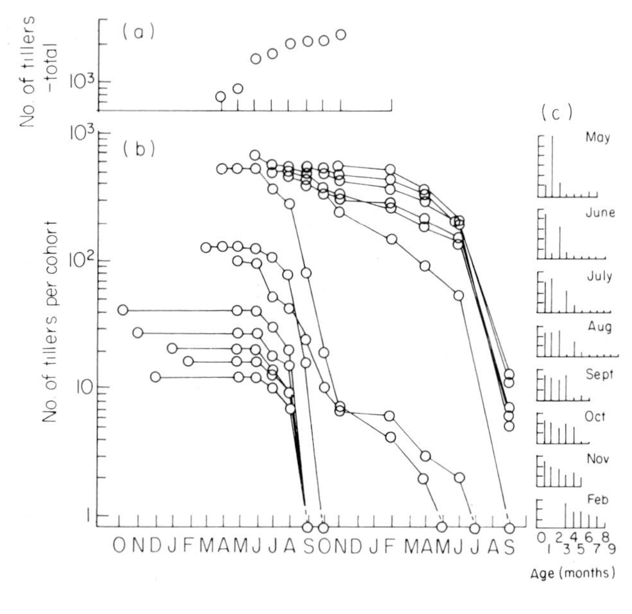

Fig. 1.18. Flux in tillers in *Phleum pratense*: (a) the total number of living tillers; (b) survivorship curves of successive monthly cohorts of tillers; (c) tiller age structure at successive monthly periods (each ordinate division is 10%). Data were gathered from forty plants each grown from seed in soil in separate 20-cm diameter pots. (from Langer 1956, after White 1980).

growth in an unrestricted environment once plants have started flowering. Death of meristems, whether vegetative or flowering, is absent and only occurs when flowers (the products of generative meristems) senesce.

Flux in modules, however, can be much more noticeable in other species. Langer (1956) followed the fate of grass tillers on individual timothy plants (*Phleum pratense*) over a period of two years. His data (Fig. 1.18) illustrate that tiller births *and deaths* are an intrinsic feature of the life of an individual plant. Whilst the size of cohorts recruited each month was seasonally dependent, the pattern of mortality in cohorts was remarkably similar, following a type 1 survivorship curve. This mortality mainly resulted from the 'monocarpic' nature of tillers (the production of an inflorescence on a tiller is followed inevitably by tiller death) but also occurred amongst non-generative tillers when recruitment of tillers was at its highest in June/July. Tillers formed in April and May were either annual (flowering in the following August and September) or biennial, remaining vegetative over winter and flowering the next year. This resulted in the stepped survivorship curves seen in Fig. 1.18.

In conclusion then it is clear that in describing a population we must carefully define the individual. This in part will depend on the nature of the scientific enquiry, but more often than not it will be determined by the growth form of the organism.

Chapter 2 Intraspecific Competition

2.1 The nature of intraspecific competition

In order to examine further the way in which the properties of individuals determine population dynamics, we will have to consider a proposition which we have not mentioned explicitly so far, but which is generally taken for granted: that each individual within a population affects, and is affected by, other individuals within the population. Consider, for instance, a thriving population of grasshoppers (all of the same species) in a field of grass. Adult males attract and court adult females by 'stridulating': they rub the insides of their hind legs against the outsides of their hardened forewings to produce a species-specific 'song'. If a male manages to attract, court and inseminate a female, he will have made some contribution to the next generation; and the more females he manages to inseminate, the greater this contribution will be. The most successful or *fittest* males within a population are those which make the greatest contribution. A solitary male amongst many females in the population might eventually inseminate every one of them. But if there are several males in the population, then they will be *competing* with one another for the females' attentions, and each will inseminate fewer females than he would have done had he been alone. The more males there are, the more intense this intraspecific competition will be; and the general effect will be to reduce the males' contributions to the next generation.

Subsequently the inseminated grasshoppers will have eggs to lay. For this they require bare soil, which may, in a grassy field, be quite rare. More to the point, they require bare soil not already occupied by another female. They can increase their contribution to the next generation by increasing the number of eggs they lay. But the more *competing* females there are, the longer it will take each one to find an appropriate site, and the fewer eggs she will lay per unit time. Moreover, along with this increased expenditure of time will go an increased expenditure of energy. This will lead to a decrease in the energy available for egg-development, and also a decrease in general viability, leading to a possible shortening of total lifespan. These, in their turn, will lead to a decrease in the number of eggs laid; and the more competing females there are, the greater this decrease will be.

Of course, in order to live, grasshoppers (male and female) must consume food (grass) to provide themselves with energy, but they must also expend energy in the process of finding and consuming the food. Each grasshopper will frequently find itself at some spot where there had previously been a palatable blade of grass—before, that is, some other grasshopper ate it. Whenever this happens the first grasshopper must move on; it must expend more energy than it would otherwise have done before it takes in food. Once again, this increased energy expenditure will lead on the one hand through increased mortality, and on the other hand through decreased rates of development, to a decreased contribution to the next generation. So the more competitors there are, the more 'moving on' each grasshopper will have to do, and the greater the decrease in contribution will be.

Considering the same hypothetical ecosystem, we can turn now to the grass itself. (We will assume, for simplicity, that it is all of one species, although in practice this is very unlikely to be so.) The contribution of an individual grass plant to the next generation will be dependent on the number of its progeny which eventually develop into reproductive adults themselves. An isolated seedling in fertile soil will have a very good chance of developing to reproductive maturity, and will also be likely to reproduce vegetatively, consisting (as a result) of multiple copies of the simplest plant form. However, a seedling which is closely surrounded by neighbours (shading it with their leaves and depleting its soil with their roots) will be very unlikely to survive at all, and will almost certainly be small and simple. The more competing individuals there are, the more likely it is that seedlings will find themselves in the latter, rather than the former, situation. Increases in density will, therefore, lead to decreases in the contributions of individuals to the next generation.

2.2 Three characteristics of intraspecific competition

Certain common features of intraspecific competition have obviously emerged. The first of these is that the *ultimate effect* of competition is a decreased contribution of individuals to the next (or, in fact, to all future) generations; a decrease, that is, from the potential contribution that the individual would have made had there been no competitors.

In some cases—stridulating males competing for females, for instance—the connection between competition and contributions to future generations is obvious and direct. With grass seedlings competing for growth resources, however, or with grasshoppers competing for food, the connection is slightly less direct, since competition leads to decreases in survivorship and/or fecundity. Nevertheless, in terms of ultimate effects, male grasshoppers competing for females, seedlings competing for light and grasshoppers competing for food are all essentially equivalent. Intraspecific competition acts more or less directly on either survivorship or fecundity, or on both, but in all cases it decreases contributions to future generations.

From a practical point of view, however, this is not quite enough. Competition must not only be likely; it must also manifest itself in measurable decreases in survivorship, fecundity or some other, less direct characteristic. Only then have we the right to conclude that it is occurring.

The second common feature of intraspecific competition is that the resource for which the individuals are competing must be in limited supply. Oxygen, for instance, although absolutely essential, is not something for which grasshoppers or grass plants need compete. Nor *necessarily* is space, food or any of the other resources we have discussed so far. They are only competed for if they *are* in limited supply.

The third feature of intraspecific competition is reciprocity. In other words the competing individuals within a population are all essentially equivalent (in contrast to the situation of a predator eating its prey, in which the predator is inherently the inflictor of the adverse effect and the prey inherently the receiver.) Of course, in any particular case intraspecific competition may be relatively one-sided: the strong early seedling shading the stunted late one; the 'resident' egg-laying grasshopper causing the later arrival to move on. However, despite this, because the early/late or resident/non-resident roles might easily be reversed, the competing individuals are *inherently* equivalent.

2.3 Density-dependence: a fourth characteristic

The fourth and final feature of intraspecific competition is that the effect of competition on any individual (i.e. the *probability* of an individual being adversely affected) is greater, the greater the number of competitors there are. The effects of intraspecific competition are, therefore, said to be *density-dependent*. Not surprisingly, they can be contrasted with density-*independent* effects, yet the point of contrast is very often confused. This can be avoided by reference to Fig. 2.1a–d (after Solomon 1969). In Fig. 2.1a the *number* of deaths is dependent on density (and, indeed, increases with density) in each of the four lines, but of these four, only three show density-dependent effects. In the fourth, the *proportion* of the population dying (or the *probability* of an individual dying) remains constant, even though the number dying increases with density: the *rate* of mortality is density-independent. Fig. 2.1b, which portrays precisely the same situations as Fig. 2.1a, makes this abundantly clear. Of course, in reality, the points leading to the density-independent plots will not all lie exactly on the straight lines. They may, in fact, be very widely spread on either side of them. However, this would not alter their essential feature: with density-independent mortality, there is *no tendency* for the mortality-*rate* to increase with increasing density. The analogous situation for fecundity is illustrated in Figs 2.1c and 2.1d.

Intraspecific competition and density-dependence are obviously bound closely together; whenever there is intraspecific competition, its effect—whether on survival, fecundity or a combination of the two—is density-dependent. However, not all density-dependent effects

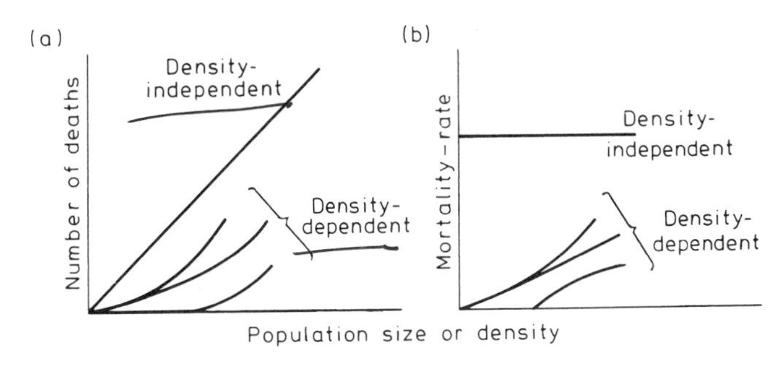

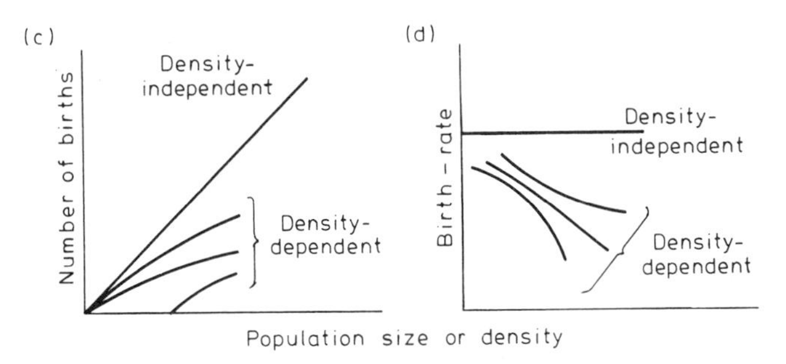

Fig. 2.1. Density-dependent and density-independent death (a) and (b), and birth (c) and (d). The vertical axes in (a) and (c) are numbers; those in (b) and (d) are rates. (After Solomon 1969.)

are the result of intraspecific competition. Chapter 4 on interspecific competition, and Chapter 5 on predation, parasitism and herbivory will make this quite clear. Nevertheless, all density-dependent effects do share a *tendency to regulate population size*.

We have already suggested (in Chapter 1) that such regulation is extremely widespread, and the subject will be examined in greater depth in Chapter 7 (Population Regulation); but three pertinent comments can be made now. First, 'regulation' refers to the ability to decrease the size of populations which are above a particular level, but to allow an increase in the size of populations below that level. This particular population level will, therefore, be a point of equilibrium. Populations below it increase, populations above it decrease, and populations actually on it neither increase nor decrease: population size is subject to negative feedback (Fig. 2.2). In the case of the effects of intraspecific competition, this equilibrium level is often called the 'carrying-capacity' of the population. In reality, however, no single carrying-capacity can ever characterize a natural population: most aspects of its environment are far too variable, and its own behaviour is never wholly predictable. For this reason 'regulation' may, more reasonably, be taken as the ability to act on a very wide range of starting densities, and bring them to a much narrower range of final densities.

Second, the word 'tendency' is used advisedly. If a density-dependent effect is not operative at all densities, or is not operative under all environmental conditions, is weak, or happens after a time-delay, then the effect—although density-dependent—may not *actually* regulate population size. Similarly, if there are several density-dependent factors acting on a population, then each factor, alone, may be incapable of regulating the population, even though each *tends* to do so.

Third, all density-dependent *effects* are the result of a density-dependent *factor* acting through a density-dependent *process*. Until now only one density-dependent process has been considered (intraspecific competition), invoked by various density-dependent factors: food, space and so on. It should be remembered, however, that the effects of intraspecific competition can easily be discussed without specifying the factor involved; and these effects—on either mortality or fecundity—can even be measured. Yet if, ultimately, we are to understand the dynamics of a population, we must identify the density-dependent factor itself.

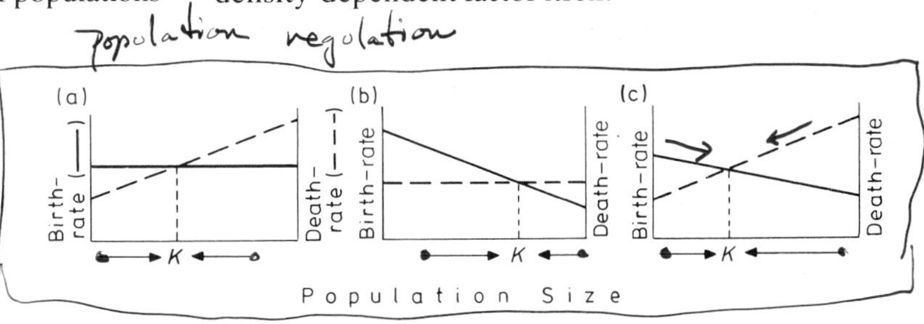

Fig. 2.2. Population regulation with (a) density-independent birth and density-dependent death, (b) density-dependent birth and density-independent death, and (c) density-dependent birth and density-dependent death. Population size increases when birth-rate exceeds death-rate below the carrying capacity, K; and decreases when death-rate exceeds birth-rate above K. K is, therefore, a stable equilibrium.

2.4 Scramble and contest

The density-dependent effects of intraspecific competition are of central importance in the dynamics of natural populations, but so far they have only been outlined. We have still to describe the precise effects that intraspecific competition can have on the quantity *and* quality of individuals within a population. Of course, competitive interactions do not all conform to precisely the same pattern. On the contrary, there is a whole spectrum of interactions, varying in their underlying biological causes, and in their effects on the quantity and quality of individuals; but in order to appreciate this variety, it will be useful to have certain standards against which actual examples can be matched. The most appropriate standards are the extreme forms of competition described by Nicholson (1954b): 'scramble' and 'contest'.

The essential features of scramble and contest are illustrated in Figs 2.3 and 2.4 respectively (adapted from Varley *et al.* 1975). It is particularly important to note that Figs 2.3b and 2.4b make use of the 'k-values' described in Chapter 1. In both cases there is no competition at all at low densities: all individuals have as much resource as they need, and all individuals need and get the same amount (% mortality = 0; $A = B$; $k = \log_{10} B/A = 0$). Above a threshold density of T individuals, however, the situation changes. In scramble competition (Fig. 2.3), all the individuals still get an equal share, but this is now less than they need, and as a consequence they all die. The slope, b, of Fig. 2.3b therefore changes suddenly from 0 to ∞ as the threshold, T, is passed. In contest competition (Fig. 2.4), on the other hand, the individuals fall into two classes when the threshold is exceeded. T individuals still get an equal and adequate share of the resource, and survive; all other individuals get no resource at all, and therefore die. There are always just T survivors in contest, irrespective of the initial density, because mortality compensates exactly for the excess number of individuals. In Fig. 2.4b

Fig. 2.3. Scramble competition. Mortality relationships (a) in terms of numbers surviving and percentage mortality, (b) in terms of k plotted against the logarithm of density.

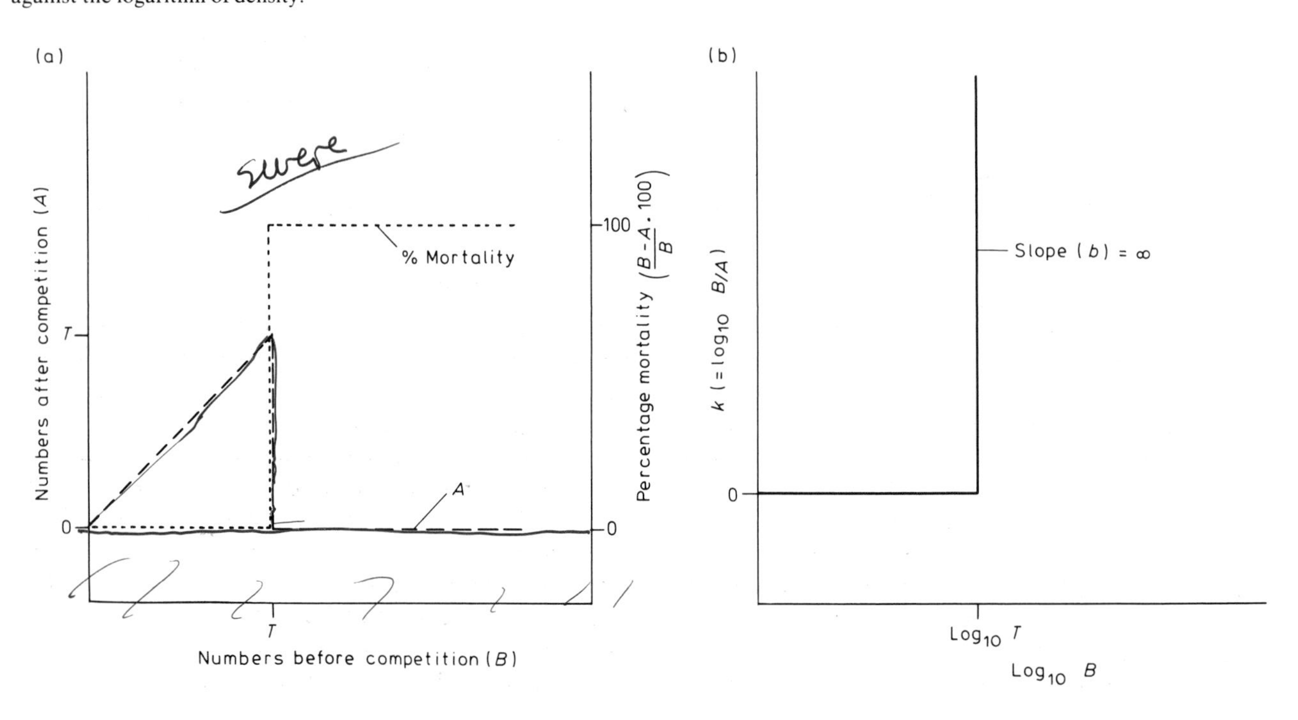

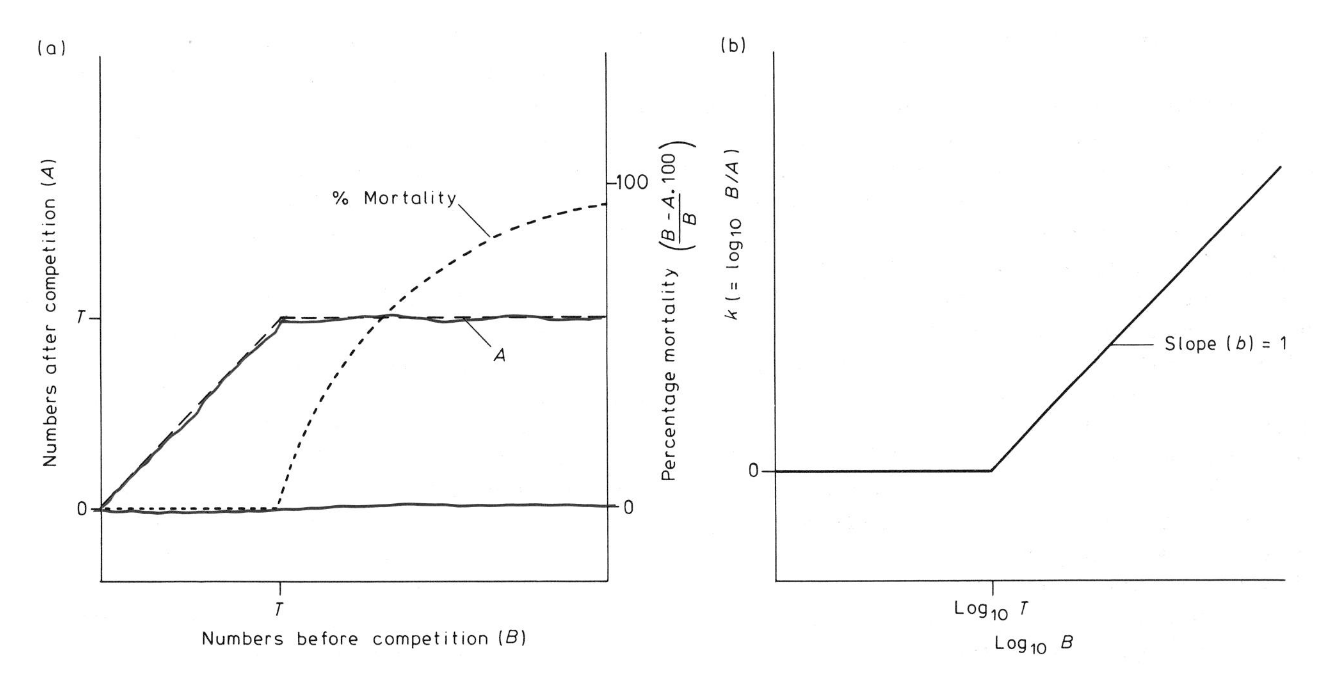

Fig. 2.4. Contest competition. Mortality relationships (a) in terms of numbers surviving and percentage mortality, (b) in terms of k plotted against the logarithm of density.

the slope changes at threshold from 0 to 1; *this b-value of 1 is indicative of the exact compensation of contest competition.*

Scramble and contest can also be seen in terms of fecundity. Below the threshold there is no competition, and all individuals produce the maximum number of offspring. Above the threshold, scramble leads to the production of no offspring whatsoever; while contest leads to T individuals producing the maximum number of offspring and the rest producing none at all.

Our 'standards' of competition have now been defined. In terms of the quantity of individuals, neither scramble nor contest are effective below some threshold; but above this threshold scramble reduces numbers to zero, while contest maintains a numerical constancy. In terms of quality, both scramble and contest allow only two classes of individual: those getting all they require and therefore surviving (or producing the maximum numbers of offspring), and those getting less than they need and therefore dying (or producing no offspring at all). The difference is that in scramble all individuals move suddenly from the first class to the second at threshold; while in contest there are still T individuals in the first class, even when threshold is exceeded.

2.5 Actual effects of intraspecific competition

2.5.1 Palmblad's data

Having described these hypothetical extremes, it is appropriate to examine some actual examples. Studies of plants indicate the inadequacies of 'scramble' and 'contest' particularly clearly.

Palmblad (1968) undertook an experimental study on the effects of intraspecific competition on several species of weed. Some of his results are summarized in Table 2.1 and Fig. 2.5. They refer to two annual species: *Capsella bursa-pastoris* (Shepherd's purse) and *Conyza canadensis* (Canadian fleabane) and the perennial, *Plantago major* (plantain). Palmblad's procedure was simply to sow seeds of each species under controlled conditions at a range of densities (1, 5, 100 and 200 seeds per pot), and then keep a careful record of their subsequent progress. As Table 2.1 shows, he was able to compute the percentage of seeds that actually germinated and produced seedlings ('germination' in Table 2.1), the percentage that subsequently died before setting their own seed at the end of the summer, and the percentage that remained alive but failed to reproduce ('vegetative' in Table 2.1). The number of seeds produced by each individual was then counted, and the 'mean number of seeds per reproducing individual' and the 'total number of seeds per pot' computed. Finally, the total dry weight of plants, both reproductive and vegetative, was measured for each pot.

The results are also summarized in Fig. 2.5, where the k-values of the various processes are plotted against the $\log_{10}$ of the sowing density (as they were in Figs 2.3b and 2.4b). $k_{germination}$, $k_{mortality}$ and $k_{vegetative}$ are all self-explanatory; $k_{fecundity}$ refers to the reduction in the number of seeds produced per individual, $\log_{10}$ (max. seeds/actual seeds); while k_{total} refers to the reduction in the total number of seeds produced, but is also the sum of all the other k-values.

The first point to note is that, almost without exception, the fifteen plots in Fig. 2.5 show k increasing with density. The density-dependent nature of the various responses of these plants to intraspecific competition is, therefore, immediately confirmed. It is also apparent, however, that the sudden threshold, characteristic of scramble and contest, is generally lacking in these real examples. Instead, as density increases, the slopes tend to increase gradually. This, not surprisingly, is characteristic of many real examples: as density increases, so the *intensity* of competition increases. One reason for this is that, because the plants were not spaced with total regularity, different plants experienced different degrees of crowding. Another is that the plants themselves were inherently (i.e. genetically) different. Other reasons will soon become apparent.

In all three species, intraspecific competition exerted its density-dependent effects on the proportions germinating, surviving and remaining vegetative, and in each case the plants fell, as a result, into one of two categories: those that 'did', and those that 'did not'. With reproduction, however, the situation was far more complex. The density-dependent effects of intraspecific competition were no less obvious, but the response was very far from being all-or-none. Instead, the mean number of seeds produced per individual varied continuously throughout an almost 200-fold range in *Plantago major*, and an approximately 100-fold range in *Capsella bursa-pastoris* and *Conyza canadensis*. This plasticity of response—admittedly exemplified by seed production in plants—is common throughout the plant and animal kingdoms. Intraspecific competition leads not only to quantitative changes in the numbers surviving in populations, but also to qualitative changes in those survivors; and these progressive decreases in quality as density increases contribute significantly to the increasing intensity of competition.

In Palmblad's experiments these qualitative changes were not confined to average seed production. Despite the considerable variation in the density of surviving plants, the total dry weight for each species, after an initial rise, remained remarkably constant with increasing density. In other words, at higher densities individual plants were smaller. There was 'compensation' so that the final 'yield' remained largely unchanged. This plasticity in the growth response of plants to intraspecific competition is so common that a term has been coined to describe the consequences of it: 'the law of constant final yield' (Kira *et al.* 1953).

Of course, the qualitative changes in dry weight and seed production are closely connected: smaller plants produce fewer seed. This, as Table 2.1 shows, leads to a comparative constancy in the total number of seeds produced. Thus, the regulatory tendencies of intraspecific competition are amply illustrated; despite a 200-fold range of sowing densities, the range of seed output is only 1.4 in *Capsella bursa-pastoris,* 2.9 in *Plantago major* and 1.7 in *Conyza canadensis.* That such regulation does indeed occur is illustrated in another way in Fig. 2.5. Remember that in contest competition there was an absolute constancy of output illustrated by a slope (b) of 1. In Fig. 2.5 the slopes of the three graphs for k_{total}, taken over the whole range, are also close to 1, indicating the near-constancy of output already noted. The resemblance to contest, however, is only superficial. In contest, constancy is achieved by a constant number of survivors all producing the same number of

Table 2.1 Aspects of Palmblad's (1968) experimental data on intraspecific competition in three species of weed

	Capsella bursa-pastoris					*Plantago major*					*Conyza canadensis*				
Sowing density	1	5	50	100	200	1	5	50	100	200	1	5	50	100	200
% germination	100	100	83	86	83	100	100	93	91	90	100	87	56	54	52
% mortality	0	0	1	3	8	0	7	6	10	24	0	0	1	4	8
% reproducing	100	100	82	83	73	100	93	72	52	34	100	87	51	42	36
% vegetative	0	0	0	0	2	0	0	15	29	32	0	0	4	8	8
Dry weight (g)†	2.01	3.44	4.83	4.51	4.16	8.05	11.09	13.06	13.74	12.57	12.7	17.24	17.75	16.66	18.32
Mean number of† seeds/reproducing individual	23 741	6102	990	451	210	11 980	2733	228	126	65	55 596	13 710	1602	836	534
Total number of seeds	23 741	30 509	40 311	37 196	30 074	11 980	12 670	8208	6552	4420	55 596	59 625	40 845	35 264	38 376
$k_{germination}$	0	0	0.08	0.07	0.08	0	0	0.03	0.04	0.05	0	0.06	0.25	0.27	0.28
$k_{mortality}$	0	0	0.01	0.02	0.04	0	0.03	0.03	0.05	0.13	0	0	0.01	0.03	0.08
$k_{vegetative}$	0	0	0	0	0.02	0	0	0.08	0.19	0.29	0	0	0.03	0.08	0.08
$k_{fecundity}$	0	0.59	1.38	1.72	2.05	0	0.64	1.72	1.98	2.27	0	0.61	1.54	1.82	2.02
k_{total}	0	0.59	1.47	1.81	2.19	0	0.67	1.86	2.26	2.74	0	0.67	1.83	2.20	2.46

†These data are averages of three individual pot values.

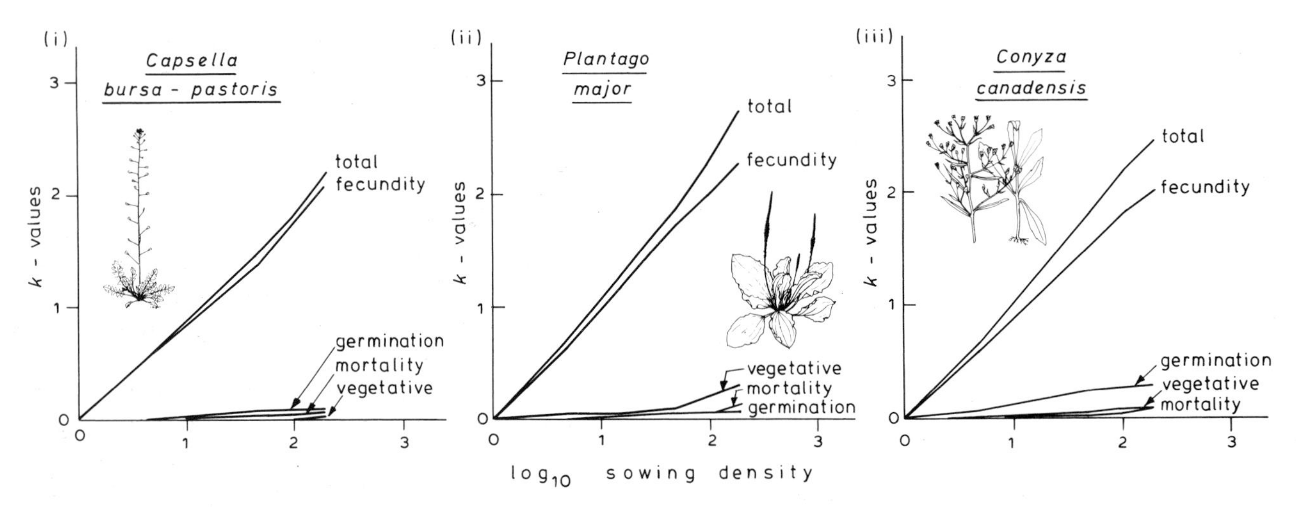

Fig. 2.5. The varied effects of intraspecific competition: experiments on populations of three species of weed. (Data from Palmblad 1968). For further discussion, see text.

offspring. In Palmblad's experiments, as in real examples generally, the near-constancy of output was achieved by a variable number of survivors producing a variable number of offspring.

Looking at these graphs more closely, we see that at lower sowing densities—below 50 in *Capsella,* and below 5 in *Conyza* and *Plantago*—the slopes of the k_{total} graphs are, in fact, less than 1. This indicates *under*-compensation (i.e. less than the *exact* compensation of $b = 1$); although there is a reduction in individual output, this is not enough to compensate for the increasing density, and the total output increases (Table 2.1). At higher densities, on the other hand (with the exception of *Conyza* between 100 and 200), there is *over*compensation; the reduction in individual output more than compensates for the increased density, and total output decreases. This is indicated by a slope greater than 1. In each species, therefore, the yield of seed reaches a peak towards the middle of the density-range. Its precise position is indicated by the point on the graph where the slope equals 1 exactly. Thus, not only do the graphs in Fig. 2.5 show the degree of compensation resulting from competition, they also indicate the sowing density that would maximize the final yield of seed. For commercially valuable crops, such graphs may be of considerable importance.

Finally, it is apparent from Fig. 2.5 that the relative importance of germination, mortality, etc., in regulating output is different in the three species. This is made particular clear by the fact that k_{total} (the total effect) is the sum of all the other k-values. In *Capsella bursa-pastoris* the effects of competition are almost entirely on the growth, and therefore the seed-production, of surviving individuals. In *Conyza canadensis* seed production is also of primary importance, but a substantial proportion of the total effect is the result of reduced rates of germination. In *Plantago major,* on the other hand, the tendency to remain vegetative and (to a lesser extent) mortality play an important role; this is no doubt associated with the perennial habit. Thus, although the end-results are similar in the three species, the ways in which they are achieved are rather different.

In summary, then, we have learnt a great deal that is of general relevance from this limited example. Typically, intraspecific competition affects not only the quantity of survivors, but their quality as well, which becomes more and more affected as density increases. This, combined with the variability of both environment and individuals, means that there is usually no sudden threshold for competition in nature. Rather, it increases gradually over an extended range. Palmblad's experiment also reiterates that the ultimate effect of intraspecific competition, acting through survival and fecundity, is on the contributions to future generations;

that individuals are affected reciprocally; that intraspecific competition tends to regulate populations; and that the effects can be measured without the unequivocal identification of the resource in limited supply.

2.5.2 *Competition in plants: a deeper look*

Palmblad's experiments also illustrate two important inter-linked events that can occur when plant populations are grown in resource-limited environments. The first is that the size of individuals (as measured for instance by seed production) is reduced; the other is that ultimately mortality may ensue. These events occur along a spectrum of competitive effects on the growth of individual plants. The extremities are the death of the individual (growth and maintenance ceased) and unconstrained growth (growth rate = maximum for that environment). The 'in between' is a reduced growth rate (< maximum) which is reflected at some point in time (harvest of the plant) in a reduced plant size.

Evidence for the density-dependent effects of competition can be examined at various levels: at the level of the population itself, amongst individuals within the population, and within individual plants. We will follow this progression, which is, perhaps not surprisingly, the way our scientific understanding has historically developed.

Fig. 2.6 shows the two general forms of 'yield–density' relationship that have emerged from the wealth of studies conducted by agronomists and ecologists. The figures relate population density *before* the action of competition—often the number of plants sown as seed or planted—to population density (yield) *after* the action of competition, always on a constant area basis. This yield may be measured in a variety of ways either as a direct fitness component (e.g. seed produced) or less directly as biomass, either of the total plant or some of its constituent parts. The form of response is either (i) *asymptotic* (Fig. 2.6a) where yield per unit area levels off with increasing density (i.e. perfect compensation; constant final yield) or (ii) *parabolic* (Fig. 2.6b), where a maximum yield is reached at an intermediate density before falling at high densities (i.e. overcompensation). From these examples, we can see (as we might expect) that the influence of adding more resources for growth (by fertilizer) increases the size of the population, measured as the height of the asymptote or peak of the parabola.

The asymptotic constant yield response, which develops progressively, can best be explained by looking at the changes in the population with time and the performance of individuals. Soybeans grown over a thousandfold density range show this particularly well (Fig. 2.7a). At sowing (day 0), yield per unit area and density are directly proportional to one another: the yield is the weight of the seed sown! With time this linear proportionality disappears, as plants grow to sizes at which they interfere with one another, the interference occurring first at the highest density. The yield curves therefore display a successively pronounced shoulder as the linear proportionality disappears, but this in its turn is replaced (after 119 days) by a horizontal yield curve when final yield is independent of original population

Fig. 2.6. Illustrative yield–density relationships in plants. (a) *Bromus uniloides* at three nitrogen fertilizer levels (from Donald 1951).

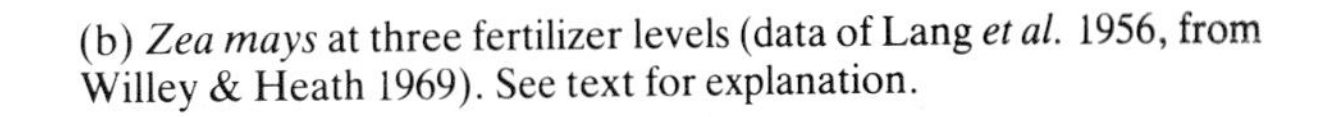
(b) *Zea mays* at three fertilizer levels (data of Lang *et al.* 1956, from Willey & Heath 1969). See text for explanation.

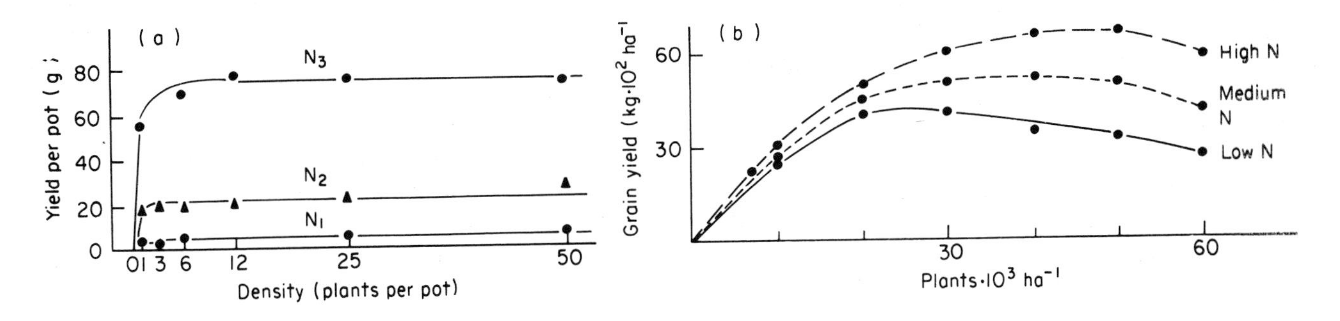

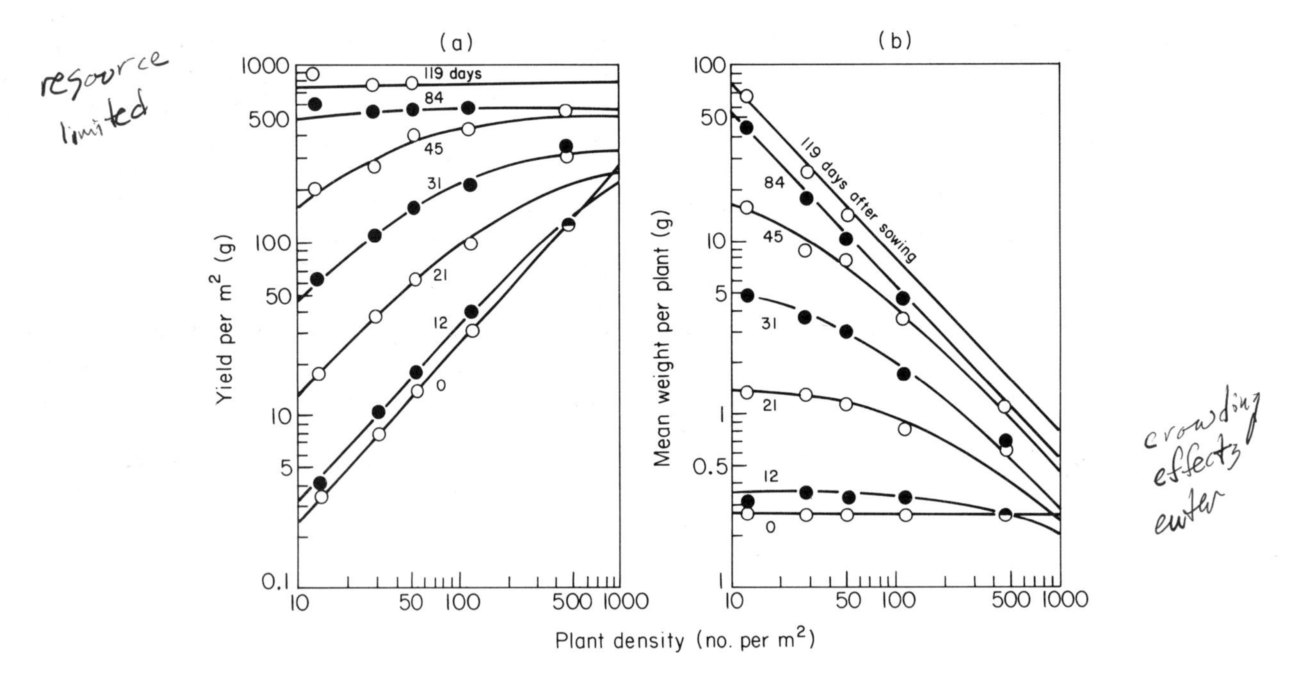

Fig. 2.7. Yield–density relationships in soybean, *Glycine max*: the progressive changes with time (a) on a unit area basis; (b) per plant. (From Shinozaki & Kira 1956, after Harper 1977.)

size. Compensation by individuals on this sort of scale (here 10 and 1000 plants m^{-2} yielding the same biomass after 119 days) is a direct reflection of the enormous *plasticity* that species with modular growth possess. Such plasticity has been shown to occur in a wide range of species including pine trees, grasses and herbs. Re-examining the same data, but from the viewpoint of an average individual soybean plant, shows us that compensation through adjustment of the performance of individuals is indeed occurring. From being initially independent of density (at day 0), the average weight of a plant becomes increasingly related to it (Fig. 2.7b). Plants at low density (10 m^{-2}) achieved a final mean weight of nearly 70 g whereas those in populations of 50 m^{-2} were only 14 g. Moreover, it is important to realize that these adjustments occurred in the absence of mortality.

How then can we explain situations in which the law of constant final yield is not obeyed (e.g. Fig. 2.6b), and yield at high densities declines from an intermediate optimum? This question cannot be fully answered for all cases, but two cogent possible explanations are (i) plant mortality does occur during the course of competition, but the surviving plants are unable to exploit fully resources freed by the death of other members of the population; and (ii) species respond *differentially* according to their developmental stage. To appreciate these explanations we must first give further consideration to two features of plant growth.

The first is that the production of flowers and hence seeds requires the differentiation of floral meristems. These may arise from the permanent conversion of vegetative meristems, or they may be borne laterally on an axis that retains a vegetative meristem at its tip. This difference leads to a dichotomy in growth form. Species may show *indeterminate* growth, bearing flowers at nodes laterally to growing axes and thus retaining the potential for indefinite vegetative extension. Contrastingly, species may show *determinate* growth, where the conversion of vegetative meristems into floral structures prohibits further vegetative development. Conversion may be triggered by changes in daylength and/or temperature (vernalization), or may be genetically fixed.

The second feature that we must consider is how yield is formed. In the case of seed, it requires the production of a range of plant parts, i.e. the *components of yield*: e.g. stems, flower-bearing branches, flowers, pods and seeds. All components may respond to density but as successive components are produced we might expect the type of density response to change.

Populations of cultivated sunflowers (*Helianthus annus*) provide us with an illustration of one type of yield–density response (Fig. 2.8a) in which there is overcompensation. In this sunflower, plants usually bear a single large capitulum (flower head). This develops from the terminal meristem when the plant is young, before the influence of competition has become marked. Thus, the response to density-induced resource limitation cannot be a reduction in capitulum number. Instead it is etiolation of the plant, and reduction both in the proportion of flowers within the capitulum that set mature achenes (seeds) and the size of those achenes. In the field bean (*Vicia faba*) on the other hand (Fig. 2.8b) responses to density are seen at all developmental stages in the production of seeds. Part of the response is a reduction in the number of stems per plant and flowers per stem, and part is the abscission of flowers and pods.

Fig. 2.8. Yield–density responses in: (a) cultivated sunflowers (data replotted from Clements, Weaver & Hanson 1929); (b) field bean (data from Hodgson & Blackman 1956).

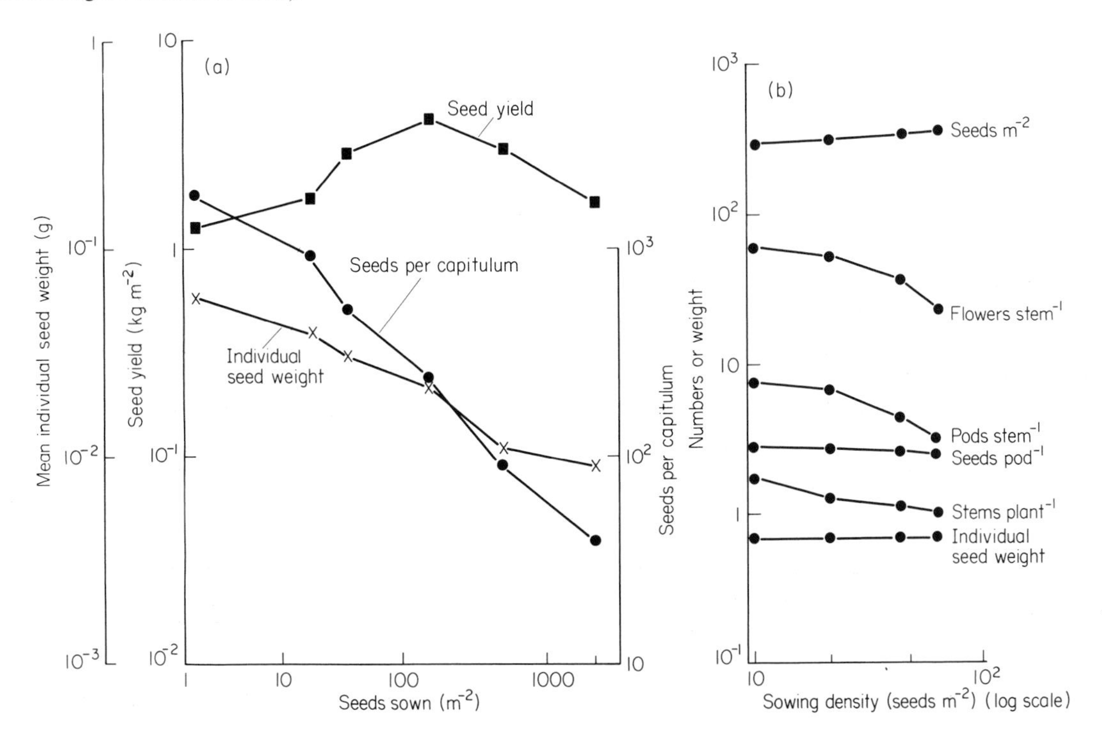

The overall outcome of this plasticity is near constancy in seed population size per unit area. One further contrast may be drawn from this example. In *Helianthus*, individual seed weight declined markedly with plant density, whereas in *Vicia* it remained constant. The former response, however, is the exception to the generally observed phenomenon of homeostasis in seed size in response to density. (Wild progenitors of *Helianthus* absorb density stress by reducing the numbers of branches and capitula, preserving constancy of seed weight; Bradshaw 1965).

Sunflowers and field beans provide us with clear illustrations of the responses of determinate and indeterminate growth forms to intraspecific competition. However, many annual grasses, as exemplified by wheat (*Triticum aestivum*), exhibit a density response that is of a combined form. Seed production in wheat is the product of the number of fertile tillers per plant and the number of grains (seeds) per ear: a fertile tiller bears only one ear which may vary in size and thus in floret and seed number. Prior to flowering and ear formation, density stress is reflected in reduced vegetative biomass and plant parts, but afterwards it is reflected in the size of the ears. This is shown (Fig. 2.9) by a field experiment of Puckridge & Donald (1967) who grew wheat over a thousandfold density range and followed the course of grain yield development. After 14 weeks growth, plants had received stimuli for flowering. At this time the number of tillers per plant was strongly density dependent (Fig. 2.9A). During ear formation (17 weeks) and subsequently through to maturity (26 weeks) the number of tillers changed little and the number of fertile tillers was a constant fraction of the total. However, the effects of density were reflected in changes in ear weight (Fig. 2.9, lines D and E). These two phases in density adjustment, therefore, resulted in a constant individual seed weight at harvest and in an asymptotic yield response per unit area up to 100 plants m^{-2}. The depression of yield at the highest sowing density in this experiment, on the other hand, was a consequence of density-induced mortality (note the end-points of lines C, E and G are displaced to the left). This mortality occurred during flowering, and clearly if it had been greater a more pronounced parabolic yield curve would have occurred.

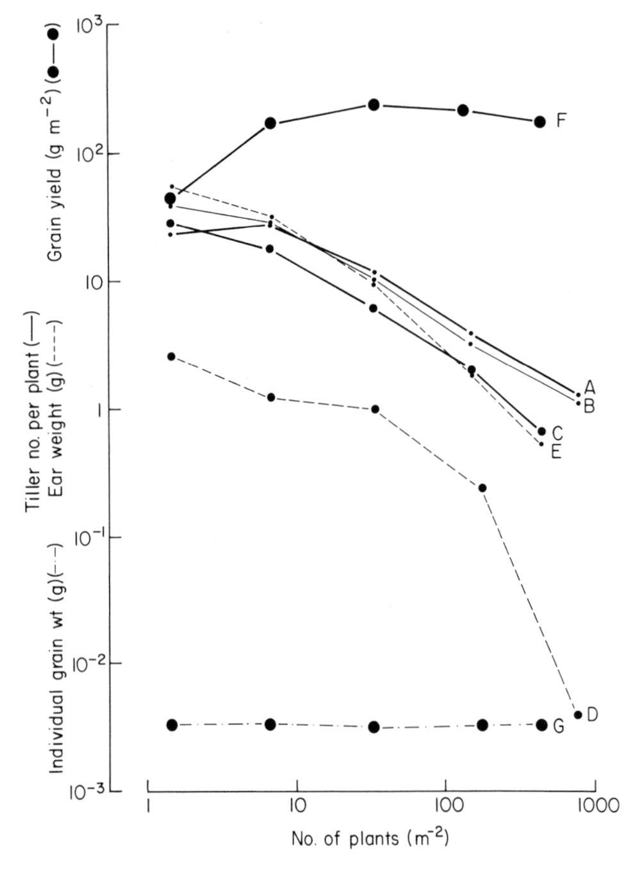

Fig. 2.9. Yield–density relationships in *Triticum aestivum*. Data from Puckridge & Donald (1967): A, tiller number per plant after 14 weeks; B, tiller number per plant after 17 weeks; C, number of fertile tillers per plant after 26 weeks; D, individual ear weight after 17 weeks; E, individual ear weight after 26 weeks; F, total grain yield at 26 weeks; G, individual grain weight after 26 weeks.

2.5.3 *Individual variability*

Characterizing plant populations by an average response as we have done up to now masks the mechanism by which compensation actually occurs. We must now consider the fates of *individuals* within the population.

Obeid *et al.* (1967) sowed *Linum usitatissimum* (flax) at three densities and harvested at three stages of development, recording the weight of each plant individually. Fig. 2.10 makes it quite clear that the frequency distribution of weights is skewed towards the left and that the skewing is increased by the passage of time and by an increase in density. The resulting populations display a few large individuals and a great many small ones. Similar distribution patterns have been observed in the growth of single species populations of crop plants, mixed tree plantations and in animals (Begon 1984).

There are two partially interrelated causes of skewness in the size distributions of populations in the absence of death of plants. Even in plant populations sown at the same time, individuals will germinate at slightly different times, perhaps due to variation in the 'local' environment for germination or because of differences in seed size. These subtle differences in time of birth and size at birth will then become exaggerated as the growth of individuals proceeds. In addition, however, competition may further exaggerate skewness. Larger individuals which will often be those that emerge first will be comparatively unaffected by interference from smaller (and later) neighbours. As a consequence they will grow quickly. Small plants and late emergers on the other hand, not only have to compete with a number of other, generally larger individuals, but have to do so on unequal terms. Intraspecific competition accentuates

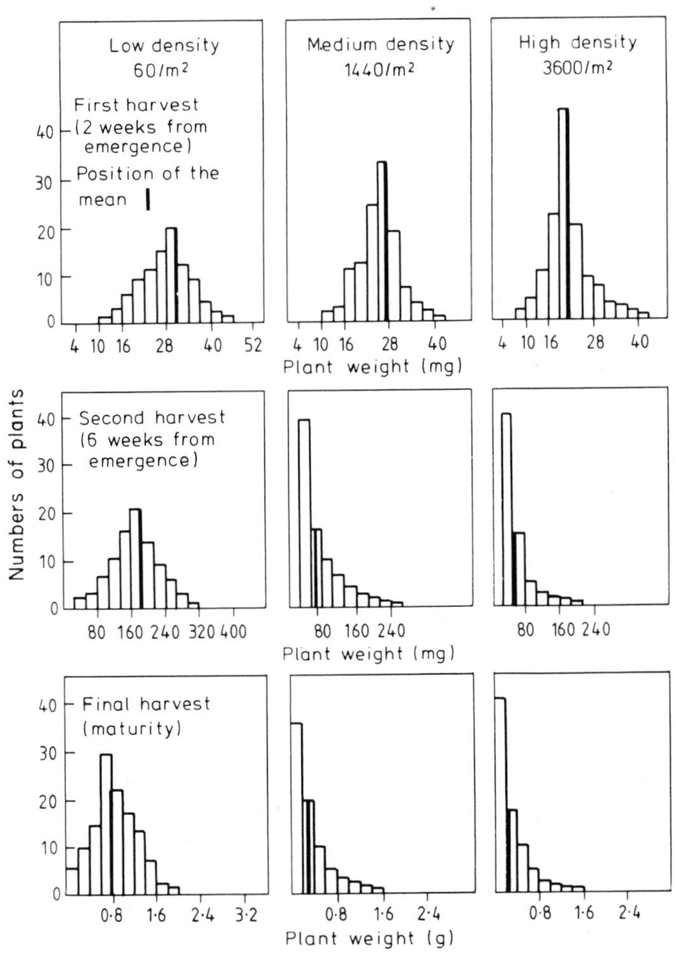

Fig. 2.10. Frequency distributions of plant weights in populations of flax, *Linum usitatissimum*, sown at three densities. (Data from Obeid *et al.* 1967, after Harper 1977.)

initial differences in size: large (early) individuals are least affected by competition and grow larger still; small (late) individuals are most affected, and lag further and further behind. Competition, then, serves to exaggerate size differences that initially may be determined by environmental chance or variation in initial starting capital: seed size.

The inevitable consequence of this process is the development of a population in which size differences will be exaggerated and a left-handed skew will develop. A 'hierarchy of exploitation' will result in a few dominant individuals with high growth rate and disproportionate share of resources whilst 'the most common type of plant in experimental (and natural) plant populations is the suppressed weakling' (Harper 1977). Experimental support for this view comes from the work of Ford (1975) on sitka spruce (*Picea sitchensis*) (Fig. 2.11). The greatest relative growth rate occurred in trees with the largest girth size, these trees being in the minority.

Finally we can consider the influence of limiting resources at the level of the modular growth of an organism. Porter (1983b) grew single *Fuchsia* plants in three different soil volumes and examined the accumulation of modules with time. His results (Fig. 2.12) show that whilst growth was initially rapid in all volumes, the onset of a stationary plant size—the plateau phase in the curves—was determined by the amount of soil available. Limiting the resources available influences the size of the module population and hence individual plant size. If this limitation arises through individuals interfering with resource gathering by one another (intraspecific competition) then differential reductions in relative growth rate amongst plants will occur and be reflected in the absolute size distribution of the population.

2.5.4 *Self-thinning in plants*

Certain aspects of these results on *Linum* are frequently repeated: the effects of intraspecific competition on a growing plant population often become accentuated with the passage of time. These effects most commonly concern plant density (which decreases with time), and mean plant weight (which increases with time); these two parameters appear to be closely related. 'Self-thinning' refers to the dynamics of density-dependent mortality that occurs progressively in cohorts as individuals grow in size. It has been studied most often in plant populations in monocultures. The process is well illustrated in an experiment by Kays & Harper (1974) in which they sowed a series of populations of *Lolium*

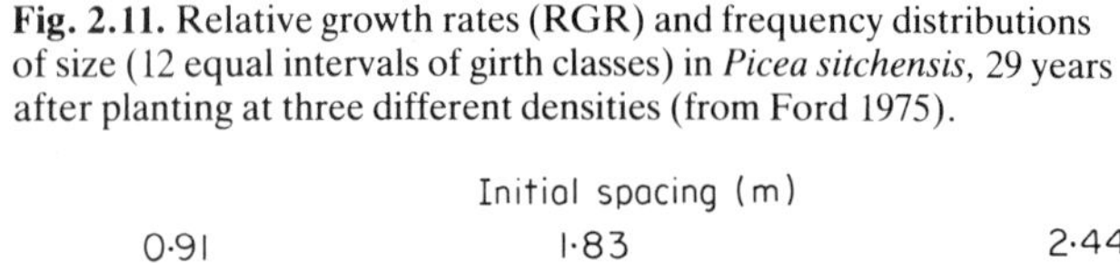
Fig. 2.11. Relative growth rates (RGR) and frequency distributions of size (12 equal intervals of girth classes) in *Picea sitchensis*, 29 years after planting at three different densities (from Ford 1975).

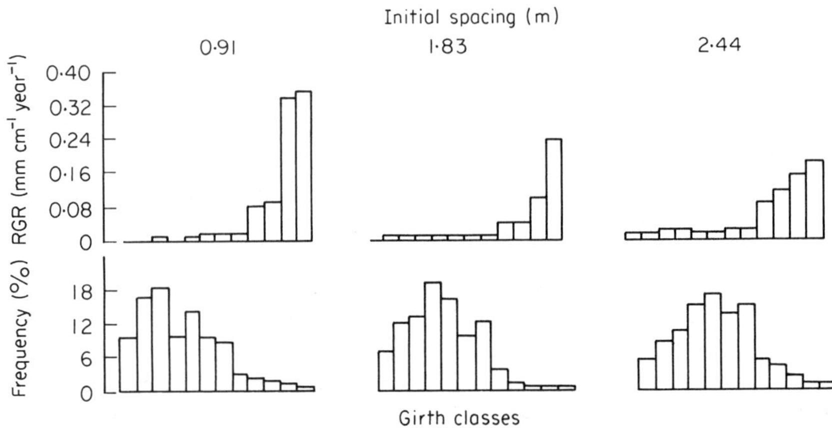

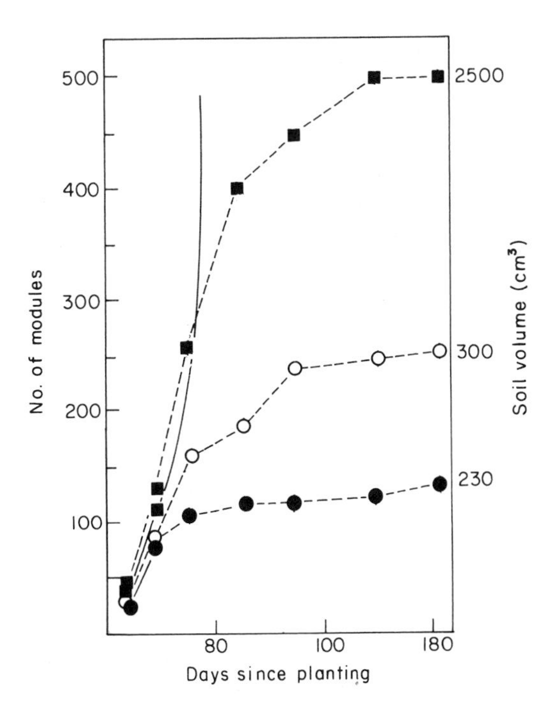

Fig. 2.12. The production of modular units by *Fuchsia*, c.v. Royal Velvet, in different soil volumes. The solid line shows the total number of modules assuming unrestricted exponential growth rate (from Porter 1983b). Plants received abundant water but no additional nutrients.

perenne, perennial ryegrass, at densities ranging from 330 to 10000 seeds m^{-2} and harvested their populations at subsequent occasions over a period of 180 days. Growth in this species occurs through the accumulation of ramets (shoots or tillers) on the genet (the established plant). Throughout the experiment (Fig. 2.13a) there was continual death of genets. Initially, this was most marked at the highest sowing density but latterly it assumed a rate independent of sowing density. During the early part of the experiment the number of ramets per unit area increased but then declined to give a similar number per m^2 after 125 days regardless of initial genet starting density. This constancy was repeated after 180 days but at a lower density. Compensatory processes within the population resulting in this constancy of tiller number derive from differential mortality of genets and differential rates of birth and death of tillers on them (Fig. 2.13a). Plotting the average genet size (mean above-ground plant biomass) against the density of surviving genets (Fig. 2.13b) we can examine the process of plastic response intertwined with mortality. The essence of studying self-thinning is to examine not a range of initial densities all at one time, but the *time-course* of a single initial density as the individuals grow in size. The time-courses show that the 'average' genet in a population grows, i.e. increases in biomass (points move upwards in the graph) at a rate diminishing with density (dotted lines curve down), until at a critical size, depending on genet density, further biomass increases can only be achieved with a concomittant loss of genets (points in the diagram shift to the left). Eventually, the time-courses appear to progress along a *boundary* delimiting the range of densities and sizes which can occur within the population. It has been described as the '−3/2 thinning line' (Yoda *et al.* 1963) since there is an inverse (negative) relationship between biomass and density with a gradient of 1.5. Perhaps somewhat surprisingly, very similar boundaries have been found to exist for a great many plant species of widely differing overall size and form (Fig. 2.14). They have also been shown true for ramet populations of ryegrass as well as for seaweeds. However, they appear not to operate in well established populations of colonial herbs (Hutchings 1979).

The −3/2 line describes a boundary, beneath which any combination of biomass and density can occur, but above which there are no permissible combinations. It is also a trajectory which the various time-courses approach and then follow.

Undoubtedly (Fig. 2.14) the slope of the self-thinning boundary is remarkably constant for a wide range of species. The precise reason for this is uncertain (see White 1981 but also Whittington 1984) but a plausible explanation is as follows. Mean plant biomass will be proportional to plant volume. The volume that a plant occupies will in turn be proportional to the spatial area in which it is growing. The space occupied by a plant in a crowded population is inversely related to density. Biomass will therefore be inversely related to density according to a function determined by the volume/area ratio. Since volume is a linear dimension cubed and area a linear dimension squared, the power is 3/2. Thus, as individuals grow in a crowded population, their 'areas' increase by a power 2, their densities therefore decrease by a power 1/2, but their mean biomass increases by a power 3. On balance, therefore, the biomass of the whole population increases.

Summarizing, we have seen how work on intraspecific competition in plants has established that the responses of actual populations are far more complex than was envisaged for 'scramble' and 'contest'. There is little uniformity of response from plant to plant, and the quality, as well as the quantity, of surviving individuals is affected. As a consequence, there is no sudden threshold of response. Plants show these features particularly clearly; it is appropriate, now, to examine some zoological examples.

2.5.5 *Competition in* Patella cochlear

Branch (1975) studied the effects of intraspecific competition on the limpet *Patella cochlear*, making observations on natural populations in South Africa varying in density from 125 to 1225 m^{-2}. *P. cochlear* feeds mainly on the alga, *Lithothamnion*, which grows not only on the limpets' rocky substrate, but also on the limpet shells themselves. Thus, the total amount of food (based on the total surface area) remains approximately constant. As density rises, however, the juvenile limpets increasingly live and feed on the shells of adults, so that competition for space is largely eliminated. But there is ever more intense competition for food. Some of Branch's results are illustrated in Fig. 2.15.

Fig. 2.13. Intraspecific competition affects tillers more than genets. (a) Changes in tiller and genet density in populations of ryegrass, *Lolium perenne*, sown at a range of densities. (b) The change in genet density and mean genet weight over the course of five harvests (H_1–H_5) for the same ryegrass populations. Arrows indicate progression of time and dashed lines link populations harvested at the same time. (Data from Kays & Harper 1974, after Harper 1977.)

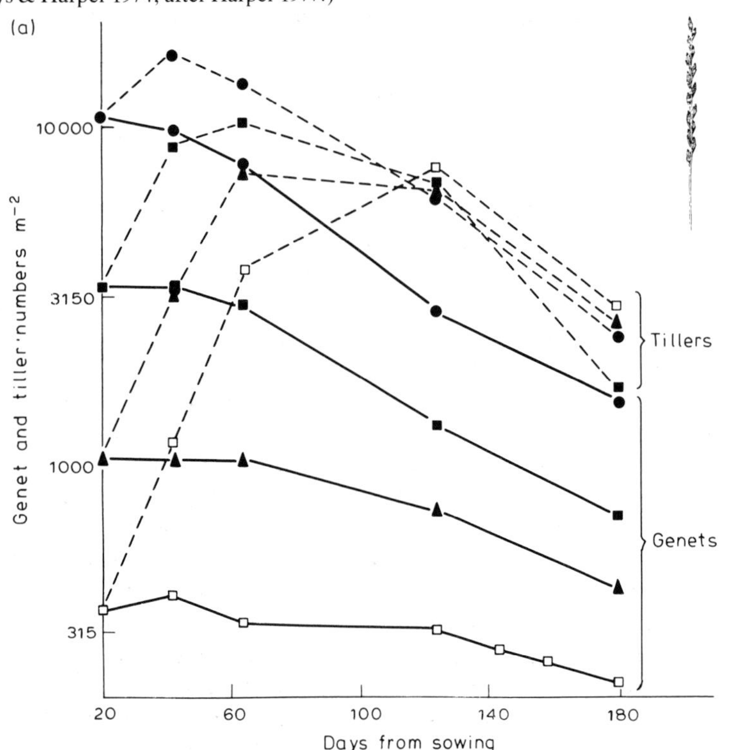

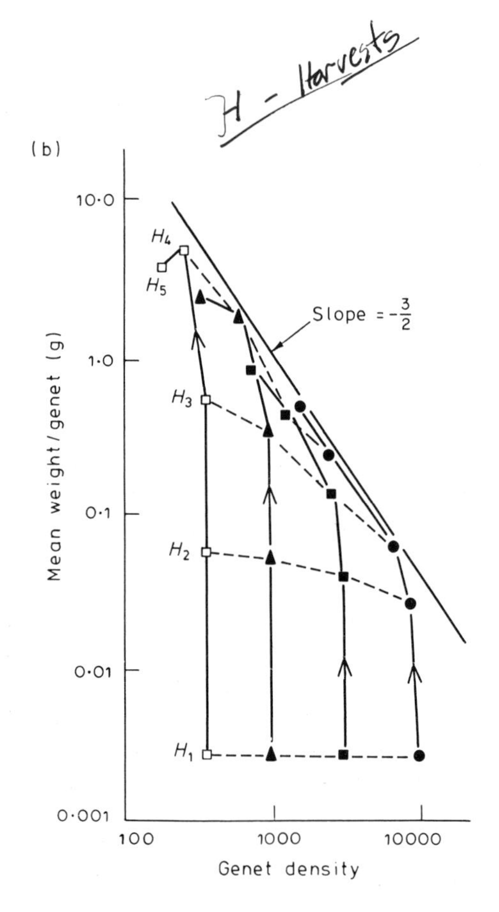

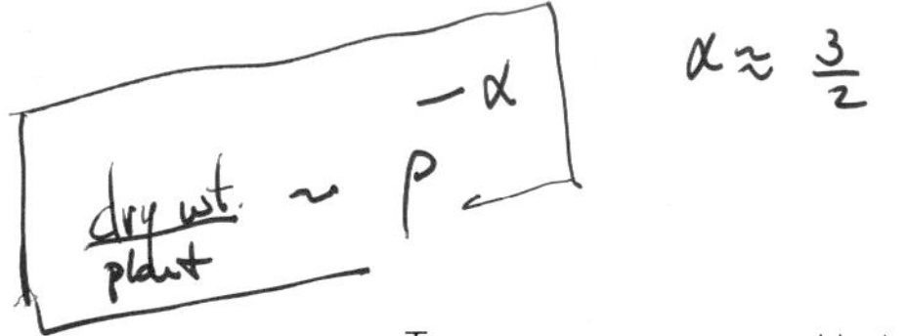

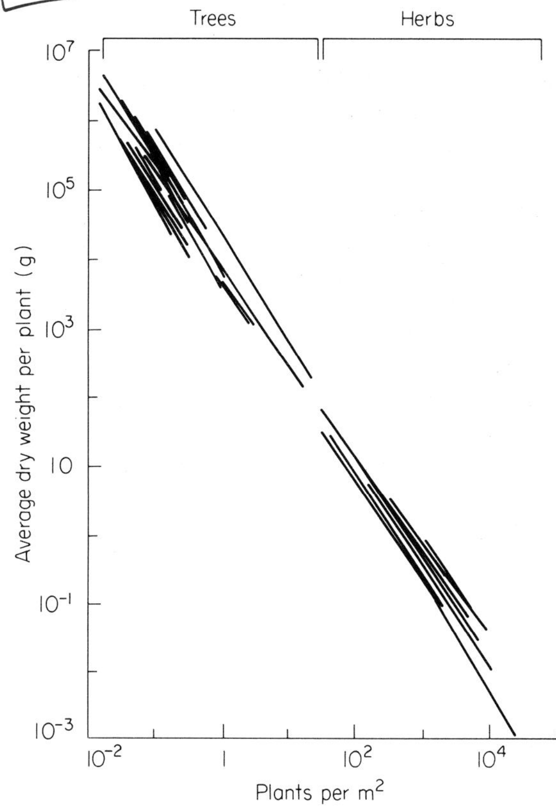

Fig. 2.14. Self-thinning in trees and herbs. Each line represents a different species in a total of 30 (from White 1980).

Fig. 2.15a shows that as density increases, there is a compensatory reduction in limpet size, leading to a stabilization of the total biomass at around 125 g m^{-2} for all densities in excess of around 450 m^{-2}. The regulatory properties of intraspecific competition and the plasticity of individual response are both readily apparent. It is clear from Fig. 2.15b, however, that consideration of only the mean or the maximum size of limpets would be very misleading; the size distribution undoubtedly alters as density increases. The reason appears to be that at low densities there is little juvenile mortality, most individuals reach the adult stage, and (large) adults come to dominate the population. As density and intraspecific competition increase, however, there is increased juvenile mortality and decreased rates of growth, so that at high densities there are comparatively few (large) adults. Once again, the effect of intense intraspecific competition is a population dominated (numerically) by 'suppressed weaklings'.

Branch had previously established that larger animals produce proportionally more gametes. Thus, the changing size-distribution of the population, despite the stabilization of biomass, would be expected to lead to a reduced gametic output as density increased. This is confirmed in Fig. 2.15c, where once again the use of k-values has proved instructive. At low densities (where there is little evidence of a sudden threshold) the reduction in size does not compensate for the increase in density ($b < 1$), and the total gametic output increases. However, at high densities there is increasing *over-compensation* ($b > 1$), and the total gametic output decreases at an accelerating rate. Once again, there is a moderate density (around 430 m^{-2}) at which $b = 1$ and the gametic output is at its maximum. The ultimate, regulatory effects of intraspecific competition are acting on the contributions of limpets to future generations.

2.5.6 *Competition in the fruit fly*

As a second zoological example, we will consider the experimental work of Bakker (1961) on competition between larvae of the fruit fly, *Drosophila melanogaster*. Bakker reared newly hatched larvae at a range of densities. Some of his results are illustrated in Fig. 2.16. As far as larval mortality is concerned, intraspecific competition appears to approach pure scramble. There is a sudden threshold at a density of around 2 larvae mg^{-1} yeast, and mortality thereafter very quickly reaches 100%. But the simplicity of this situation—and its similarity to pure scramble—is very largely an illusion. Up to the threshold density, competition has little or no effect on larval mortality; but the growth-rate of the larvae is very much affected, as is their final weight at pupation. Moreover, it is well known that in *D. melanogaster* small larvae lead to small pupae, which lead to small adults; and that these small adults, if female, produce comparatively few eggs. As Fig. 2.16 shows, with increasing larval density there is a decrease in the size of adults produced; this will lead to a decrease in the number of eggs contributed to the next generation. Hence, even below the threshold for larval mortality, intraspecific competition is exerting a density-dependent regulatory effect on the *D. melanogaster* larvae. But it is the quality, not the quantity, of larvae which is affected, and, ultimately, their contribution to the next generation which is reduced. Above the mortality threshold, the larvae are so reduced in size that they are not even large enough to pupate; when this happens they eventually die.

Clearly, in animals as in plants, intraspecific competition is very much more complicated than scramble or contest. Moreover, the complications—lack of a sudden threshold, individual variability, and so on—are very largely the same in both major kingdoms.

2.6 Negative competition

Finally, we need to consider an interaction which is beyond the spectrum of intraspecific competition dealt with so far. It is an example of a situation in which fecundity *increases* (or mortality decreases) with rising density. Birkhead (1977) studied breeding success in the common guillemot (*Uria aalge*) on Skomer Island, South Wales. The birds breed there in several sub-colonies of differing density. Female guillemots lay just one egg, and a pair of birds can be considered successful if they rear their chick until fledging. By visiting the various sub-colonies at least once a day throughout the breeding season, Birkhead was able to make careful observations on the losses of eggs and chicks, and could

Fig. 2.15. Intraspecific competition in a limpet. (a) Maximum length (o) and biomass (x) of the limpet, *Patella cochlear,* in relation to density. (b) Size-distributions of the limpets at three densities. (c) The effect of intraspecific competition on the limpets' gametic output. (Data from Branch 1975.)

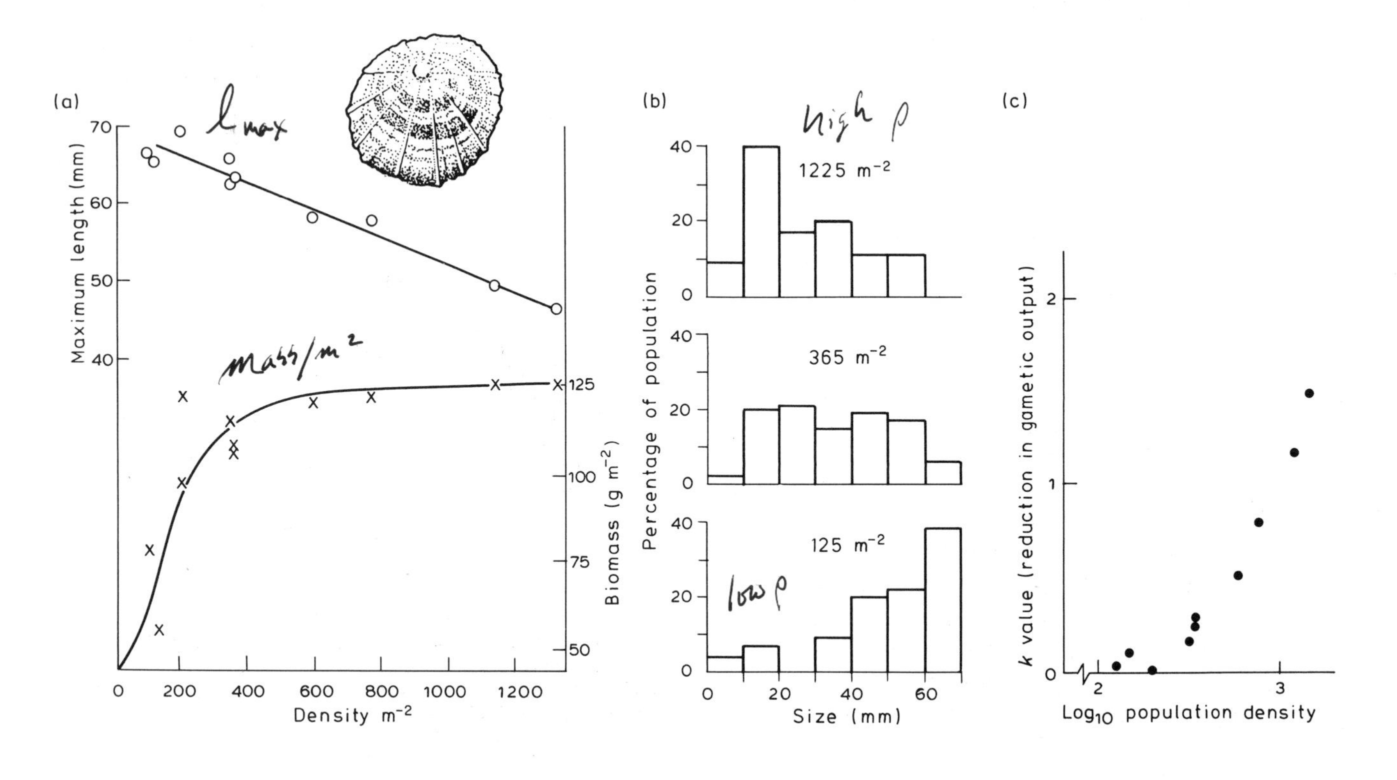

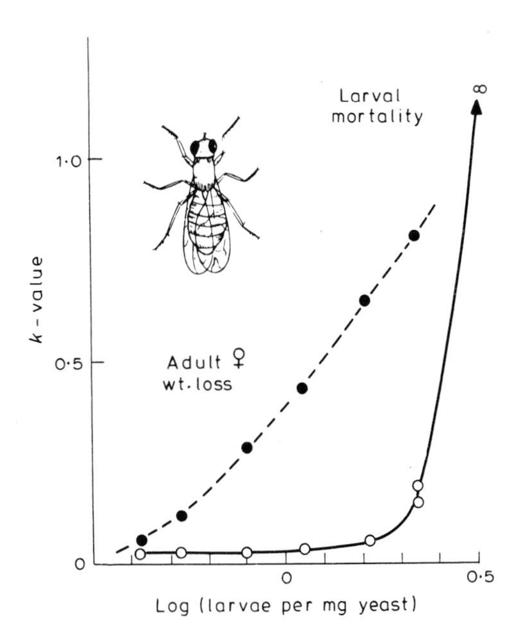

Fig. 2.16. The effects of intraspecific competition for food in the fruit fly, *Drosophila melanogaster*. For adult weight loss, a variable number of larvae competed for a constant amount of food. For larval mortality, a constant number of larvae were given a variable amount of food. (Data from Bakker 1961, after Varley *et al.* 1975.)

compute the percentage of pairs breeding successfully. His results are shown in Fig. 2.17.

It is quite clear that as density increases, breeding success, and thus fecundity, also increases. This is a case of *inverse* density-dependence. Competition is actually *negative*, and might more properly be called co-operation. Indeed, co-operation does appear to be the explanation for these results. Great black-backed and herring gulls are both important predators of the eggs and chicks of guillemots on Skomer, and denser groups are less susceptible to predation, because a number of guillemots are able (together) to deter gulls by lunging at them. Thus, our spectrum of competition (and density-dependence) must for completeness, be extended to 'co-operation' and inverse density-dependence.

Not surprisingly, just as there is a regulatory tendency associated with density-dependence, there is a destabilizing tendency associated with inverse density-dependence. This is illustrated diagrammatically for the guillemots on the left-hand side of Fig. 2.18, in which birth-rate rises with density while death-rate remains constant. Small populations get smaller still (because death-rate exceeds birth-rate), while larger populations increase in size (because birth-rate exceeds death-rate). However, it is likely that these destabilizing tendencies will disappear and then be reversed as population size increases and the carrying capacity is approached. The situation over all densities, therefore, is likely to be as

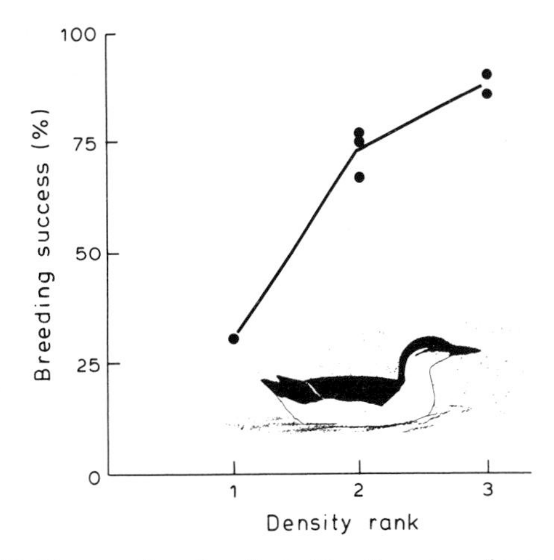

Fig. 2.17. Co-operation: the effect of density on breeding success in sub-populations of the common guillemot, *Uria aalge*, on Skomer. (Data from Birkhead 1977.)

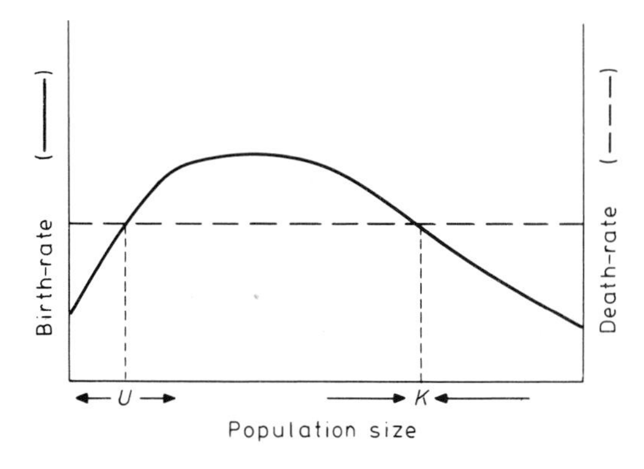

Fig. 2.18. The 'Allee effect' (Allee 1931). Density-dependent birth at moderate and high densities leads to a stable carrying capacity, *K*, but the change to inversely density-dependent birth at low densities leads to an unstable equilibrium, *U*, below which the population will decline to extinction (see also Fig. 2.2).

illustrated in the whole of Fig. 2.18, and is known as the 'Allee effect' (Allee 1931). The regulatory density-dependent effects to the right of the figure are easy to imagine. The destabilizing inversely density-dependent effects to the left of the figure could result from co-operation (as with the guillemots), or from certain other problems associated with low density (mate-finding, for instance). We shall consider the Allee effect again in Section 5.13.2.

Chapter 3 Models of Single-Species Populations

3.1 Introduction

It would be easy to follow one example of intraspecific competition with another; each would, in some way, be different from the rest. However, science progresses not by merely accumulating facts but by discovering patterns. We must, therefore, try to pin-point those features of the organism and its environment which are common to many, or even all, of the special cases. We can then concentrate on these essential features in our attempts to understand the dynamics of populations *in general*. We may even be able to discover how best to control the distribution and abundance of organisms, or how to predict the responses of populations to proposed or envisaged alterations in their environment. But in order to do this—or even make a start in doing this—we require a general conceptual framework on which each special case can be fitted. We require a system which embodies everything that the special cases have in common, but which, by manipulation of its parts, can be made to mirror each of the special cases in turn. In short, we require a model.

As Levin (1968) has suggested, our perfect model would be maximally general, maximally realistic, maximally precise and maximally simple. However, in practice, an increase in one of these characteristics leads to decreases in the others. Each model is, therefore, an imperfect compromise, and it is in this light that models of population dynamics should be seen. Mostly they sacrifice precision and a certain amount of realism, so as to retain a high degree of generality and simplicity. They are usually in the form of an equation or equations illustrated by graphs, but may, occasionally, be in the form of graphs without accompanying equations.

To be useful, and therefore successful, our model of single-species population dynamics should be:

(a) a satisfactory description of the diverse systems of the natural and experimental worlds;

(b) an aid to enlightenment on aspects of population dynamics which had previously been unclear; and

(c) a system which can be easily incorporated into more complex models of interspecific interactions.

We will begin by imagining a population with discrete generations, i.e. one in which breeding occurs in particular seasons only. Having developed a model for this situation, we can return to populations with continuous breeding to develop an analogous model in a similar fashion.

3.2 Populations breeding at discrete intervals

3.2.1 The basic equations

Suppose that each individual in one generation gives rise to two individuals in the next. If we begin with 10 individuals in the first generation, then the series of population sizes in succeeding generations will obviously be: 20, 40, 80, 160, and so on. This factor by which population size is multiplied each generation is commonly called the *reproductive-rate,* and we can denote it by R, i.e. in the above example $R = 2$.

Note immediately that we are assuming that a *single* figure can characterize reproduction for a *whole* (presumably heterogeneous) population on *all* occasions. Note also, that since 'reproduction' is equivalent to 'birth minus death', we have avoided dealing with birth and death separately. R is, therefore, a '*net* rate of increase', or 'net reproductive-rate'.

We can denote the initial population size (10 in our example) by N_0, meaning the population size when no time has elapsed. Similarly when one generation has elapsed the population size is N_1, when two generations have elapsed it is N_2 and, generally, when t generations have elapsed the population size is N_t. Thus:

$$N_0 = 10,$$
$$N_1 = 20,$$

that is:

$$N_1 = N_0 \times R$$

and, for instance:

$$N_3 = N_2 \times R = N_0 \times R \times R \times R = N_0 R^3 = 80.$$

In general terms:

$$N_{t+1} = N_t R \text{ (a \textit{difference} equation)} \quad (3.1)$$

and

$$N_t = N_0 R^t. \quad (3.2)$$

It is plain to see, however, that these equations lead to populations which continue to increase in size indefinitely. The next, obvious step, therefore, is to make the net reproductive-rate subject to intraspecific competition. To do this, we will incorporate intraspecific competition into the difference equation.

Consider Fig. 3.1. The justification for point *A* is as follows: when the population size (N_t) is very, very small (virtually zero), there is little or no competition in the population, and the net reproductive-rate (R) does not require modification. It is still true, therefore, that $N_{t+1} = N_t R$, or, rearranging the equation:

$$\frac{N_t}{N_{t+1}} = \frac{1}{R}.$$

As population size increases, however, there is more and more competition, and the actual net reproductive-rate is increasingly modified by it. There must presumably come a point at which competition is so great, and net reproductive-rate so modified, that the population can do no better than replace itself each generation. In other words, N_{t+1} is merely the same as N_t (no greater), and N_t/N_{t+1} equals 1. The population size at which this occurs, as we have already seen, is the *carrying-capacity* of the population, denoted by K. This is the justification for point *B* in Fig. 3.1.

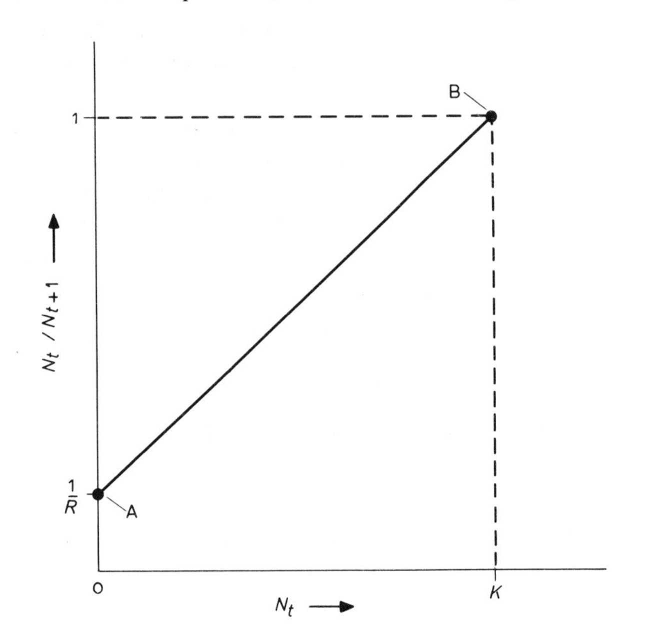

Fig. 3.1. The inverse of generation increase (N_t/N_{t+1}) rising with density (N_t). For further discussion, see text.

It is clear, then, that as population size increases from point *A* to point *B* the value of N_t/N_{t+1} must also rise. But it is for simplicity's sake, and only for simplicity's sake, that we assume this rise follows the straight line in Fig. 3.1. This is because all straight lines are of the simple form: $y = (\text{slope})x + (\text{intercept})$. The value for the 'intercept' is clearly $1/R$. That for the 'slope', considering the portion between points *A* and *B*, is $(1-1/R)K$. The equation of our straight line is, therefore.

$$\frac{N_t}{N_{t+1}} = \frac{\left(1-\frac{1}{R}\right)N_t}{K} + \frac{1}{R},$$

which, by simple rearrangement, gives:

$$N_{t+1} = \frac{N_t R}{1+\frac{(R-1)N_t}{K}}.$$

It is probably simpler to replace $(R-1)/K$ by a and remember:

$$N_{t+1} = \frac{N_t R}{1+aN_t}. \quad (3.3)$$

The *new* reproductive-rate (replacing the unrealistically constant R) is, therefore:

$$R/(1+aN_t).$$

We can now examine the properties of this equation and the net reproductive-rate within it, to see if it is satisfactory; and in some respects it certainly is. Fig. 3.2 shows a population increasing in size from very low numbers under the influence of equation 3.3. In the first place, when numbers *are* very low, the population increases in the same, 'exponential' fashion as a population unaffected by competition (equation 3.1). In other words, when:

$$N_t \text{ approaches } 0,$$

$$1+aN_t \text{ approaches } 1,$$

and $$\frac{R}{1+aN_t} \text{ approaches } R.$$

During this initial phase, therefore, the rate at which the population increases in size is dependent only on the size of the population and the potential net reproductive rate (R) of the individuals within it.

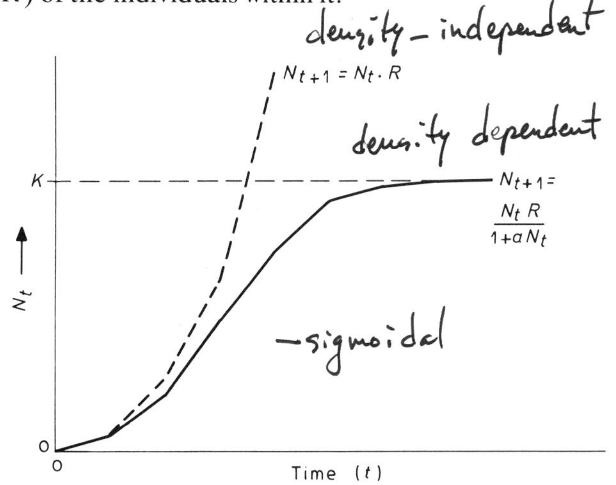

Fig. 3.2. Exponential (dashed) and sigmoidal increase in density (N_t) with time, in a population with discrete generations.

The larger N_t becomes, however, (and thus the more competition there is) the larger $1+aN_t$ becomes, and the smaller the actual net reproductive-rate ($R/[1+aN_t]$) becomes. This situation, in which the net reproductive-rate is under density-dependent control, is responsible for the 'S-shaped' or *sigmoidal* nature of the curve shown in Fig. 3.2. It is important to note, however, that while such sigmoidal increase is desirable, in as much as it shows our population gradually approaching the carrying capacity (K) first mentioned in Chapter 2, equation 3.3 is only one of many that would lead to such a pattern. We chose equation 3.3 because of its simplicity.

At the top of the curve, N_t approaches K, so that:

$$1+aN_t \equiv 1+\frac{(R-1)N_t}{K} \simeq 1+\frac{(R-1)K}{K} = 1+(R-1) = R.$$

Thus, $$N_{t+1} = \frac{N_t R}{1+aN_t} = \frac{N_t R}{R} = N_t$$

and $$N_t = K = N_{t+1} = N_{t+2} = N_{t+3}, \text{ etc.}$$

If, however, the population is perturbed such that N_t *exceeds* K, then:

$$1+\frac{(R-1)N_t}{K} > R,$$

and $$N_{t+1} < N_t.$$

In other words, the population will return to K. It will also do so if perturbed to below K. The carrying capacity is a stable equilibrium point to which populations return after perturbation. Our model, therefore, exhibits the regulatory properties classically characteristic of intraspecific competition.

3.2.2 *Incorporation of a range of competition*

It is clear, from the type of population behaviour that it leads to, that equation 3.3 can describe situations in which a population with discrete generations reproduces at a density-dependent rate. What is not clear is the exact nature of the density-dependence, or the exact type of competition that is being assumed. This can now be examined.

Each generation, the *potential* production of offspring is obviously $N_t R$, i.e. the number of offspring that *would* be produced if competition did not intervene. The *actual* production (i.e. the number that survive) is $N_t R/(1+aN_t)$. We already know, however, that the 'k-value' of any population process is given by:

$$k = \log_{10}(\text{no. produced}) - \log_{10}(\text{no. surviving}).$$

In the case of our equation, therefore:

$$k = \log_{10} N_t R - \log_{10}\{N_t R/(1+aN_t)\},$$

or

$$k = \log_{10} N_t + \log_{10} R - \{\log_{10} N_t + \log_{10} R - \log_{10}(1+aN_t)\},$$

or $$k = \log_{10}(1+aN_t).$$

We also know, from Chapter 2, that the type of intraspecific competition can be determined from a graph of k

against $\log_{10} N_t$. Fig. 3.3 shows a series of such plots for a variety of values for a and N_0. In each case the slope of the graph approaches 1. In other words, in each case the density-dependence begins by undercompensating, and approaches exact compensation at higher values of N_t.

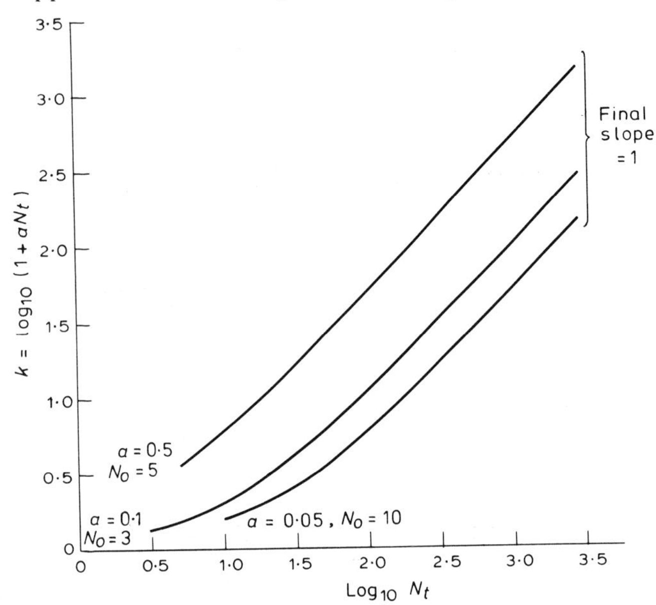

Fig. 3.3. The intraspecific competition inherent in equation 3.3. The final slope (of k against $\log_{10} N_t$) is unity (exact compensation), irrespective of the starting density N_0, or the constant $a(=[R-1]/K)$.

Our model, therefore, lacks the generality we would wish it to have. But such generality is easily incorporated. All we require is a model that can give us, potentially, b-values (slopes of the k versus $\log_{10} N_t$ graph) varying from zero to infinity.

One such general model (and there are others), was originally proposed by Hassell (1975). It is a simple modification of equation 3.3:

$$N_{t+1} = \frac{N_t R}{(1+aN_t)^b}. \tag{3.4}$$

As Fig. 3.4 indicates, plots of k against $\log_{10} N_t$ will now approach b rather than 1, irrespective of the values of a and N_0. This equation can, therefore, describe populations reacting in a whole range of ways to the effects of competition, and b is the parameter which measures this degree of under- or overcompensation. This whole

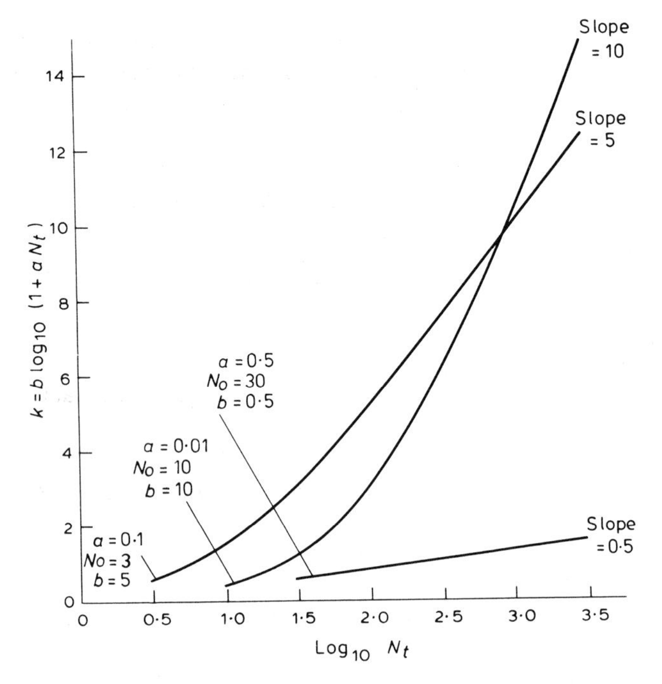

Fig. 3.4. The range of intraspecific competition which can be expressed by equation 3.4. At higher densities, the slope of k against $\log_{10} N_t$ reaches the value of the constant b in the equation, irrespective of N_0 and a.

range is only apparent at higher densities (competition gradually increasing in intensity as N_t increases); but when we remember the actual examples considered in Chapter 2, this seems, if anything, a positive advantage of the model. Note, however, that we have still not specified whether it is births or deaths or both which are being affected: to do this would require an even more complex model. Nevertheless, provided we remember that we are dealing with a net rate of increase, equation 3.4 can be a useful, simple model.

3.2.3 Models for annual plants

The same model can also be applied to annual plants with discrete generations. A convenient starting point is the law of constant final yield (Fig. 2.6), the underlying relationship between biomass and density being an inverse one (Fig. 2.7). Early workers expressed this in a variety of mathematical models (see Willey & Heath (1969) for a review) but we will concentrate on one recent development. Watkinson (1980) used a generalized expression that can describe both asymptotic and parabolic yield responses to density and also allows a biological interpretation of the constants involved. It expresses average plant in the populations, $\bar{w}$, as both a function of plant size in the absence of competition w_m and population density, N, after the action of competition (i.e. at harvest);

$$\bar{w} = \frac{w_m}{(1+aN)^b} \tag{3.5}$$

In this, there is clearly more than a passing similarity to equation 3.4. The term $(1+aN)$ here acts in a manner identical to that in equation 3.4, while b allows the incorporation of a range of compensatory responses; w_m, like NR in equation 3.4 is 'what is achieved' in the absence of competition.

To appreciate the features of this model we will explore the hypothetical data presented in Fig. 3.5a, which shows the two general forms of the yield response curve that we have already discussed (the range of units on the axes of the graph are arbitrary). Fig. 3.5b shows the corresponding data for mean individual plant size (measured perhaps as biomass) plotted against density. Note initially that the maximum competition-free plant size (w_m) is 100 units of biomass. Generally speaking, the pattern of divergence of the two types of response is determined by the values of the two parameters a and b in equation 3.5. For convenience, however, the value of a (= 0.1) is common to both curves. In Fig. 3.5, curves (1) and (2) both depart appreciably from the density-independent line of constant proportionality (dotted) at $N = 10$ and this in fact is the reciprocal of a in equation 3.5. We can see from Fig. 3.5b that it is at this density that individual plant size begins to be depressed. The reciprocal of this density (i.e. a) is then broadly an estimate of the space required for one isolated plant to grow to maximum size. In this arbitrary case, it is clearly 0.1 units of area. This area has been called the 'ecological neighbourhood area' (Antonovics & Levin 1980; Watkinson 1981).

Where exact compensation for density is occurring at high density—curve (1) in Fig. 3.5a—plant biomass is directly inversely proportional to density. The slope of the line is −1 (Fig. 3.5b), i.e. $b = 1$. Overcompensation, reflected by values of $b > 1$, results in an enhanced

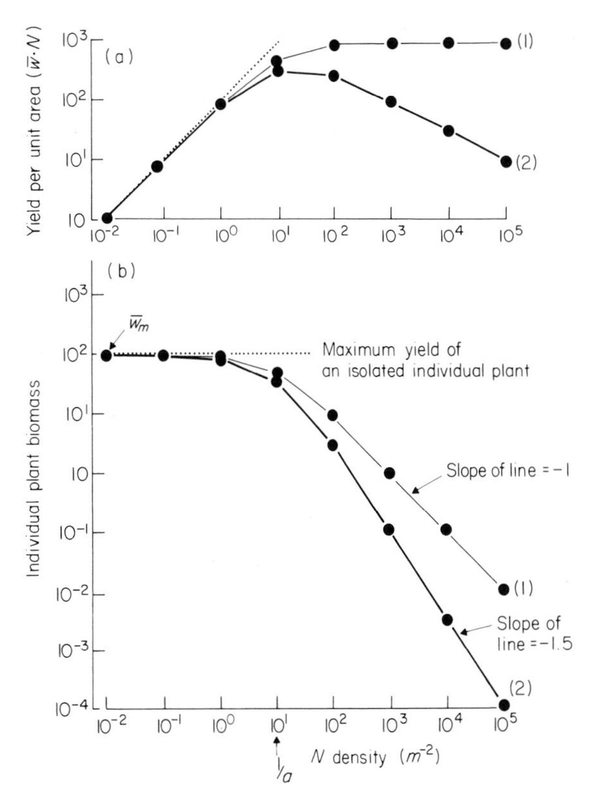

Fig. 3.5. The two general yield–density responses in plant populations expressed on a unit area basis (a) and on an individual plant basis (b), according to equation 3.5. The dotted line indicates a constant proportional relationship between yield and sowing density. (1) Exact compensation, $b = 1$; (2) overcompensation, $b = 1.5$. In both cases, $w_m = 100$ and $a = 0.1$. In (b) the values of the terms in equation 3.5 are indicated for the two yield responses. (The value of 1.5 in curve (2) has no particular significance to the power term of the self-thinning line.) Plant mortality is absent.

proportional reduction of plant size with density (the slope of the line in Fig. 3.5b is steeper) and yield per unit area is depressed at high density below a maximum occurring at an intermediate one.

The model's most conspicuous and valuable features are that it has the generality to describe the whole range of yield density responses in the absence of mortality (via the constants a and b) and that the parameters in the model are biologically interpretable.

Equation 3.5, however, does not incorporate mortality of plants. A number of studies have shown that the relationship between density before and after the action of density-dependent mortality follows the graphical form shown in Fig. 3.6. The mathematical description of this is another variant of equations 3.3 and 3.4, namely

$$N = \frac{N_i}{(1+mN_i)} \quad (3.6)$$

where N_i *is the number of plants sown,* N is the number surviving and $1/m$ is the maximum population size that can (asymptotically) be reached. We can now combine equations 3.5 ad 3.6 to give a model describing the relationship between $\bar{w}$ and N where there are effects both on mortality and growth.

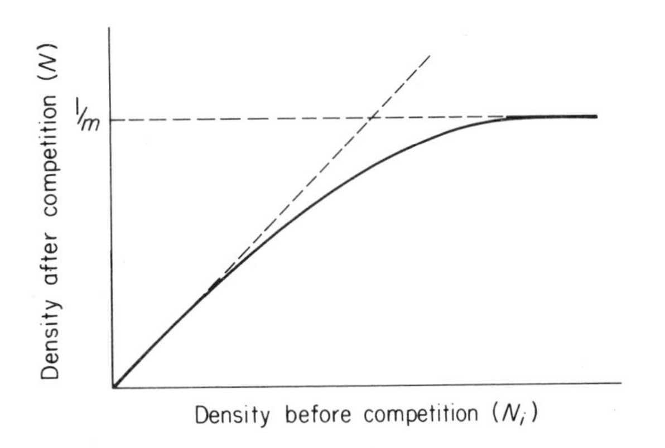

Fig. 3.6. Population sizes before (N_i) and after (N) density-dependent mortality. The curve follows the equation $N = N_i(1+mN_i)^{-1}$.

In equation 3.5

$$\bar{w} = w_m\,(1+aN)^{-b},$$

but N is the number of plants at harvest which is given by equation 3.6. Hence, by substitution,

$$\bar{w} = w_m\,(1+aN_i(1+mN_i)^{-1})^{-b} \quad (3.7)$$

Furthermore, to calculate yield per unit area at harvest we merely have to multiply $\bar{w}$, (the average plant weight) by N (the density at harvest). We can now examine graphically the behaviour of this model in Fig. 3.7, which shows the yield of a population of plants after density-dependent regulation, in relation to starting population size. In this simulation, as before, each plant when grown under isolated conditions requires 0.1 units of area (a) to achieve a maximum size of 100 biomass units (w_m). Where there is exact compensation at high densities ($b = 1$), the effect of decreasing the maximum

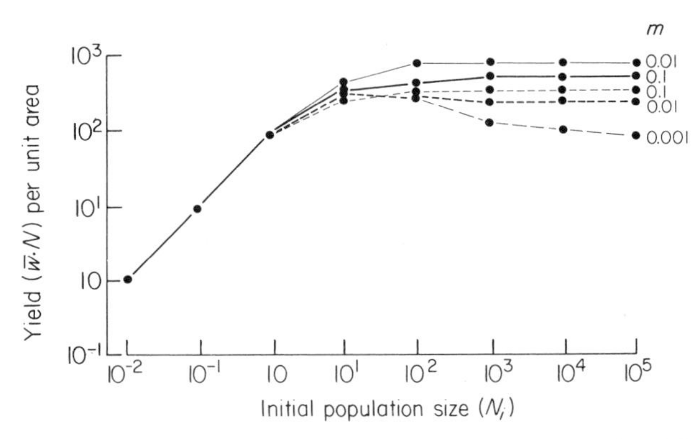

Fig. 3.7. The relationship between initial population size N_i and yield per unit area as determined by equations 3.6 and 3.7. Constants in the model are $a = 0.1$, $w_m = 10$, $b = 1.0$ (solid line), $b = 1.5$ (dotted lines). Values of m as indicated.

population size attainable ($1/m$) is to lower yield. If, however, there is overcompensation ($b > 1$) the shape of the yield–density curve depends on the intensity of density-dependent mortality (governed through m) in relation to the ecological neighbourhood area (a). Overt overcompensation is greatest when the potential maximum population size ($1/m$) is highest. Hence, the model has the generality to encompass the range of density responses which we know to occur. Furthermore, it incorporates the basic mechanisms of plasticity and mortality that are part of the response. We must bear in mind though that it is only a static description of an essentially dynamic process occurring within plant populations.

Up to now we have been concerned only with the relationship between plant size after the action of competition and initial population density (i.e. at sowing), but it is a relatively easy step to extend our model to one that describes changes from generation to generation. To do this we need to know the relationship between plant biomass at harvest, $\bar{w}$, and the number of seeds produced per plant, S. Often this may be described by

$$S = q\bar{w}^p$$

where q and p are constants describing the exact form of the relationship. We may therefore express seed number per plant at harvest as a function of seeds sown in an identical way to equation 3.7:

$$S = \lambda(1+aN_i(1+mN_i)^{-1})^{-b}$$

where λ is the number of seeds produced by a plant in isolated (non-competitive) conditions. Multiplying both sides of the equation by the number of surviving plants gives us the seed yield per unit area, and hence we have a population model relating the number of seeds sown to those harvested. Thus,

$$SN = \lambda(1+a(N_i(1+mN_i)^{-1})^{-b}N$$

Since N_i and SN are the population sizes in successive generations, we may replace them by N_t and N_{t+1}, remembering that we are dealing with populations of seeds. Upon rearrangement our model becomes

$$N_{t+1} = \frac{\lambda N_t}{(1+aN_t)^b + m\lambda N_t}$$

On inspection, this too is very similar to equation 3.4, differing only insofar as it includes an additional term for density-dependent mortality: $m\lambda N_t$. In the absence of mortality ($m = 0$), this model contracts to equation 3.4 and will of course display the properties already discussed.

3.3 Continuous breeding

We can now return to populations which breed continuously. For our purposes, the essential difference between these and ones with discrete generations is that population size itself changes continuously rather than in discrete 'jumps'. This is illustrated in Fig. 3.8. Curves A and B are both of the form $N_{t+1} = N_t R$, and represent populations increasing (exponentially) at essentially the same rate. ('Rate' in this case means 'the amount by which population size (N_t) increases per unit time (t)'.) The difference between them is that generation time is much shorter in curve B than curve A. The same process has been extended in curve C until the straight line segments are so small (infinitesimally small) that the graph is a continuous curve. Nevertheless, the rate at which the population increases has remained essentially the same. Curve C describes a situation in which, during each and every exceedingly small time interval, there is the possibility, at least, of birth and death. In other words, breeding and dying are continuous. Thus, the

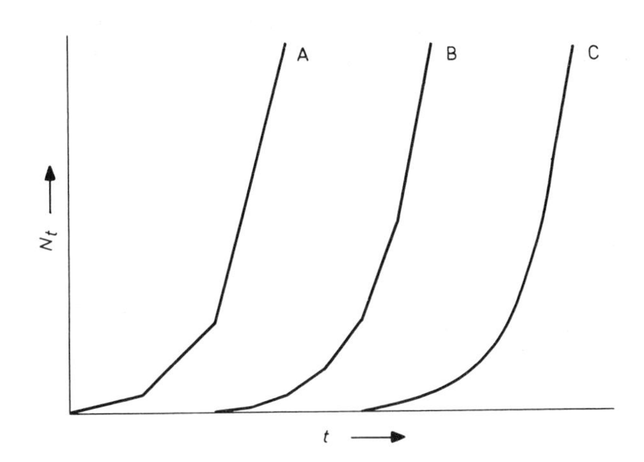

Fig. 3.8. Three populations growing exponentially at the same rate. A has long, discrete generations; B has short, discrete generations; C breeds continuously.

initial equation we require in order to build our general model of a continuously breeding population must retain the essential properties of $N_t = N_0 R^t$, but must describe a continuous curve.

Differentiation is a mathematical process which is specifically designed to deal with changes occurring during infinitesimally small intervals of time (or of any other measurement on the x-axis). Those familiar with differential calculus will clearly see (and those unfamiliar with it need merely accept) that differentiating $N_t = N_0 R^t$ by t we get:

$$\log_e N_t = \log_e N_0 + t \log_e R,$$

and

$$\frac{dN}{dt} \times \frac{1}{N} = 0 + \log_e R,$$

or

$$\frac{dN}{dt} = N \log_e R. \quad = rN$$

This is the equation we require. dN/dt is the *slope* of the curve, defining the *rate* at which population size increases with time. $\log_e R$ is usually replaced by r, 'the intrinsic rate of natural increase' or 'instantaneous rate of increase', but the change from R to r is simply a change of currency: the commodity dealt with remains the same (essentially 'birth minus death'). The *differential* equation $dN/dt = rN$, just like the difference equation $N_{t+1} = N_t R$, describes a situation of exponential population increase dependent only on the size of the population and the reproductive-rate of individuals.

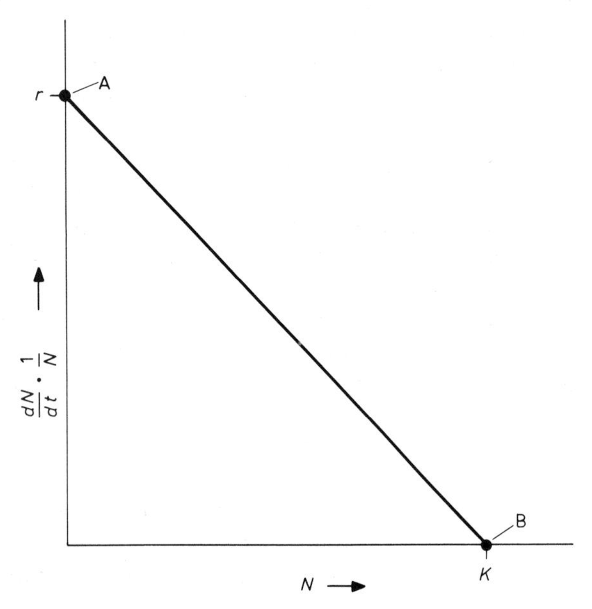

Fig. 3.9. The rate of increase per individual $dN/dt \cdot 1/N$ falling with density (N). For further discussion, see text.

We can proceed now in Fig. 3.9, exactly as we did in Fig. 3.1. The rate of increase *per individual* is unaffected by competition when N approaches zero, and is, therefore, given by r (point A). When N reaches K, the carrying capacity, the rate of increase per individual is 0 (point B). As before, we assume that the line between A and B is straight, and thus:

$$\frac{dN}{dt} \times \frac{1}{N} = \frac{-r}{K} \times N + r,$$

or

$$\frac{dN}{dt} \times \frac{1}{N} = r\left(1 - \frac{N}{K}\right),$$

and

$$\boxed{\frac{dN}{dt} = rN\left(1 - \frac{N}{K}\right).} \qquad (3.8)$$

This is the so-called logistic equation. Its characteristics are essentially the same as those of the difference equation 3.3, described previously. Minor dissimilarities, associated with the change from the difference to the differential form, will be discussed later; and, of course, it describes a continuous sigmoidal curve, rather than a series of straight lines. Note once again, moreover, that there are many other equations which would lead to sigmoidal increase; the justification of the logistic is the simplicity of its derivation. Note, too, that the logistic equation is based on exact compensation, in just the same way as its discrete generation analogue. In the case of the logistic, however, it is by no means easy to incorporate a factor, such as b, which will generalize the model to cover all types of competition. Our equation for a continuously breeding population must, therefore, remain comparatively imperfect.

Drawing these arguments together then, we are left with two equations: equation 3.4 as an apparently satisfactory general model for discrete-generation increase and equation 3.8 as a somewhat less satisfactory model for the behaviour of a continuously breeding population.

3.4 The utility of the equations

3.4.1 Causes of population fluctuations

We can now proceed to examine the models' utility. The extent to which they can be incorporated into more complex models of interspecific interactions will become apparent in Chapters 4 and 5. We shall begin here by determining whether our models can indeed throw important *new* light on aspects of population dynamics. To do so we shall examine the question of fluctuations in the sizes of natural, single-species populations. That *some* degree of fluctuation in size is shown by *all* natural populations hardly needs stressing.

The causes of these fluctuations can be divided into two groups; extrinsic and intrinsic factors. Amongst the extrinsic factors we include the effects of other species on a population, and the effects of changes in environmental conditions. These are topics which will be discussed in later chapters. For now, we will concern ourselves with intrinsic factors. Our approach will be to examine our models to see which values of the various parameters, and which minor alterations to the models themselves, lead to population fluctuations; and to see what type of fluctuation they lead to.

The method by which our discrete-generation equation may be examined has been set out and discussed by May (1975). It is a method which uses fairly sophisticated mathematical techniques, but we can ignore these and concentrate on May's results. These are summarized in Fig. 3.10. Remember that Fig. 3.10a refers to equation 3.4. It describes the way in which populations fluctuate (Fig. 3.10b) with different values of R and b inserted. [The value of a, i.e. $(R-1)/K$, determines the level about which populations fluctuate, but not the manner in which they do so.] As Fig. 3.10a shows, low values of b and/or R lead to populations which approach their equilibrium size without fluctuating at all. Increases in b and/or R, however, lead firstly to damped oscillations gradually approaching the equilibrium; and then to 'stable limit cycles' in which the population fluctuates around the equilibrium level, revisiting the same two, or four, or even more points time and time again. Finally, with large values of b and R, we have a population fluctuating in a wholly irregular and chaotic fashion.

The *biological* significance of this becomes apparent when we remember that our equation was designed to model a population which *regulated* itself in a density-dependent fashion. We can see, however, that if a population has even a moderate net reproductive-rate (and an individual leaving 100(= R) offspring in the next generation in a competition-free environment is not unreasonable), and if it has a density-dependent reaction which even moderately overcompensates, then *far from being stable,* it may fluctuate in numbers without any extrinsic factor acting. Thus, our model system has *taught* us that we need not look beyond the intrinsic dynamics of a species in order to understand the fluctuations in numbers of its natural populations. It is worthwhile stressing this. Our intuition would probably tell us that the ability of a population to regulate its numbers in a density-dependent way should lend stability to its dynamics. Yet our model shows us that if this regulation involves a moderate or large degree of overcompensation, and at least a moderate reproductive-rate, then the population's numbers may fluctuate considerably because of those very 'regulatory' processes. As May, himself, concludes: '. . . even if the natural world was 100% predictable, the dynamics of populations with "density-dependent" regulation could nonetheless in some circumstances be indistinguishable from chaos.'

In examining this general model, the special, exact-compensation case (b = 1) has itself been covered. Irrespective of R, the population will reach K without overshooting; and with R-values greater than about 10

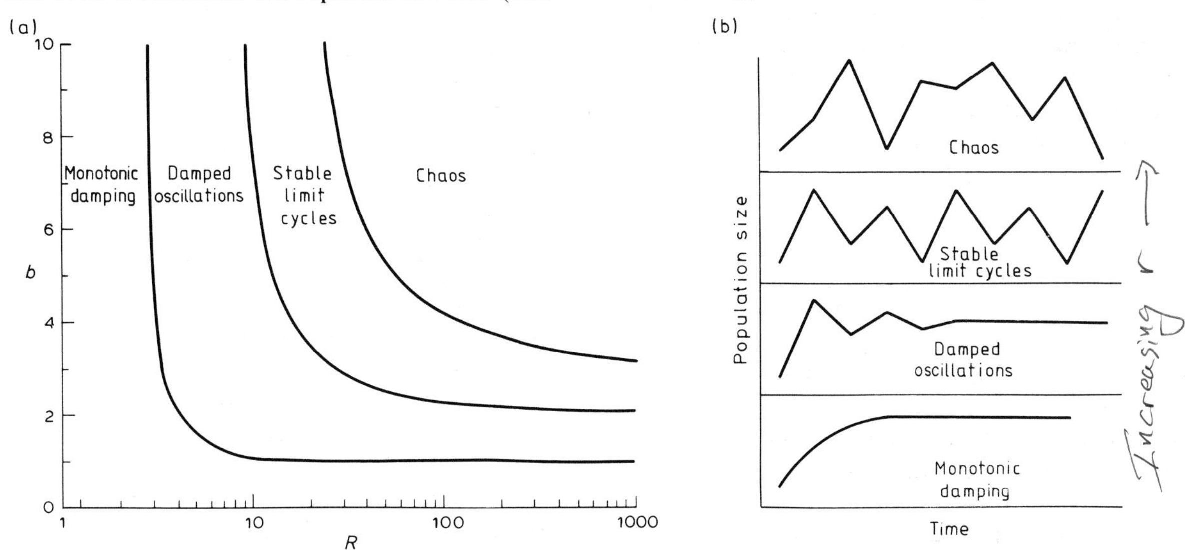

Fig. 3.10. The effect of intraspecific competition on population. (a) The range of population fluctuations (themselves shown in b) generated by equation 3.4 with various combinations of b and R inserted. For further discussion, see text. (After May 1975.)

this will take little more than a single generation. (The analogous differential equation—the logistic—behaves very similarly to this: always approaching K with exponential damping.) This special case can be modified, however, to incorporate an additional feature. We have assumed until now that populations respond *instantaneously* to changes in their own density. Suppose, on the contrary, that the reproductive-rate is determined by the amount of resource available to the population, but that the amount of resource is determined by the density of the *previous* generation. This will mean that the reproductive-rate is dependent on the density of the previous generation. Thus, since:

$$N_{t+1} = N_t \times \text{reproductive-rate}$$

$$N_{t+1} = \frac{N_t R}{1 + aN_{t-1}}. \qquad (3.9)$$

In other words, there is a *time-lag* in the population's response to its own density, caused by a time-lag in the response of its resource. The amount of grass in a field in spring being determined by the level of grazing the previous year is a simple but reasonable example of this. The behaviour of this modified equation, as shown by computer simulations, is as follows:

$R <$ approx. 1.3: exponential damping
$R >$ approx. 1.3: damped oscillations.

In comparison, the original equation, without a time-lag, led to exponential damping for all values of R. *The time-lag has provoked the fluctuations.* Once again, therefore, examination of our model has taught us which intrinsic features of a species' dynamics can lead to fluctuations in its population density.

In fact, there are other types of time-lag and they all *tend* to provoke fluctuations in density. Consider again the difference between a population with discrete generations and one with continuous breeding. In both cases, the population responds throughout each 'time-interval' to the density at the start of that time-interval. With continuous breeding the time-intervals are infinitesimally small, and the response of the population is, therefore, continually changing; with discrete generations, on the other hand, the population is still responding at the end of a time-interval to the density at its start. In the meantime, of course, this density has altered: there is a time-lag. Thus, the difference equation model is more liable to lead to fluctuations than the differential logistic, and it is the time-lag which accounts for the difference. Moreover, some organisms may respond to density at one point in their life-cycle, and actually reproduce some time later; this 'developmental time' between response and reproduction is also a time-lag. In all cases, the population is responding to a situation that has already changed, and an increased level of fluctuation is the likely result.

There are two important conclusions to be drawn from this discussion. The first is that time-lags, high reproductive-rates and highly overcompensating density-dependence (either alone or in combination) are capable of provoking all types of fluctuations in population density, without invoking any extrinsic cause. The second conclusion is that this is clear to us only because we have studied the behaviour of our model systems.

3.4.2 The equations as descriptions

Finally, we must examine the ability of our models to describe the behaviour of natural and experimental systems. There are two aspects of this. The first is concerned with the way in which single-species populations increase or fluctuate in size; the second with the precise way in which intraspecific competition affects the various facets of fecundity and survival.

A number of examples of population increase and fluctuation are illustrated in Fig. 3.11. Fig. 3.11a describes the continuous increase of a population of yeast cells under laboratory conditions (Pearl 1927): the resemblance to the logistic curve is quite striking. Fig. 3.11b, on the other hand, describes the year-by-year change in the size of the sheep population of Tasmania (Davidson 1938). There is no obvious detailed resemblance between this and the pattern predicted by any of our models. Nevertheless, the initial rise in numbers, decelerating as it approaches a plateau (albeit a fluctuating one), is reminiscent of the pattern generated by our difference equation model with b close to 1. If, furthermore, we take into account the fluctuations in environmental conditions, the interactions of the sheep

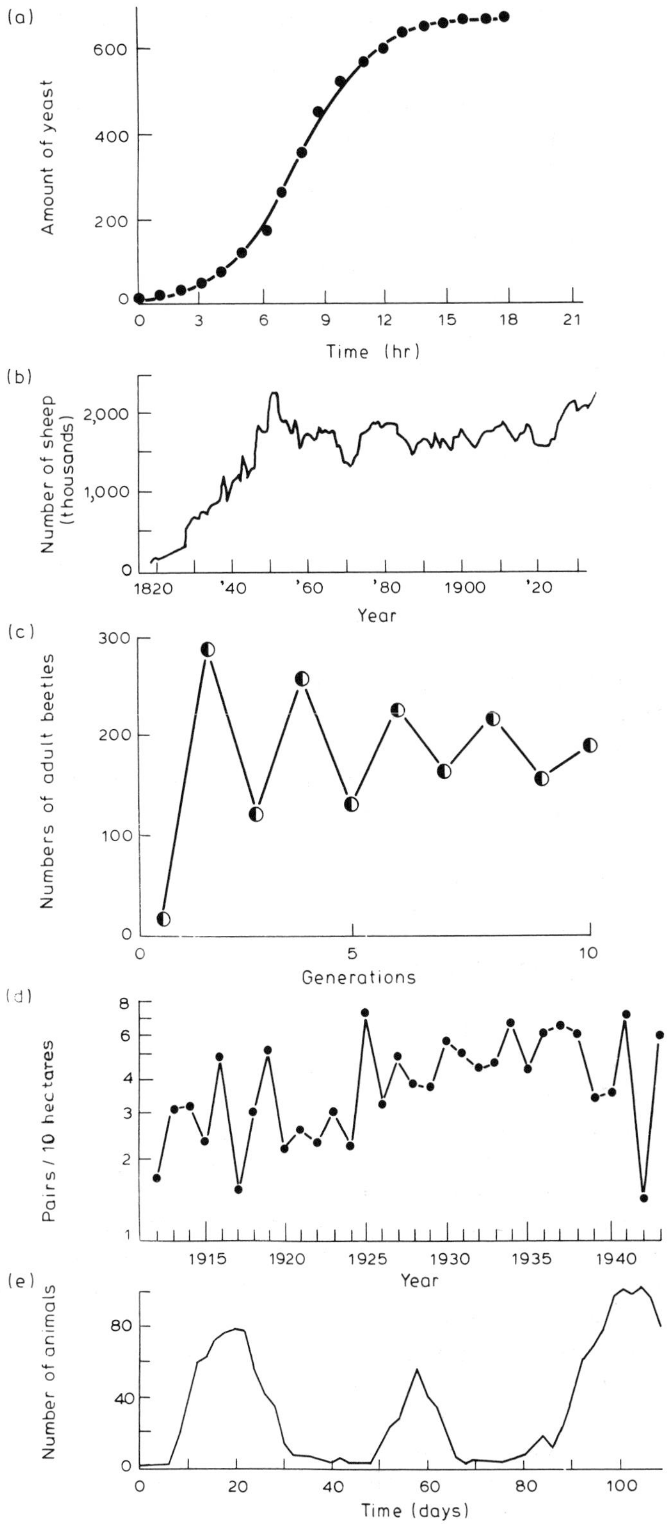

Fig. 3.11. Observed population fluctuations. (a) Yeast cells (after Pearl 1927). (b) Tasmanian sheep (after Davidson 1938). (c) The stored-product beetle, *Callosobruchus maculatus* (after Utida 1967). (d) The great tit, *Parus major* in Holland (after Kluyver 1951). (e) The water flea, *Daphnia magna* (after Pratt 1943).

with other species, and the undoubted imperfections of the sampling method, then the resemblance between Fig. 3.11b and our model begins to look more significant. This is the nub of our problem in deciding how satisfactory our models are. On the one hand, the field data, unlike some data collected under controlled conditions, do not follow our models' behaviour *exactly*. On the other hand, there is nothing in the field data to make us *reject* our models. We could claim, in fact, that our models represent a very satisfactory description of the population dynamics of sheep in Tasmania, but that the complexities of the real world tend to blur some of the edges. Similar conclusions could be drawn from Fig. 3.11c–e, although in each case the model concerned might be slightly different; perhaps a different value for a or b, or the addition of some sort of time-lag. Fig. 3.11c, for example, would be quite adequately described by our difference equation with $R = 20$ and $b = 2$; the pattern in Fig. 3.11 could be accounted for, simply, by environmental fluctuations; and the fluctuations in numbers of *Daphnia* (Fig. 3.11e) do appear, quite genuinely, to be caused by a time-lag.

The most reasonable conclusion seems to be this. Our models *seem to be* satisfactory, in as much as they are *capable* of describing the observed patterns of increase and fluctuation as long as environmental fluctuations, interspecific interactions and the imperfections of sampling are taken into account. On the other hand, the discrepancies between our models and our data could be due, quite simply, to the essential inadequacies of our models, particularly the unrealistic simplifying assumptions they make. To examine their utility more critically, we must consider their ability to describe the effects of intraspecific competition on fecundity and survival. It should also be stressed, however, that a true test of a model's ability to describe population behaviour should not consist only of a comparison of *numbers*; it should also consider the underlying *biological* similarities: the respective R-values for instance, or the existence—in both model and fact—of a time-lag.

In considering the ability of our models to describe the effects of intraspecific competition, we must confine ourselves to the discrete-generation equation. The logistic model, restricted as it is to perfect compensation, cannot be expected to describe a range of situations; its utility lies in its simplicity and the ease with which it can be understood. The capabilities of our difference equation can be critically assessed, because, for any set of data the most appropriate values for a and b—those giving the 'best fit' to the data—can be estimated using a statistical technique (following Hassell 1975 and Hassell *et al.* 1976). The curve produced by the model can then be compared with the original data points. This has been done for several examples (including two from Chapter 2), and the results are shown as plots of k-values against $\log_{10}$ density in Fig. 3.12 (experimental data) and Fig. 3.13 (field data). As originally noted by Hassell (1975), there is a tendency, particularly under laboratory conditions, for the model to be very satisfactory at high and low densities, but incapable of describing the sudden transition from a shallow to a steep slope. This may be a result of the data reflecting two superimposed density-dependent processes (one weak and one strong), while the model treats them as a single process (Stubbs 1977). Nevertheless, the model's performance overall is impressive: whether the plot is more or less straight or curved, its fit to the data is very satisfactory.

A final example of the utility of the difference equation model is shown in Fig. 3.14. The data come from a pot experiment in which *Agrostemma githago*, corncockle, was sown over a wide range of densities. We can see that there were two components of density-dependent regulation: plant mortality and plasticity. Self-thinning was noticeable above sowing densities of 1000 seeds m^{-2} (Fig. 3.14a), whilst the number of seeds borne per flowering plant declined with density over the entire range (Fig. 3.14b).

Fitting the models (equations 3.5 and 3.6) described earlier to these data by statistical means, we can see that they do indeed give a very close fit to the observed results. It appears (again) that our basic difference equation model is very satisfactory in describing reality. Since, however, we can interpret the constants in the models in biological terms we can make further statements about the *Agrostemma* population. The data suggest that the maximum population size that can be supported (if the experiment is repeated again under identical conditions) is 10 869 seed-bearing plants m^{-2} ($= m^{-1} = 1/(9.2 \times 10^{-5})$, and that the maximum seed

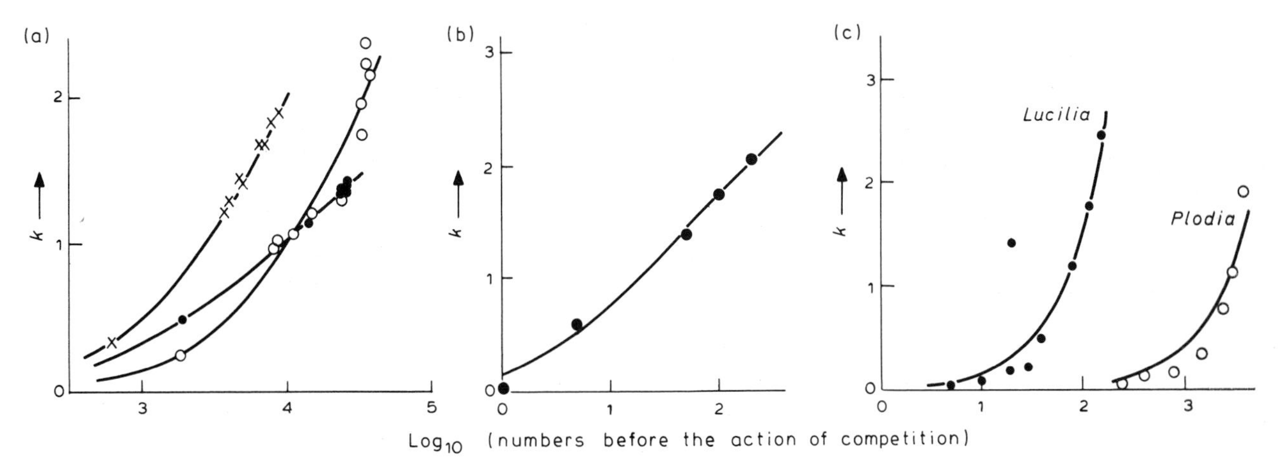

Fig. 3.12. Data from competition experiments and their description based on equation 3.4. (a) Beetles: *Callosobruchus chinensis* (●) $a = 0.00013$, $b = 0.9$ (Fujii 1968); *C. maculatus* (×) $a = 0.0006$, $b = 2.2$ (Utida 1967); *C. maculatus* (O) $a = 0.0001$, $b = 2.7$ (Fujii 1967). (After Hassell *et al.* 1976). (b) Shepherd's purse, *Capsella bursa-pastoris*: $a = 0.377$, $b = 1.085$ (Palmblad 1968). (c) The blow-fly, *Lucilia cuprina* (Nicholson 1954b); the moth, *Plodia interpunctella* (Snyman 1949). (After Hassell 1975.)

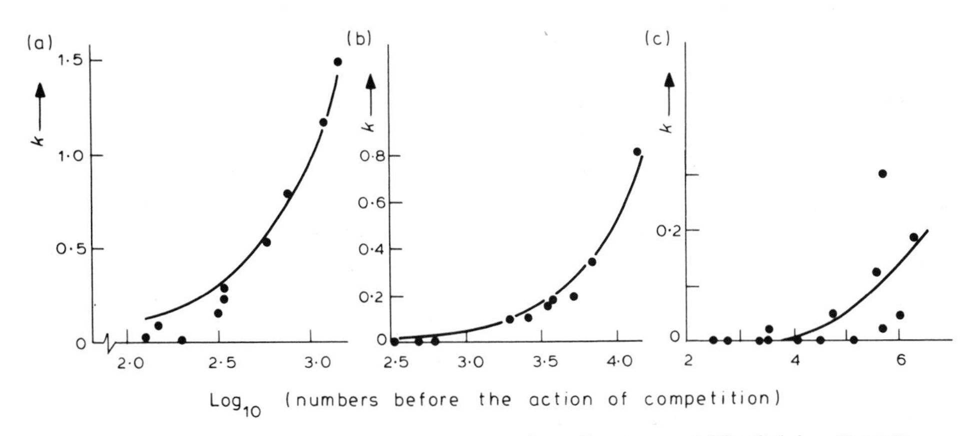

Fig. 3.13. Field data on intraspecific competition and their description based on equation 3,4 (a) The limpet, *Patella cochlear*: $a = 4.13\times10^{-5}$, $b = 51.49$ (Branch 1975). (b) The Colorado beetle, *Leptinotarsa decemlineata*: $a = 3.97\times10^{-6}$, $b = 30.95$ (Harcourt 1971). (c) Larch tortrix, *Zeiraphera diniana*: $a = 1.8\times10^{-5}$, $b = 0.11$ Auer 1968). (After Hassell 1975.)

yield of an isolated plant, λ, is 3685 seeds. Also, to grow to maximum size, a plant required 325 cm² ($a = 0.0325$) of space. The value of b (1.15) in equation 3.6 was significantly greater than unity, indicating overcompensation at high density. This is evident when the components of seed yield are examined (Fig. 3.14c). The effects of density did not fall equally on all parts of the plant. Most density stress was absorbed by the number of capsules per plant which declined at a constant rate with density as did seed weight. The number of seeds per capsule, however, fell sharply at high density reflecting the overcompensatory density response. The difference-equation model, despite its simplicity, does indeed appear to encapsulate the essential characteristics of the behaviour of single-species populations.

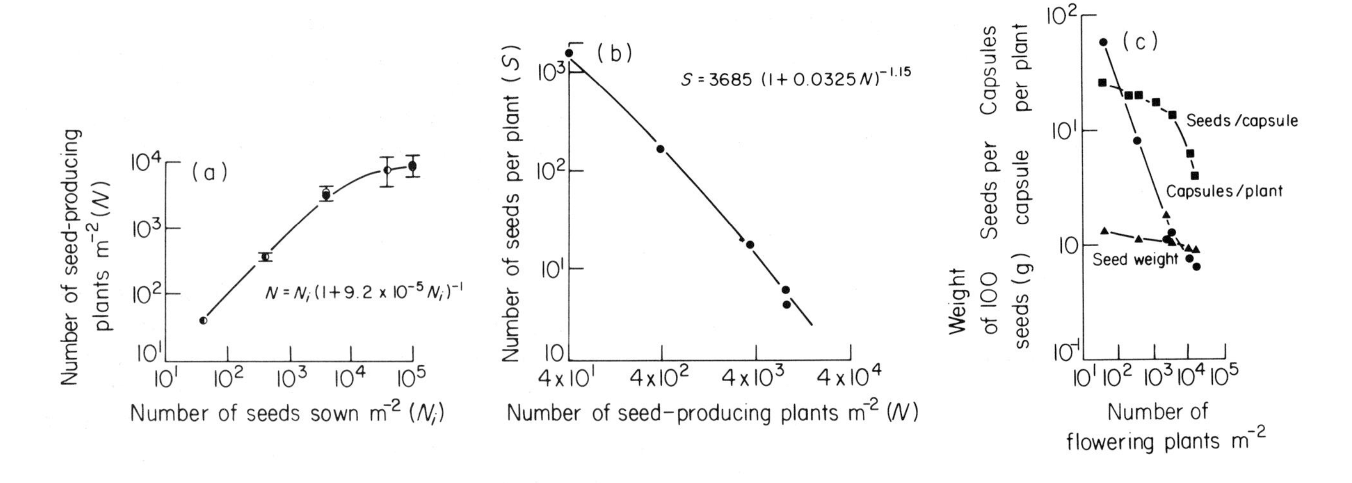

Fig. 3.14. Density dependent regulation in *Agrostemma githago*: (a) survival to reproduction; (b) seed production; (c) yield components. (From Watkinson 1981.)

3.5 Incorporation of age-specific fecundity and mortality

We have seen from the life-table data for annual meadow grass and red deer (Chapter 2) that mortality and fecundity are often age-specific. The models we have developed so far, however, do not include these important features, and to this extent they are deficient. It may seem a formidable task to attempt to model a population of overlapping generations in these terms, but by the application of appropriate mathematical techniques it can be achieved fairly easily.

The best starting point is the diagrammatic life-table for overlapping generations (Fig. 1.6), expanded into a more complete form (Fig. 3.15). Here we suppose that a population can be conveniently divided into four age groups: a_0, a_1, a_2 and a_3; a_0 representing the youngest adults and a_3 the oldest. In a single time step, t_1 to t_2, individuals from group a_0, a_1 and a_2 pass to the next respective age group; each age group contributes new individuals to a_0 (through birth); and the individuals in a_3 die. This clearly rests on the assumption that the population consists of discrete age groupings and has discrete survivorship and birth statistics, in contrast to the reality of a continuously ageing population.

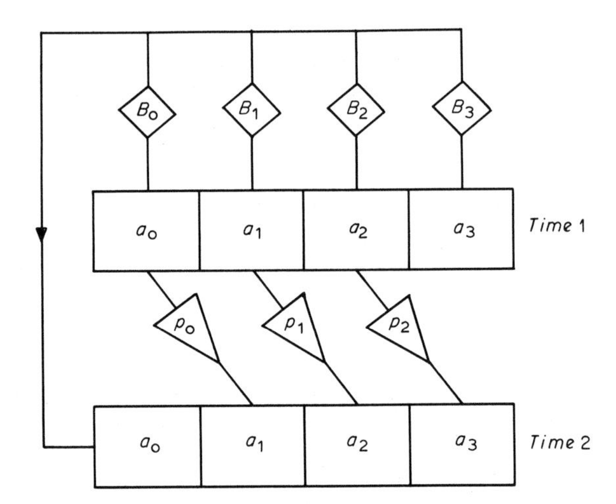

Fig. 3.15. The diagrammatic life-table for a population with overlapping generations: a = numbers in different age-groups, B = age-specific fecundities and p = age-specific survivorships.

We can write a series of algebraic equations to express the changes that are occurring in Fig. 3.15. These are:

$$ {}_{t_2}a_0 = ({}_{t_1}a_0 \times B_0) + ({}_{t_1}a_1 \times B_1) + ({}_{t_1}a_2 \times B_2) + ({}_{t_1}a_3 \times B_3) \quad (3.10) $$

$$ {}_{t_2}a_1 = ({}_{t_1}a_0 \times p_0) \quad (3.11) $$

$$ {}_{t_2}a_2 = ({}_{t_1}a_1 \times p_1) \quad (3.12) $$

$$ {}_{t_2}a_3 = ({}_{t_1}a_2 \times p_2) \quad (3.13) $$

where the numbers in the age group are subscripted t_1 or t_2 to identify the time period to which they refer. There are four equations because there are four age groups, and they specifically state how the numbers in age groups are determined over the time-step t_1 to t_2.

Suppose that our population at t_1 consists of 2000 individuals distributed among the age groups as follows:

$$ {}_{t_1}a_0 = 1750 \quad {}_{t_1}a_1 = 100 \quad {}_{t_1}a_2 = 100 \quad {}_{t_1}a_3 = 50. $$

Suppose, further, that the age-specific birth- and survival-rates are as follows:

Age	*B*	*p*
0	0	0.1
1	5	0.6
2	15	0.3
3	10	0.0.

Then, for instance, no offspring are born to adults aged a_0, whilst adults in age group a_2 are the most fecund. Also, only 10% of the individuals aged a_0 survive the next time-step and become incorporated into a_1, whereas none of the individuals in a_3 survive ($p = 0$). The numbers at time t_2 will be:

$$ \begin{aligned} {}_{t_2}a_0 &= (1750\times0)+(100\times5)+(100\times15)+(50\times10) &= 2500 \\ {}_{t_2}a_1 &= 1750\times0.1 &= 175 \\ {}_{t_2}a_2 &= 100\times0.6 &= 60 \\ {}_{t_2}a_3 &= 100\times0.3 &= 30 \\ & & \overline{2765} \end{aligned} $$

During this time-step our population has increased by 765 individuals and our age distribution has changed:

from			
	$a_0 =$	1750 to	2500
	$a_1 =$	100	175
	$a_2 =$	100	60
	$a_3 =$	50	30.

It is important to realize that these equations state that *individuals reproduce before they die.* Taking age group a_2, for instance, the figure 100 was used in equation 3.10 to compute the number of births ($100\times15 = 1500$), but *then* used in equation 3.13 to compute the number of a_2 survivors into age group a_3 ($100\times0.3 = 30$).

3.5.1 *The matrix model*

The equations, 3.10 to 3.13 are called 'linear recurrence equations', but it is clear that in this form our model is rather cumbersome, consisting as it does of as many equations as there are age groups. One way of expressing our model in a much more compact form is by the use of matrix algebra. In essence, matrix algebra is a tool designed to manipulate and store large sets of data. We will certainly not need to cover all aspects of matrix algebra to understand how our model is constructed in matrix terms, but it is vital that the basic concepts are clearly understood.

A matrix is simply a group, or table, or array of numbers. For instance, the numbers for B and p envisaged above could be set out as the matrix:

$$ \begin{bmatrix} 0 & 0.1 \\ 5 & 0.6 \\ 15 & 0.3 \\ 10 & 0.0 \end{bmatrix} $$

and we signify that it *is* a matrix by surrounding the numbers by square brackets. Conventionally, matrices are symbolized by a letter in bold face, say **X**.

The number of rows and columns that comprise a matrix may vary, and we may have matrices consisting of only a single column. We can, in fact, write the initial age structure of our previous population in this way:

$$ \begin{bmatrix} 1750 \\ 100 \\ 100 \\ 50 \end{bmatrix} $$

and symbolize it as ${}_{t_1}\mathbf{A}$. Our age distribution at t_2, ${}_{t_2}\mathbf{A}$, can obviously then be written as

$$ \begin{bmatrix} 2500 \\ 175 \\ 60 \\ 30 \end{bmatrix} $$

The two matrices ${}_{t_1}\mathbf{A}$ and ${}_{t_2}\mathbf{A}$ are technically called *column vectors,* indicating that our matrices are in fact just one column of figures.

We have seen how our population changes from ${}_{t_1}\mathbf{A}$ to ${}_{t_2}\mathbf{A}$ through the medium of the recurrence equations. Now, to complete our matrix model, we have to construct a suitable matrix to enable ${}_{t_1}\mathbf{A}$ to become ${}_{t_2}\mathbf{A}$ by multiplication. The construction of this matrix is determined by the rules of matrix multiplication, as we shall learn below, and in consequence it appears as:

$$\begin{bmatrix} 0 & 5 & 15 & 10 \\ 0.1 & 0 & 0 & 0 \\ 0 & 0.6 & 0 & 0 \\ 0 & 0 & 0.3 & 0 \end{bmatrix} = \mathbf{T}.$$

Note that matrix **T** is square, and that we have entered into it all our age-specific survival and birth statistics in particular positions, writing all the other numbers (or matrix *elements* as they are called) as zero.

The multiplication of our initial, age-distributed population occurs according to the rules of matrix algebra, as follows. Take, in turn, each *row* of elements in **T**. Each individual element in the row is multiplied with the corresponding element in ${}_{t_2}\mathbf{A}$: the first with the first, the second with the second, and so on. These pairwise multiplications are then summed, and the sum entered as the appropriate element in a new column vector. Thus, the first row of the square matrix leads to the first element of the new column vector, the second leads to the second, and so on. This may be illustrated as:

$$\begin{bmatrix} 0 & 5 & 15 & 10 \\ 0.1 & 0 & 0 & 0 \\ 0 & 0.6 & 0 & 0 \\ 0 & 0 & 0.3 & 0 \end{bmatrix} \times \begin{bmatrix} 1750 \\ 100 \\ 100 \\ 50 \end{bmatrix} = \begin{bmatrix} (0)(1750)+(5)(100)+(15)(100)+(10)(50) \\ (0.1)(1750)+(0)(100)+(0)(100)+(0)(50) \\ (0)(1750)+(0.6)(100)+(0)(100)+(0)(50) \\ (0)(1750)+(0)(100)+(0.3)(100)+(0)(50) \end{bmatrix} = \begin{bmatrix} 2500 \\ 175 \\ 60 \\ 30 \end{bmatrix}$$

We can observe now that the positions of the zeros in the matrix **T** are critical, because they reduce terms in each row of the multiplication to zero, giving rise in ${}_{t_2}\mathbf{A}$ to the age-structure that we have already seen.

This matrix model was first introduced to population biology by P. H. Leslie in 1945 and is often known as the Leslie matrix model. In general form, for n age groups, it is:

$$\begin{bmatrix} B_0 & B_1 & B_2 & \dots & B_{n-1} & B_n \\ p_0 & 0 & 0 & \dots & 0 & 0 \\ 0 & p_1 & 0 & \dots & 0 & 0 \\ 0 & 0 & p_2 & \dots & 0 & 0 \\ \cdot & \cdot & \cdot & & \cdot & \cdot \\ \cdot & \cdot & \cdot & & \cdot & \cdot \\ \cdot & \cdot & \cdot & & \cdot & \cdot \\ 0 & 0 & 0 & \dots & p_{n-1} & 0 \end{bmatrix} \times \begin{bmatrix} {}_{t_1}a_0 \\ {}_{t_1}a_1 \\ {}_{t_1}a_2 \\ {}_{t_1}a_3 \\ \cdot \\ \cdot \\ \cdot \\ {}_{t_1}a_n \end{bmatrix} = \begin{bmatrix} {}_{t_2}a_0 \\ {}_{t_2}a_1 \\ {}_{t_2}a_2 \\ {}_{t_2}a_3 \\ \cdot \\ \cdot \\ \cdot \\ {}_{t_2}a_n \end{bmatrix} \quad (3.14)$$

which may, alternatively, be written:

$$\mathbf{T} \times {}_{t_1}\mathbf{A} = {}_{t_2}\mathbf{A}. \quad (3.15)$$

T is called the *transition matrix,* which, when *post-multiplied* by the vector of ages at t_1, gives the age distribution at t_2. (In matrix algebra, in contrast to conventional algebra, $\mathbf{T}\times\mathbf{A}$ is different from $\mathbf{A}\times\mathbf{T}$ (cf. $x \times b = b \times x$), and we distinguish between the two by referring to post- and pre-multiplication.)

By now it must be clear that our matrix model allows us to condense the complexities of age-specific schedules into a simply written but explicit form. It might also seem that to use the model requires endlessly repeated multiplications and additions. This disadvantage has been removed, however, by the widespread use of computers, which are ideally suited (not to say designed) to perform such iterative procedures at speed. Our model is, therefore, of great value.

The specific rules and procedures of matrix algebra are numerous. We have only dealt here with those which are necessary in this particular context. If further explanations are needed, reference may be made to one of the several books on matrix algebra for biologists (e.g. Searle 1966).

3.5.2 *Using the model*

If we now wish to compute the changing size of our population, on the assumption that the birth and mortality statistics are constant from one time to the next, we can write:

$$\mathbf{T}\times{}_{t_1}\mathbf{A} = {}_{t_2}\mathbf{A}; \quad \mathbf{T}\times{}_{t_2}\mathbf{A} = {}_{t_3}\mathbf{A}; \quad \mathbf{T}\times{}_{t_3}\mathbf{A} = {}_{t_4}\mathbf{A},$$

and so on. This is a process of iterative premultiplication of the successive age-groups by the transition matrix, giving:

$$\mathbf{T}\times\begin{bmatrix}1750\\100\\100\\50\end{bmatrix}=\begin{bmatrix}2500\\175\\60\\30\end{bmatrix}; \quad \mathbf{T}\times\begin{bmatrix}2500\\175\\60\\30\end{bmatrix}=\begin{bmatrix}2075\\250\\105\\18\end{bmatrix};$$

and

$$\mathbf{T}\times\begin{bmatrix}2075\\250\\105\\18\end{bmatrix}=\begin{bmatrix}3005\\207\\150\\32\end{bmatrix}$$

Over these three time-steps the age-distribution is changing and the population is increasing in size; the numbers of individuals in all age-groups are oscillating. If we were to continue the iteration, however, these oscillations would disappear (Fig. 3.16), and after 17 and 18 multiplications the vectors would be

$$\begin{bmatrix}24845\\2110\\1070\\272\end{bmatrix} \text{ and } \begin{bmatrix}29320\\2484\\1266\\321\end{bmatrix} \text{ respectively.}$$

Note that the population has grown enormously, but that the ratios of ${}_{t_{18}}a_x : {}_{t_{17}}a_x$ are all equal (to two significant figures at least),

$$\text{i.e.} \quad \frac{29320}{24845} = \frac{2484}{2110} = \frac{1266}{1070} = \frac{321}{272} = 1.18.$$

This means that we have reached a situation in which the age-distribution is proportionally constant or *stable* as the population increases. In other words, $a_0 : a_1 : a_2 : a_3$ is the same in all subsequent generations. In fact, this particular stable age-distribution is an attribute of the transition matrix, and would have been reached irrespective of the initial column vector; and in general,

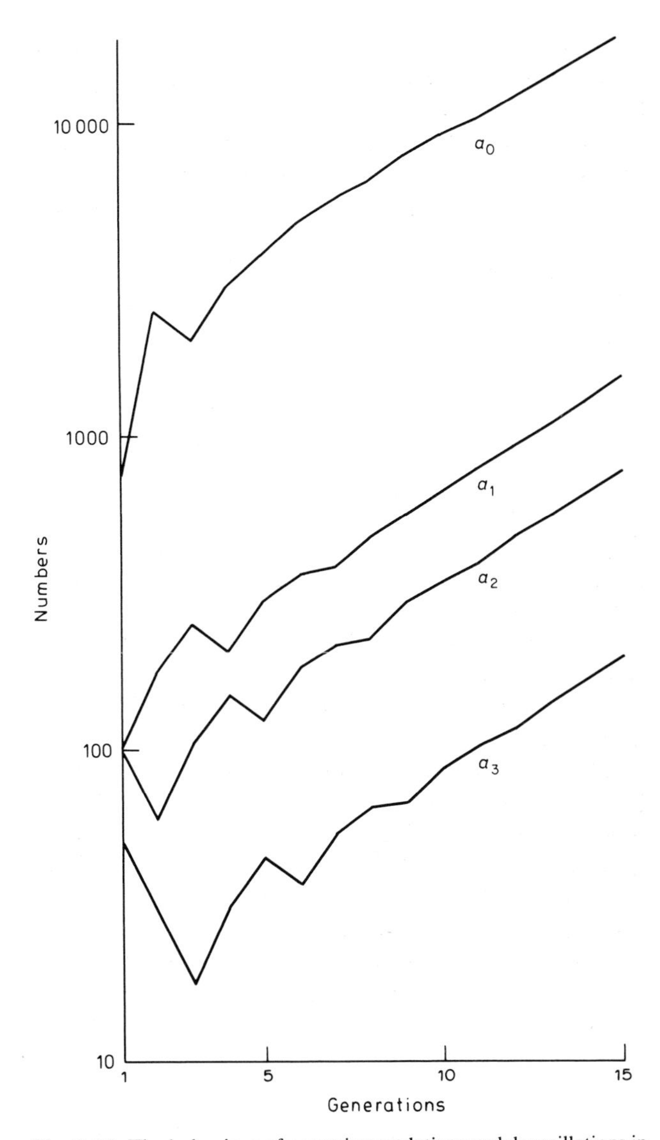

Fig. 3.16. The behaviour of a matrix population model: oscillations in an age-distributed population as a stable age distribution is approached. See text for details.

populations, when subjected to repeated pre-multiplication by the same transition matrix, achieve the stable age-distribution characteristic of that matrix.

Moreover, as Fig. 3.16 shows, the ratios from one generation to the next (1.18 in this case) are also constant: the population (as well as each of the age-classes within it) is increasing at a constant rate. This rate is also a reflection of the transition matrix, and is, in fact, the net reproductive-rate or finite rate of increase per unit time, R. Repeated pre-multiplication is, therefore, the mathematically simplest means of determining R from an age-specific schedule of births and deaths.

We should not, however, simply dismiss the matrix model as a device for only handling sets of recurrence equations in an iterative manner. A transition matrix, stating age-specific data as it does, is a precise description of the population statistics and can be subjected to algebraic analysis. We can, for instance, find R by calculating 'the dominant latent root' of the transition matrix, which is itself equal to R, although such a calculation requires some further knowledge of matrix algebra (see for instance Usher 1972), and is beyond the scope of this book. A further usage of transition matrices comes in harvesting theory (Chapter 5), which allows us to predict the level of 'cropping' that can occur without driving a population to extinction. All that need be appreciated at this stage is that matrix modelling can lead to sophisticated analyses of populations and their behaviour.

Our computations of R and the stable age-distribution rest on the assumption that the elements of the transition matrix are *constant over time* and *independent of population density*. In these respects our matrix model, in its present form, is very much divorced from the biological reality of natural populations. We can incorporate temporal changes in fecundity and survival in a very straightforward manner, by simply changing our transition matrix with every time-step. Thus, whereas we have assumed that:

$$\mathbf{T}\times{}_{t_1}\mathbf{A} = {}_{t_2}\mathbf{A} \quad \text{and} \quad \mathbf{T}\times{}_{t_2}\mathbf{A} = {}_{t_3}\mathbf{A},$$

so that

$$\mathbf{T}\times\mathbf{T}\times{}_{t_1}\mathbf{A} = {}_{t_3}\mathbf{A},$$

we can assume instead that, in general terms, for r time periods:

$$\mathbf{T}_r\times\mathbf{T}_{r-1}\times\mathbf{T}_{r-2}\ldots\mathbf{T}_2\times\mathbf{T}_1\times{}_{t_1}\mathbf{A} = {}_{t_r}\mathbf{A} \quad (3.16)$$

Furthermore, we can incorporate the idea that the birth and death statistics are density-dependent by varying the fecundity and survival elements in the transition matrix in relation to population size. To achieve this, we need simply derive the relevant elements of the successive transition matrices from equations which relate fecundity and survivorship to the population size, which is itself the sum of the elements of the most recent age vector.

3.5.3 *A working example:* Poa annua

Both of these additions to the model are well illustrated in practice by the work of Law (1975) on the annual meadow grass, *Poa annua*: a particularly successful colonizer of open habitats, which responds strongly to intraspecific competition. In modelling populations of this species, Law envisaged that the life cycle included four ages of plants besides seeds (Fig. 3.17), and that the span of each age group was approximately 8 weeks. The transition matrix appropriate to this life cycle is given in Table 3.1 and, although most of the elements are

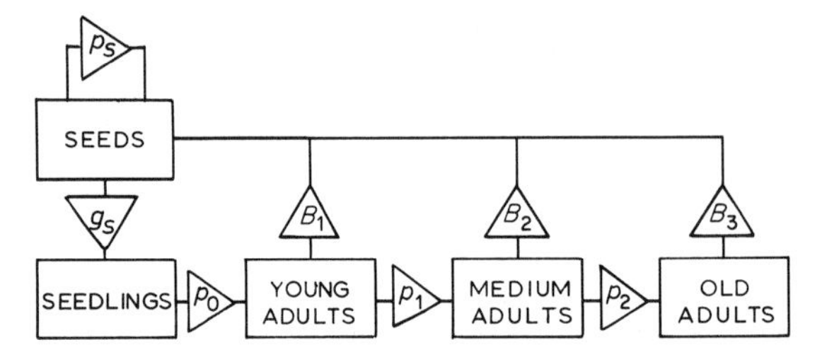

Fig. 3.17. The life cycle for annual meadow grass, *Poa annua*. (After Law 1975.) Proportions (ranging from 0 to 1): p_s = seeds surviving; g_s = seeds germinating; p_0 = seedlings surviving to become young adults; p_1 = young adults surviving to become medium adults; p_2 = medium adults surviving to become old adults. Fecundities (0 or > 0): B_1, B_2 and B_3 = seed produced by young, medium and old adults respectively. All events occur over the time period t to $t+1$.

Table 3.1 A matrix model of the life cycle outlined in Fig. 3.17.

$$\begin{bmatrix} p_s & 0 & B_1 & B_2 & B_3 \\ g_s & 0 & 0 & 0 & 0 \\ 0 & p_0 & 0 & 0 & 0 \\ 0 & 0 & p_1 & 0 & 0 \\ 0 & 0 & 0 & p_2 & 0 \end{bmatrix}$$

familiar to us, the incorporation of a seed bank into the model requires that the first element of the matrix (p_s) is not a fecundity but the probability of a seed surviving in the bank if it does not germinate.

Law was able to develop curves of both seedling survivorship to young adults, and age-specific seed output per plant in relation to overall population density. These are shown in Fig. 3.18 together with the equations that describe these relationships. The transition matrix which incorporates these density-dependent functions of survival and reproduction takes the form shown in Table 3.2. Between successive time intervals, 0.2 of the seeds in the banks remain dormant and survive, while 0.05 of them germinate. The proportion of individuals that survive from the seedling age-class, $p_0(N)$, is initially 0.75 at very low population density, but declines according to the function shown in Fig. 3.18. The proportion of individuals that survive from the young adult age-class onward, however, is fixed at 0.75. This represents the specific assertion that the survivorships of individuals in these age-classes are not subject to density-dependence. The maximum seed production of young and old adults at very low population density is 100 seeds per plant, but, with increasing density, this decreases in a negative exponential fashion. Similarly, the maximum seed production of medium adults is 200 seeds per plant, which also decreases with density. This reflects *P. annua*'s schedule of fecundity in which seed output peaks in the medium adult age-class (see Fig. 1.8).

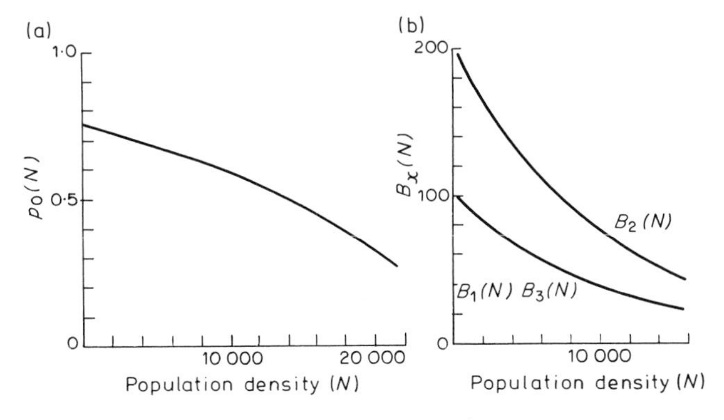

Fig. 3.18. (a) The proportion of seedlings of *Poa annua* surviving to become young adults as a function of population density:
$p_0(N) = 0.75 - 0.25 \exp(0.00005\,N)$ if $N < 27726$
$p_0(N) = 0$ if $N > 27726$.
(b) Age-specific seed production per plant for young (B_1) medium (B_2) and old (B_3) adults of *Poa annua* as a function of population density:
$B_1(N) = B_3(N) = 100 \exp(-0.0001\,N)$
$B_2(N) = 200 \exp(-0.0001\,N)$. (From Law 1975.)

Table 3.2 A transition matrix, **P**, for a *Poa annua* population. (After Law 1975.)

$$\begin{bmatrix} 0.2 & 0 & B_1(N) & B_3(N) & B_2(N) \\ 0.05 & 0 & 0 & 0 & 0 \\ 0 & p_0(N) & 0 & 0 & 0 \\ 0 & 0 & 0.75 & 0 & 0 \\ 0 & 0 & 0 & 0.75 & 0 \end{bmatrix}$$

The matrix model,

$$\mathbf{P} \times {}_{t_r}\mathbf{A} = {}_{t_{r+1}}\mathbf{A}$$

can now be used to simulate the behaviour of the population, which is depicted in Fig. 3.19. We can see that

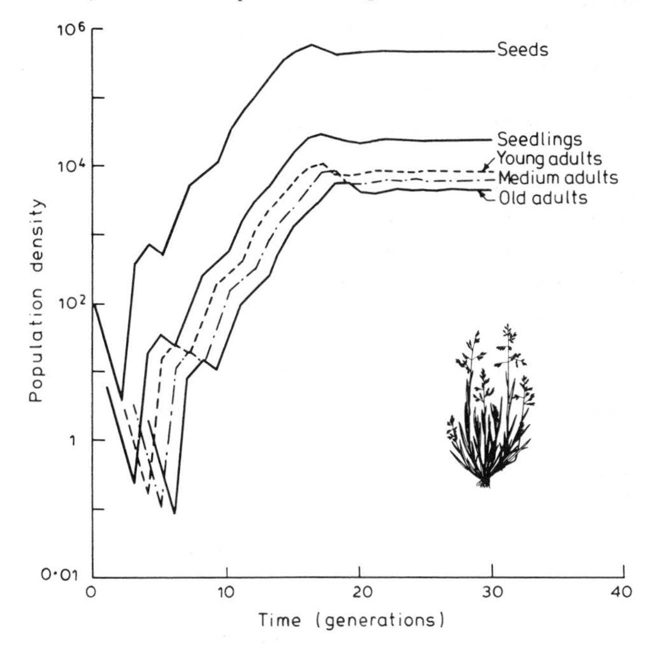

Fig. 3.19. Application of matrix model: simulation of the density-regulated changes in density of five age-classes in a population of *Poa annua*. (From Law 1975.)

each group of individuals undergoes oscillations in numbers, but when 18 time periods have elapsed the size of each has stabilized and population increase has been halted by intraspecific competition. (Note that the age-structure stabilization after 18 time periods both here and in Section 3.5.2 is purely coincidental.)

Seed germination and seed production in *P. annua* vary considerably with the time of year, and to model this realistically it would be necessary to introduce a different transition matrix for each month of the year. The number of seeds per individual plant in each age-class would then be a function of population density and time during the season. Although we do not need to examine the workings of this sophistication, we should realize that the facility of changing elements in successive transition matrices is essential for biological reality.

Finally, it is interesting to note that the net reproductive-rate, R, can still be estimated, even from this more sophisticated form of the model. R is, by definition, the rate of increase in the *absence* of competition. If we, therefore, consider the transition matrix **P** with its elements unaffected by density, we can derive a figure for R as previously described. Moreover, by considering a variety of transition matrices, based on a range of densities, we can examine the density-dependent effects on the *actual* reproductive-rate. This is illustrated in Fig. 3.20.

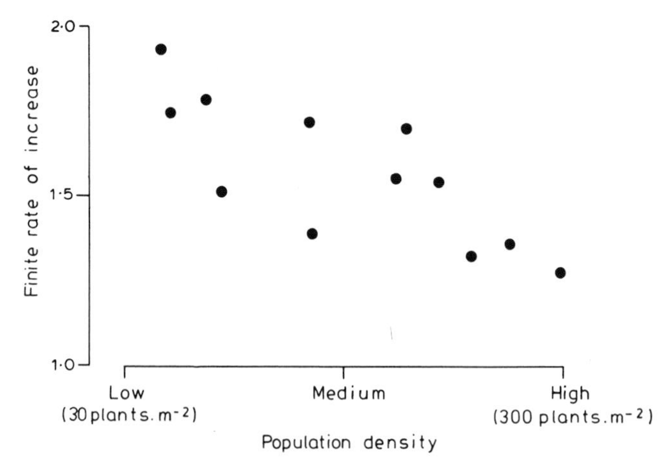

Fig. 3.20. Net reproductive-rate of *Poa annua* growing at a range of population densities. (Data from Law 1975.)

PART 2 INTERSPECIFIC INTERACTIONS

Chapter 4 Interspecific Competition

4.1 The nature of interspecific interactions

In Chapters 2 and 3 we examined the consequences of individuals of the same species competing with one another. Natural communities, however, are usually assemblages of species, and in many habitats the 'neighbours' of any one individual may well be of a different species to the individual itself. Yet, although the *process* of interaction between species might occur in a variety of ways (fighting in animals, shading in plants, etc.), there are only three basic *effects*. One species may cause (or be the cause of) increases in the survival, growth or fecundity of another species, or it may cause decreases, or may have no effect at all. Placing these effects in symbolic form in Table 4.1, we can see that there are six possible outcomes, in five of which at least one species is affected.

The most obvious example of the '+ −' type of interaction is a predator–prey relationship in which one species is eaten by the other; but there are other situations in which one species provides a food-source or growth-requirement for a second species at the first species' expense. Herbivores 'prey' on vegetation but rarely destroy entire plants by grazing; and parasites, while reducing the vitality or fecundity of their host, do not usually cause its immediate death. All aspects of the '+ −' interaction will be considered in detail in Chapter 5.

Table 4.1 The effects of species 1 on survival, growth or fecundity in species 2. The effects of species 2 on survival, growth or fecundity in species 1.

		The effects of species 1 on fitness in species 2		
		Increase	Neutral	Decrease
The effects of species 2 on fitness in species 1	Increase	+ +		
	Neutral	0+	00	
	Decrease	− +	−0	− −

We can only make passing reference to the '+0', '−0' and '+ +' interactions. This is not to belie their biological significance, but more to indicate the lack of consideration they have received in terms of population analysis. Commensalism (+0) is the state in which prerequisite conditions for the existence of one species are maintained or provided by a second, but in which there is no associated adverse effect for the second species. Saprophytism between fungi and higher plants might enter this category, as would 'parasites' that have no measurable effect on their host. The reverse of commensalism is amensalism (−0), an often-cited example of which is allelopathy between plants: toxic metabolite production by one species causing growth-reduction in another. Allelopathy is difficult to investigate in controlled laboratory experiments, and its role in the field is nothing but uncertain (see Harper 1977).

Mutualism (+ +) completes this group of three 'minor' categories, and although it represents a uniquely fascinating topic in evolutionary terms, it has received relatively little attention from population ecologists. Nevertheless, a single example (Janzen 1966) will illustrate the essential continuity between mutualism and other interactions. The bullhorn acacia of Central America (*Acacia cornigera*) gets its name from the pairs of large, swollen, hollow spines it bears on its trunk. The spines have a patch of thin tissue on one side, and small, aggressive, stinging ants (*Pseudomyrmex ferruginea*) perforate this tissue and nest inside the spine. The ants also feed on nectar produced at the base of the bullhorn acacia leaves, and on protein-rich bodies produced at the leaf-tips, and they are, therefore, able to complete their whole life cycle on *A. cornigera*. In addition, however, the ants attack any insects that attempt to eat the acacia leaves, and they cut the shoots of any other plants which come into contact with the acacia and may shade it. As Table 4.2 shows, therefore, the ants, apart from obtaining food and a protected place to live from the acacia, also cause measurable *improvements* in the fecundity and the survivorship of the acacia itself by

Table 4.2. Effects of the ant *Pseudomoyrmex ferruginea* on the bullhorn acacia, *Acacia cornigera*. (After Janzen 1966.)

	Ants present	Ants removed
Weight of suckers (g)	41 750	2900
Number of leaves	7785	34600
Average growth in 45 days (cm)	72.6	10.23
Percentage mortality	28	56
Percentage of shoots with other insects: day	2.7	38.5
night:	12.9	58.8
Mean number of insects per shoot: day	0.039	0.881
night	0.226	2.707

protecting it from predators and competitors. Fecundity and survivorship are clearly the common currency linking mutualism with every other ecological interaction.

4.2 Interspecific competition

The final interaction in Table 4.1 is interspecific competition, and the '− −' symbolism stresses its essential aspect: that the two species cause *demonstrable* reductions in *each other*'s survival, growth or fecundity. Nevertheless, having stressed this, it must also be emphasized that whenever two species compete there will be some circumstances in which one species will be very much more affected than the other. In such 'one-sided' cases of competition it may even be impossible to discern any measurable detrimental effects on the stronger competitor. These cases will *appear* to be amensal. (The relationship between amensalism and interspecific competition is a subtle one which we shall consider in more detail in Section 4.11.)

As with intraspecific competition, we can expect that the detrimental effect of interspecific competition will act through some combination of fecundity and survivorship; that the interaction will be *essentially* reciprocal; that competition will be for a resource which is in limited supply; and that the effects will be density-dependent.

We can, however, expect important differences between the precise nature of interspecific competition in animals and plants. All animals obtain their food from the growth, reproduction and by-products of other living organisms, and they commonly compete for this food. The factors required for plant growth, on the other hand, and for which they compete, are neither self-sustaining nor the products of reproductive processes. Moreover, because plants are sessile organisms, they will, once rooted and fixed in position, interfere mainly with their *neighbours*' growth and reproduction (also true of sessile animals, of course). Amongst most animals, on the other hand, there is rarely a continuous struggle between two, or even a few, individuals. Moreover, whereas two plants in close proximity to one another may immediately suggest the possibility of competition for a limited resource (perhaps soil nitrogen, or light in a canopy of leaves), many animal species may never even encounter their competitors (because of differences in foraging strategy, feeding times and so on). In assessing the nature of competitive interactions, therefore, particularly amongst animals, very detailed observation and exacting experimentation are required. Nevertheless, despite these difficulties and differences, there is, as we shall see, a coherent view of interspecific competition which applies to both animals and plants.

4.3 A field example: granivorous ants

Desert environments are of interest to physiologists because of the opportunities they present for studying animals and plants adapted to the extremes of water-shortage; but because they are rather simple environments, there are also many instances in which ecological studies in deserts have been very instructive. One example is the work of Davidson (1977a, b, 1978) and Brown & Davidson (1977) on interspecific competition in the seed-eating ants and rodents living in the deserts of the south-western United States. In examining this example, we shall take the approach which we try to follow throughout the present chapter. First we shall discover whether or not interspecific competition occurs, *then* we shall examine the form it takes and its consequences.

Seeds play a major role in desert ecology, since they constitute a dormant, resistant stage in the life-histories of plants, allowing them to survive the long, unfavourable intervals between short periods of growth. But these seeds are also a food-source for several, distantly related taxa (including ants and rodents), which feed as specialized granivores. It is well established that in arid regions mean annual precipitation is a good measure of productivity; and this productivity will determine the size of the seed resource available to these granivorous animals. Thus, the graphs in Fig. 4.1 (Brown & Davidson 1977) of numbers of common ant and rodent species against mean annual precipitation, indicate that species number is correlated with the size of the seed resource. This suggests that for both granivorous guilds—ants and rodents—the size of the food-resource limits the number of common species, and probably also the total number of individuals (a 'guild' is defined as a group of species exploiting the same resource in a similar fashion; Root 1967). Moreover, by actually taking seeds from foraging ants and rodents, Brown and Davidson were also able to show that the two guilds overlapped considerably in the sizes of seeds they ate (Fig. 4.2), suggesting not only that the ants and rodents are limited by their food, but also that the two guilds compete with one another for this limiting resource.

These are only suggestions, however: plausible deductions from field correlations. Realizing this, Brown and Davidson performed an experiment in which

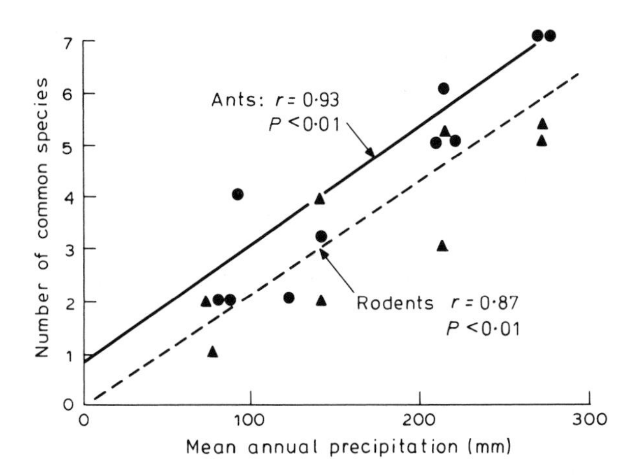

Fig. 4.1. Patterns of species diversity of seed-eating rodents (▲) and ants (●) inhabiting sandy soils in a geographic gradient of precipitation and productivity. (After Brown & Davidson 1977.)

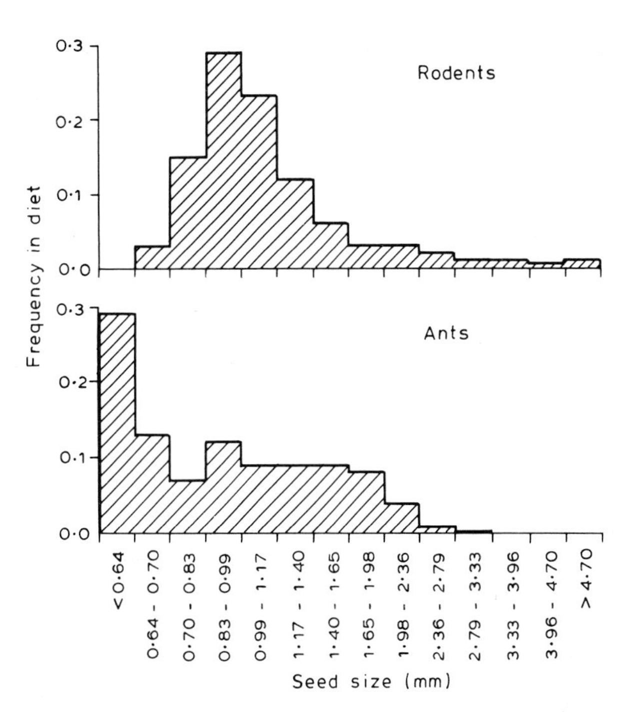

Fig. 4.2. The foods of ants and rodents overlap: sizes of native seeds harvested by coexisting ants and rodents near Portal, Arizona. (After Brown & Davidson 1977.)

four types of 36-m diameter plots were established in relatively level, homogeneous desert scrub. In two plots, rodents were excluded by trapping residents and fencing to preclude immigration; in another two plots, ants were removed by repeated insecticide applications; in two further plots, both rodents and ants were removed and excluded; and, finally, two plots were reserved as unmanipulated controls. The results are shown in Table 4.3, and constitute positive evidence that the two guilds compete interspecifically with one another. When either

Table 4.3 Competition affects competitors and the resource competed for. Responses of ants, rodents and seed density to ant and rodent removal. (After Brown & Davidson 1977.)

	Rodents removed	Ants removed	Rodents and ants removed	Control
Ant colonies	543	—	—	318
Rodents numbers	—	144	—	122
Seed density relative to control	1.0	1.0	5.5	1.0

rodents or ants were removed, there was a statistically significant increase in the numbers of the other guild; the reciprocally depressive effect of interspecific competition was clearly shown. Moreover, when rodents were removed, the ants ate as many seeds as the rodents and ants had previously eaten between them, as did the rodents when the ants were removed; only when both were removed did the amount of resource increase. In other words, under normal circumstances both guilds eat less and achieve lower levels of abundance than they would do if the other guild was absent. This clearly indicates that the rodents and ants, although they coexist in the same habitat, compete interspecifically with one another. It also suggests strongly that the resource for which they compete is seed.

Davidson (1977a, b) went on from this to examine the various species of ants more closely. She was particularly interested in two facets of the ants' feeding ecology. The first of these was the relationship between a species' worker body length and the size of the seeds which the species harvested. Some of her data are illustrated in Fig. 4.3 (Davidson 1977a), which represents the results of an experiment in which eight species of granivorous ants were presented with an artificially produced range of seeds and seed fragments of various sizes. Workers and the seeds they were carrying were then sampled and measured. The points in Fig. 4.3 refer to mean values, and therefore fail to illustrate the fact that species overlap considerably in the sizes of seed that they take. Nevertheless, it is clear from Fig. 4.3 that seed size and body size are strongly correlated, and that each species tends to *specialize* in seeds of a particular size depending on its own size.

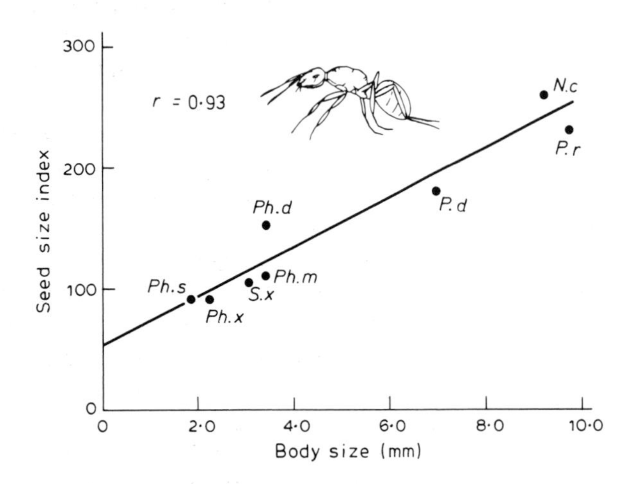

Fig. 4.3. The relationship between worker body length and seed size index for experiments with eight species of seed-eating ants near Rodeo, New Mexico. Species designations are as follows: *N. c.* = *Novomessor cockerelli. Ph. d.* = *Pheidole desertorum; Ph. m.* = *Pheidole militicida; Ph. s.* = *Pheidole sitarches; Ph. x* = *Pheidole xerophila; P. d.* = *Pogonomyrmex desertorum; P. r.* = *Pogonomyrmex rugosus; S. x.* = *Solenopsis xyloni.* All species except *Ph m.* coexist at Rodeo. (After Davidson 1977a.)

The second facet studied by Davidson was the species' foraging strategy (Davidson 1977b), of which there were essentially two types: 'group' and 'individual'. Workers of group-foraging species tend to move together in well-defined columns, so that, at any one time, most of the searching and feeding take place in a restricted portion of the area surrounding the nest. By contrast, in colonies of individual foragers, workers search for and collect seeds independently of one another, and, as a result, all of the area surrounding the colony is continuously and simultaneously searched. From a series of observations and experiments, Davidson was able to show that group foraging was more efficient than individual foraging when seed densities were high and when the distribution of seeds was clumped; but the relative efficiencies were reversed at low seed densities and when the seeds were more evenly distributed. The situation with regard to foraging is therefore directly comparable with that regarding size. Group and individual foragers show a considerable potential for overlap in the seeds which they harvest, but each specializes in a particular arrangement of the resource. In fact, the specialization resulting from foraging strategy tends to be a temporal one. Group foragers have marked peaks of activity coinciding with periods of high seed density, and pass less favourable periods in a 'resting' state. Individual foragers, although very active at high seed densities, retain intermediate levels of activity even during the less favourable periods.

Bearing these observations on size and foraging strategy in mind, we can turn now to some of Davidson's (1977a) results concerning the occurrence of various ant species at a range of sites. These are illustrated in Fig. 4.4. It will be convenient, initially, to restrict our

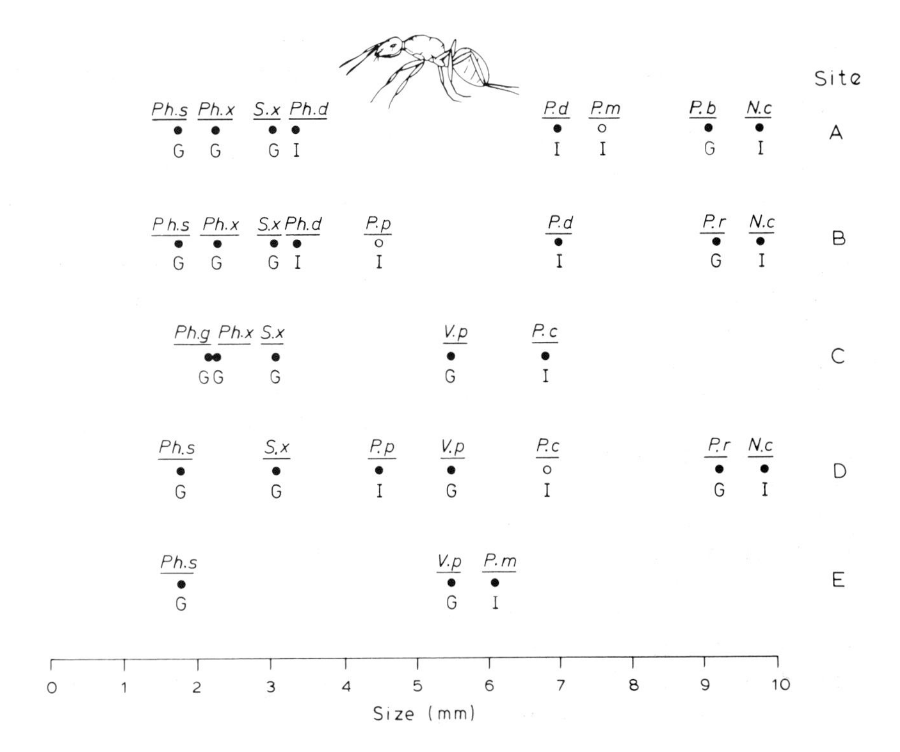

Fig. 4.4. Mean worker body lengths of seed-eating ants at five sites: A = Rodeo 'A', New Mexico; B = Rodeo 'B', New Mexico; C = Casa Grande, Arizona; D = Ajo, Arizona; E = Gila Bend, Arizona. Species designations as for Fig. 4.3 plus *Ph. g.* = *Pheidole gilvescens; P. b.* = *Pogonomyrmex barbatus; P. c.* = *Pogonomyrmex californicus; P. m.* = *Pogonomyrmex maricopa; P. p.* = *Pogonomyrmex pima; V. p.* = *Veromessor pergandei.* G = group forager ; I = individual forager; species designated by open circles occur only rarely. (After Davidson 1977a.)

discussion to the ant species with mean worker sizes exceeding 3 mm. If we do this, there are several important conclusions we can draw. Almost without exception, when species of similar size coexist at a site they differ in foraging strategy, and when species of similar foraging strategy coexist at a site they differ in size. The only apparent exception, in fact, is the coexistence of *Pogonomyrmex desertorum* and *P. maricopa* at site A, and of these the latter occurs only rarely. Certainly this should not prevent us from drawing the general conclusion that when several species of granivorous ant coexist at a single site, each specializes in a different way in its utilization of the food-resource.

Conversely, when the species-compositions of different sites are compared, it is apparent that the species similar in both size *and* foraging strategy act as 'ecological replacements' for one another. Thus, of the two group foragers exceeding 9 mm in length, *P. barbatus* and *P. rugosus,* only one ever inhabits a site; and of the three individual foragers between 6 mm and 7.6 mm in length, *P. californicus, P. desertorum* and *P. maricopa,* there is never more than one that is common. Given the numbers of species and sites, this is very unlikely to be mere coincidence. Overall, therefore, it appears that although amongst the guild of granivorous ants there is overlap in resource-utilization and interspecific competition for food, there is coexistence only between species that differ in size, or foraging strategy, or both. Species that do not differ in at least one of these respects are apparently unable to coexist.

The data for ant species less than 3 mm in length, however, do not conform to this pattern. It could be claimed that *Pheidole sitarches* and *Ph. gilvescens* are ecological replacements for one another, but, overall, it is apparent that these small species coexist without any differentiation in size or foraging strategy. There are three possible explanations. The first is that the mode of coexistence of the larger species, described above, is not a general phenomenon: these small species compete for a resource and coexist, even though they utilize the resource in the same way. The second explanation is that the *Pheidole* species do utilize the seed-resource in different ways, but at present the basis for the differentiation—some third facet of their feeding ecology—is unknown to us. Finally, the third explanation is that these species are not limited, and do not compete, for the seed resource because they are limited in some other way; perhaps by some other resource, or by a predator that keeps their densities so low that there is no competition.

The most important aspect of these three alternative explanations is that they illustrate some very basic methodological problems in the study of interspecific competition. The second and third explanations are based on our ignorance concerning the ants' ecology; but since we are always likely to be ignorant of a species' ecology to some extent, these explanations can never be discounted. The first explanation is, therefore, left as a last resort: to be used only when we are confident of our own infallibility! Furthermore, these second and third explanations are essentially dependent on there being some 'other' important differences between the *Pheidole* species. Yet the mere discovery of differences between the species cannot, *in itself*, support either of the explanations. The differences must also be shown to reflect differential utilization of a resource for which the species *do* compete. Coexistence requires differences; but there are likely to be many differences that have nothing to do with coexistence. There are, therefore, some very real difficulties in studying interspecific competition (to which we shall return in Section 4.11). Yet, as we shall see below, concerted effort often confirms that when competing species coexist they do so by differential utilization of the resource for which they are competing.

Finally, Davidson (1978) looked closer still at just one of these and species (*Veromessor pergandei*) that differs in size, and size-variation, from site to site. Some of her results are illustrated in Figs 4.5 and 4.6. In this case Davidson used mandible length as a measure of size, and, as Fig. 4.5 shows, the variability of this measurement decreased significantly as the diversity of potential competitors at a site increased. In other words, *V. pergandei* is apparently more of a size-specialist at those sites in which interspecific competition is most likely, and less of a specialist where it is least likely. This is also apparent from Fig. 4.6, which suggests, in addition, that size itself is strongly influenced by those species which are most similar to *V. pergandei* in mandible length. The effect is particularly noticeable at Ajo, where *P. pima* (mean mandible length, 0.64 mm) coexists with *V. pergandei* (0.81 mm). Note that, because of the nature of the data, we cannot conclude with absolute certainty that *V. pergandei* competes with the other ant species for

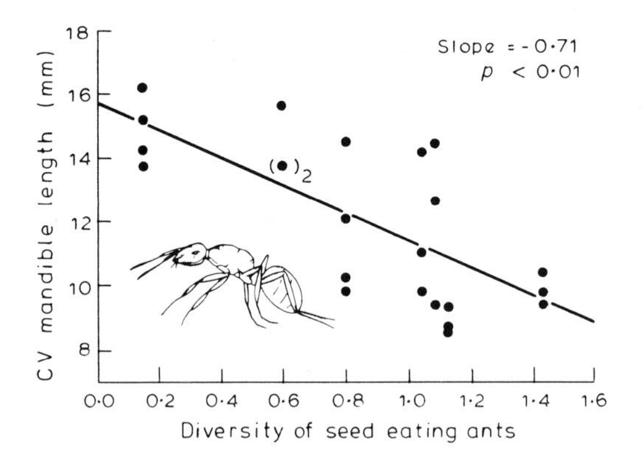

Fig. 4.5. The relationship between the within-colony coefficient of variation (CV) in mandible length for *Veromessor pergandei* and the species diversity of seed-eating ants in the community. (After Davidson 1978.)

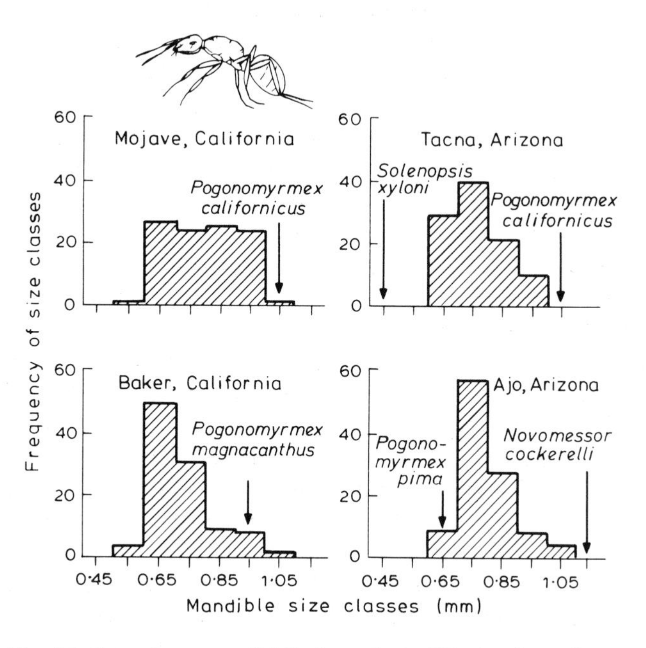

Fig. 4.6. Some frequency distributions of mandible size classes for *Veromessor pergandei*, with the mean mandible lengths of competitors most similar in size indicated by arrows. (After Davidson 1978.)

food; but we can say that, at each site, a form of *V. pergandei* has evolved which competes less with the other species than it might have done.

Thus, this single, elegant example illustrates four important points.

(a) Species from distantly related taxa can compete with one another for a limited resource.

(b) Competition need not lead to exclusion: competitors can still coexist.

(c) Coexisting species tend to differ in at least one respect in the way they utilize the limited resource, and species which utilize the resource in the same way tend to exclude one another from a site.

(d) The precise nature of a species, and thus the precise way in which it utilizes the resource, can itself respond to the species' competitive *milieu*.

4.4 Competition between plant species: experimental approaches

Plants are eminently suitable for sophisticated and detailed types of experimentation to determine whether or not competition is occurring, and also its outcome, primarily because they are easy to handle. Deliberate manipulation of the number of plants in a pot allows the density and proportion of species in mixture to be changed at will, and, since the plants are together in the same pot, there will, at suitable densities, necessarily be root competition for finite (nutrient and water) resources in the soil, and shading amongst adjacent leaves resulting in competition for incident radiation within the canopy. However, although the execution of a competition experiment may appear straightforward, its design is not. Consider a pot containing a mixture of plants: 100 of species A and 50 of species B. Addition of 50 plants of species B to this pot has *two* immediate effects. Firstly, it changes the *overall density*, increasing it by a third; but it also alters the *proportion* of species B to A in the mixture from ⅓ to ½. In other words, *additive experiments* like this confound two important variables which should, ideally, be clearly separated. One solution to this problem in design is to deliberately maintain the overall plant density constant, but to vary the proportions of the mixture by substitution. The inception and development of the design and analysis of these *substitutive* competition experiments has largely been the work of de Wit and his colleagues in the Netherlands (de Wit 1960). The basis of the experimental design is the 'replacement series' in which seeds of two species are sown to constant overall density, but the proportions of both species are varied in the mixture from 0 to 100%. At a density of 200 plants we may have 100 A and 100 B, 50 A and 150 B, 0 A and 200 B and so on, representing a set of mixtures all at constant density. Such a series may be repeated at different densities if required. The overall density is chosen to ensure competition between species, and the yield of both is measured after a suitable growth period. Results are presented in a 'replacement diagram', in which the yields of both species in monoculture may be compared with those in mixture.

However, if we consider the invasion of an already occupied natural habitat by a newcomer, both the proportion of newcomer to indigenous species *and* the total density will change. Overall, therefore, both types of design can provide instructive data on the nature of competition between plants.

4.4.1 *Additive experiments*

Crop–weed mixtures have long been the subject of additive competition experiments, partly because of agricultural interests, but also because of the relative ease of conducting them. Typically, both species are sown at the same time and germinate more or less synchronously, so that competitive effects may be examined in even-aged mixtures of species. Trenbath & Harper (1973) investigated the competitive ability of several *Avena* species, including an examination of the relative aggressiveness of the weed species *A. fatua* and *A. ludoviciana* towards the crop species *A. sativa*. They used an additive design, sowing the crop species in plots at a constant density and adding the weed species at two sowing densities to represent levels of weed infestation. For comparative purposes the experiment also included 'additions' of the crop itself, leading to monocultures of the crop at comparable densities. The experiment was conducted in an unheated greenhouse in summer, seeds being sown directly into the ground.

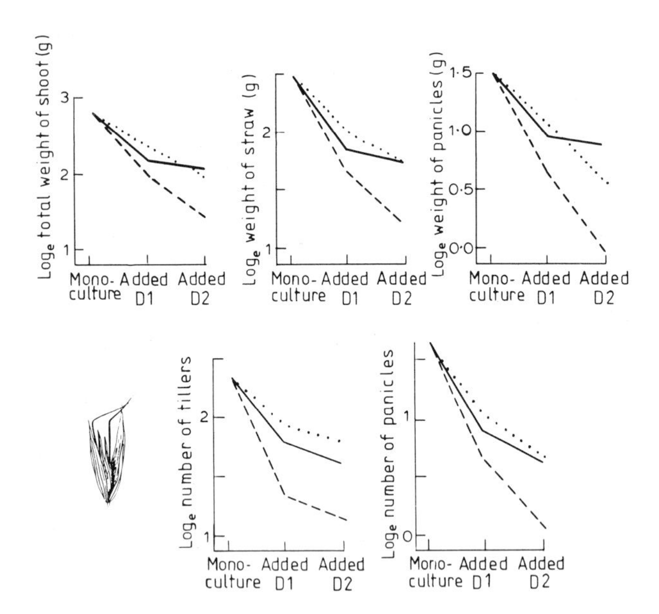

Fig. 4.7. The effect of competition on the oat *Avena sativa* as revealed by an additive experiment. *A. sativa* itself (——), *A. fatua* (– – –) and *A. ludoviciana* (. . . . .) were added at two densities to a monoculture of *A. sativa*. D1 = 38 plants m^{-2}; D2 = 76 plants m^{-2}. Lines are drawn for ease of interpretation; they do not imply a continuous relationship with density. (After Trenbath & Harper 1973.)

The results are shown in Fig. 4.7. They are plotted in a way that shows the depression in yield in *A. sativa* when grown with additions of the other species and of itself: the magnitude of depression can be taken as a general measure of the intensity of inter- and intra-specific competition. It is immediately apparent that *A. fatua* is the most vigorous competitor, depressing both shoot, straw and panicle (flowerhead) weight, as well as reducing the number of tillers and panicles. In contrast, *A. ludoviciana* caused the least depression, the relative ranking of aggressiveness being *A. fatua* > *A. sativa* > *A. ludoviciana*, when considered overall.

Whilst reduction in the weight and number of plants is strongly suggestive of interspecific competition, it does not fully establish whether there was a reduction in the reproductive output of the competing adults. Even though the number of panicles is reduced in the presence of added species, the number of seeds borne per panicle could increase to maintain a constant seed number per plant. The data for the components of seed yield, however, do not bear this out (Table 4.4). The number of seeds per panicle remains relatively constant, regardless of the added species in the mixture, and the output of plants is thus directly related to the number of panicles per plant. Further evidence of an overall reduced output is indicated by the weights of individual seeds. The reduction in weight of seeds of *A. sativa*, when in association with *A. fatua*, clearly suggests that these seeds will be at a relative disadvantage at germination to those produced in monoculture.

Additive experiments of this type only go part way to providing a clear illustration of interspecific competition. In particular they do not illustrate the influence of competition over the full range of densities that may occur when two species are grown together. To examine this, Firbank & Watkinson (1985) grew wheat, *Triticum aestivum* and corncockle, *Agrostemma githago* together in mixture combinations covering a range of total sowing densities from 4 to 1200 m^{-2}. They were then able to examine the mean yield per plant of both species as a 'response surface' according to the densities of the components in the mixture (Fig. 4.8). Comparison of the two surfaces reveals that in general wheat was more sensitive to increases in plant density than corncockle—lines across the surface are generally steeper for wheat. More importantly, the curves illustrate a second feature of interspecific competition, namely that there is an

Table 4.4 The components of seed yield of *A. sativa* in monoculture and with competitors in mixture. (After Trenbath & Harper 1973.)

	Number of panicles per plant	Number of seeds per panicle	Number of seeds per plant	Weight per seed (mg)
Monoculture	3.3	33.9	111.9	35.3
Mixture + *A. sativa*	1.5	44.8	67.2	34.4
Mixture + *A. fatua*	1.0	43.8	43.8	17.4
Mixture + *A. ludoviciana*	1.5	36.1	54.2	28.9

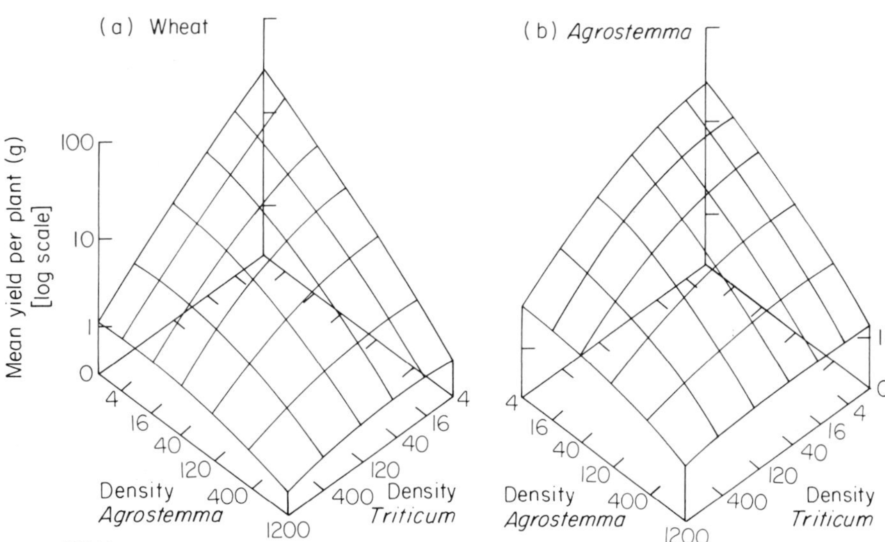

Fig. 4.8. Calculated density responses in a mixture of *Triticum aestivum* and *Agrostemma githago*: yields per m² of (a) *Triticum* and (b) *Agrostemma*. After (Firbank & Watkinson 1985.)

intergrading of the effects of inter- and intraspecific competition across the total plant density range. When represented in mixture at low density, both species were sensitive to an increase in the density of the other component, a sensitivity which diminished with increasing density of the mixture. Indeed, regardless of sowing density, wheat had very little effect on the biomass of corncockle plants when they themselves were sown at a density of 1200 m^{-2}. We may infer from this that the most important competitive interactions at high density tend to be intraspecific ones. Nevertheless, it is clear that interspecific competition does influence corncockle performance at lower corncockle density. To turn to the means of disentangling inter and intraspecific competition we must now look at substitutive experiments.

4.4.2 Substitutive experiments with wild oats

In the annual grasslands of California two wild oat species, *Avena fatua* and *A. barbata*, occur naturally together, and Marshall & Jain (1969) undertook an experimental analysis of the competitive interaction between them, in order to elucidate the extent and pattern of their cohabitation. The two oats were grown together from seed at four densities—32, 64, 128 and 256 plants per pot—and five frequencies: 0, 12.5, 50.0, 87.5 and 100% of the total sown. After 29 weeks growth in a greenhouse, the yield of spikelets per pot was assessed for each species. To measure the intensity of *intra*-specific competition across the range of densities in the mixtures, both species were also grown in monoculture (pure stand). Fig. 4.9 shows that the yield per pot

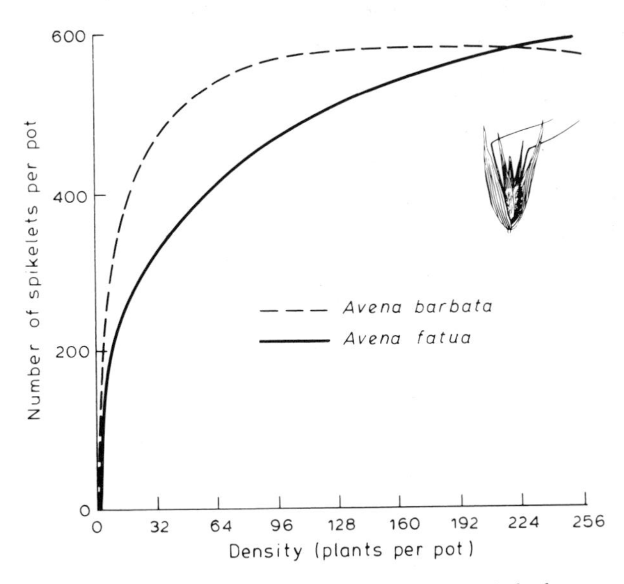

Fig. 4.9. Intraspecific competition in *Avena fatua* and *A. barbata*. (From Marshall & Jain 1969.)

in monoculture finally became independent of initial sowing density. This is the background of intraspecific competition against which we must assess the reaction of each species grown in mixture.

Fig. 4.10 shows the replacement diagrams for each of the four densities, with yields per pot plotted against the proportion in mixture. The dotted lines in these diagrams are the appropriate yield responses from Fig. 4.9 of each species grown on its own; solid lines show the responses of each in mixture. Comparison of these lines for each species thus allows us to gauge the effect of interspecific competition. At the lowest total density (32 plants per pot), the yield responses of *A. fatua* in mixture and pure stand are almost identical, suggesting that this species is not responding to the presence of *A. barbata* in mixture. For *A. barbata,* on the other hand, yield is depressed so that at the equiproportional mixture it is reduced to only 44%. The interaction appears to be amensal.

Examining the higher planting densities, however,

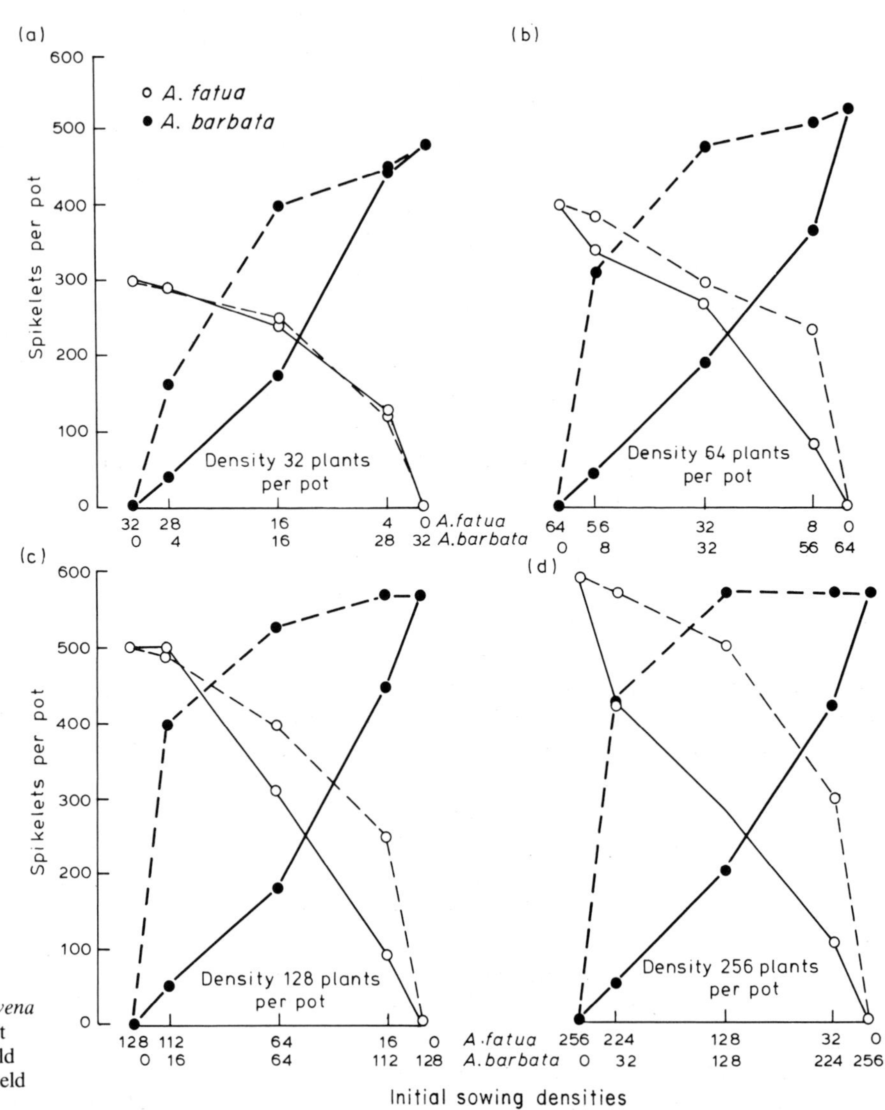

Fig. 4.10. De Wit replacement diagrams for *Avena fatua* and *A. barbata* when grown in mixture at four overall densities. Dotted lines are the yield expectations in monoculture, solid lines the yield in mixture. (After Marshall & Jain 1969.)

(Fig. 4.10b–d) reveals a truly competitive interaction between the species. The monoculture yield responses are all convex, indicating yield limitation by intraspecific competition. Yet in mixture the yield responses of *A. barbata* are all concave, and with increasing total plant density they become substantially depressed below the expected yield in pure stand. The same is also true for *A. fatua,* since if we examine this species' response in mixture over the four densities, spikelet yield departs more and more from the monoculture expectation. There is, therefore, mutual depression resulting from interspecific competition. However, the shape of *A. fatua*'s response curve changes from convex (Fig. 4.10a), through linear (Fig. 4.10c), to almost concave (Fig. 4.10d). So, whilst suffering interspecific competition, *A fatua* is performing relatively better than *A. barbata* in mixture: it has a competitive advantage over *A. barbata*; and for both species the intensity of interspecific competition increases with overall density.

Interpretation of replacement diagrams, then, involves two considerations:

(a) an assessment of what a species might do when growing on its own at densities equivalent to the densities it experiences in the replacement series, and

(b) a measurement of the departure from this response when grown in mixture.

A response in mixture that is concave necessarily means that a species is suffering interspecific competition, since the yield–density response in monoculture can only be linear or convex (intraspecifically limited). However, a convex response does not mean that a species is not experiencing interspecific competition. We can only assess this with the additional knowledge of the equivalent pure stand response. Our important conclusions from this experiment then, are, firstly, that interspecific competition between two plant species may affect the performance of both, but to different extents; and secondly, that the intensity of competition is dependent on the density at which the interaction takes place.

If we wished to know the outcome of prolonged competition between the two species *at this particular density*, we could repeatedly resow (each successive generation) the seeds collected from mixtures of *A. fatua* and *A. barbata* growing together. Yet while this is perfectly feasible, it is in fact unnecessary; we can deduce the outcome using the data from our replacement diagram. If we calculate the ratio of the spikelets sown (or input) to the ratio of those collected (or output) for a particular mixture, and the ratios are equal, then obviously the proportions in the mixture remain stable on repeated sowings. If, however, one species succeeds at the expense of the other, then there is inequality between the input and output ratios, and the successful species will gain proportionally in the mixture. We may explore these consequences by plotting, logarithmically, the ratios of input to output from the replacement series (Fig. 4.11). The solid line with a slope of 1 indicates that whatever ratio is sown (input) is also collected at harvest (output): the values of A and B in Fig. 4.11 are equal. If, however, the relationship departs from this line, for instance the dashed line in Fig. 4.11, then an input ratio C yields the ratio D, a mixture in which species 1 has increased in proportion to species 2. Resowing of D (i.e. at E, equal to D) yields F and the proportion of species 1 to 2 changes again. Repeated sowing 'steps up' the proportion of species 1 to 2 in Fig. 4.11 (dotted lines), each step indicating a generation of competition. Similarly, for a line relating input to output that occurs *below* the equality input/output line, repeated sowing will 'step

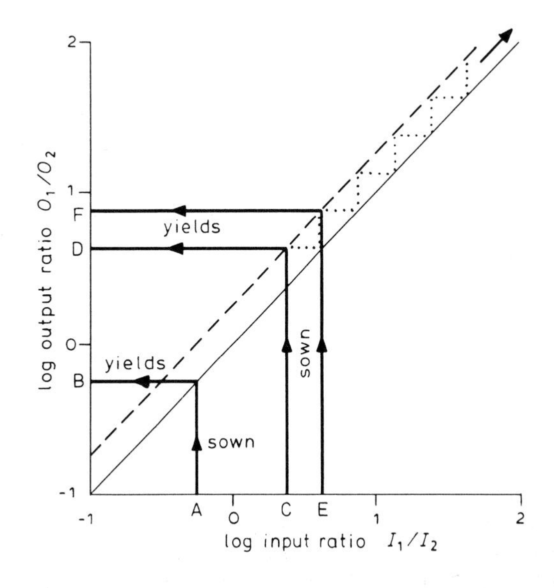

Fig. 4.11. The de Wit ratio diagram. See text for explanation.

down' the graph, leading to a steady increase of species 2 in the mixture.

Plotting Marshall and Jain's data in the form of such a *ratio diagram* immediately tells us what the outcome of continued competition will be at each of their densities (Fig. 4.12). At the lowest density (Fig. 4.12a) the line relating inputs and outputs crosses the line of unit slope at a proportion, per hundred, of 78 *A. fatua* plants to 22 of *A. barbata*. This point is in fact an *equilibrium point*, since any mixtures to either side of it will move towards it by the stepping-up or stepping-down processes described previously in Fig. 4.11 (dotted lines). This may seem somewhat surprising. However, *A. barbata* outyields *A. fatua* in pure stand by 175 spikelets, but suffers interspecific competition to a much greater extent than *A. fatua*. The competitive aggressiveness of *A. fatua*, therefore, seems to reduce the fecundity advantage of *A. barbata* to a level at which they both coexist; and at the higher densities, equilibrium points of coexistence are present in mixtures which are even more dominated by *A. fatua*. Yet Marshall and Jain were unable to establish the underlying mechanisms which lead to this.

4.4.3 *Substitutive experiments with a grass–legume mixture*

To see the processes by which this coexistence can be achieved, we must turn to another substitutive experiment (de Wit *et al.* 1966). *Panicum maximum* and *Glycine javanica* form a grass–legume mixture in some Australian pastures. *Panicum*, the grass, acquires its nitrogen from the soil; while *Glycine* acquires part of

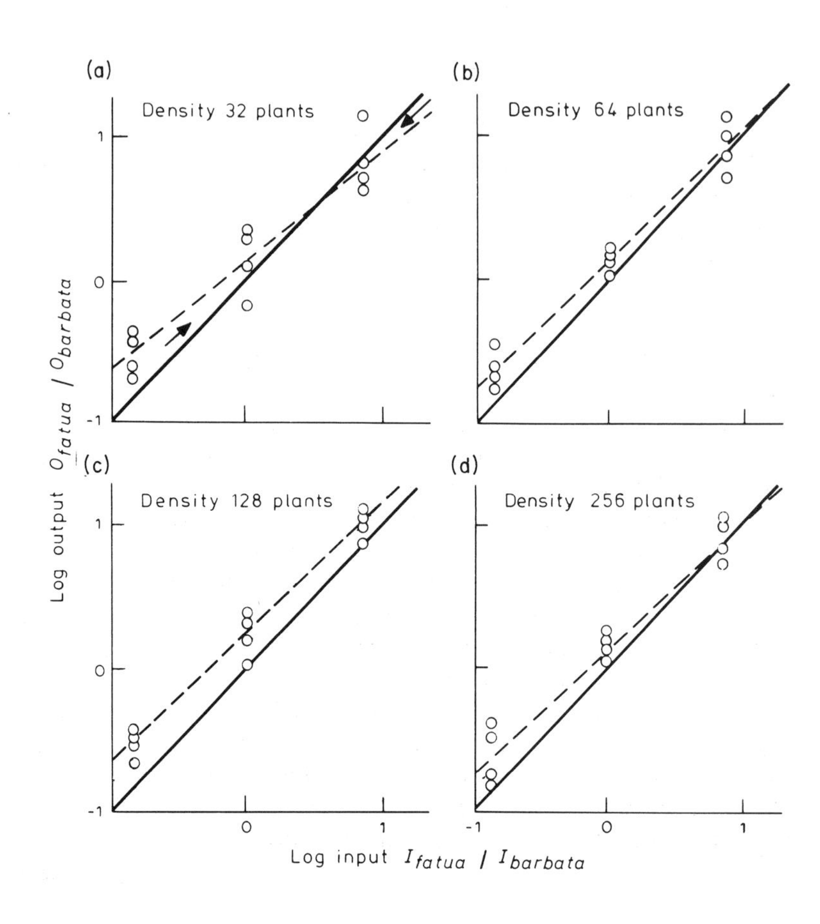

Fig. 4.12. Ratio diagrams for mixtures of *Avena fatua* and *A. barbata* at four densities. (From Marshall & Jain 1969.)

its nitrogen from the air, by nitrogen fixation, through its root association with the micro-organism *Rhizobium*. De Wit and his colleagues grew both species in a replacement series with and without an inoculation of *Rhizobium*; the results (Fig. 4.13) are given as replacement and ratio diagrams and as 'relative yield totals'. The relative yield of a species in a given mixture is the ratio of its yield in mixture to its yield in monoculture in the replacement series. Calculating yield in this way removes any absolute yield differences that may exist between species and refers both yields to the same scale. The relative yield *total* (RYT) is then the sum of the two relative yields. RYTs of 1 indicate that both species are competing for the same environmental resources, in that the relative gain of one species (in numbers *and* resource-acquisition) is exactly balanced by the relative loss of the other. Values greater than 1, therefore, signify differing resource demands; one species' gain is accompanied by a comparatively minor loss by the other. Fig. 4.13 shows that in the absence of *Rhizobium*, *G. javanica* is suppressed when grown in mixture with *P. maximum*, the yield response curve being concave. The RYTs for both dry matter yield and nitrogen in the plant are very close to 1 at all of the mixtures. On the other hand, when inoculated with *Rhizobium*, the legume's yield response curve in the replacement diagram becomes convex, and the RYTs exceed 1. The ratio diagrams illustrate the long-term consequences of growth together: coexistence when *Rhizobium* is present, exclusion of the legume when it is not.

The clear inference from this experiment is that when both grass and legume compete for the same nitrogen source (soil nitrogen), the legume will experience such intense competition from the grass that it will ultimately disappear from the mixture; but if the legume is able to utilize a different source, namely nitrogen from the air fixed by *Rhizobium*, then the two species will coexist.

These experiments on plants confirm the findings and interpretations of the studies on ants. Two species may compete with one another and coexist, even though the competition has a detrimental effect on both species. Moreover, the conclusion has also been reinforced that for two species to coexist they must in some way avoid making identical demands for limited environmental resources.

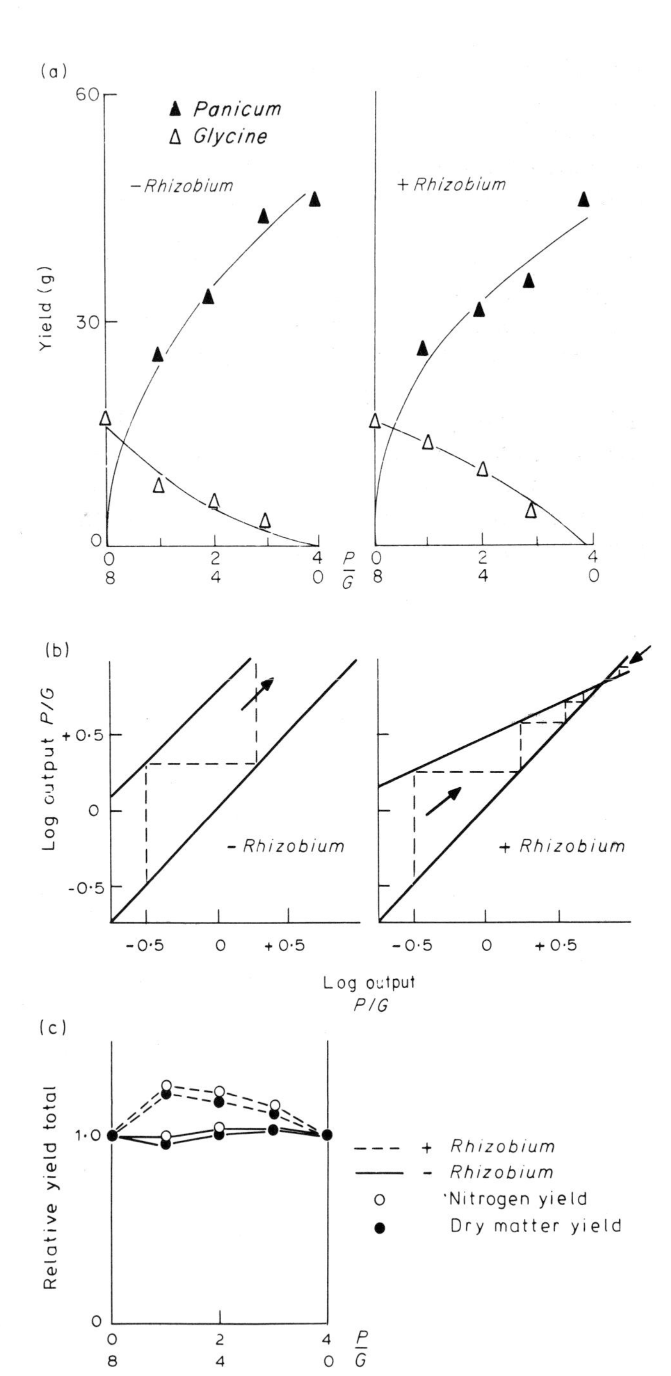

Fig. 4.13. Interspecific competition between *Panicum maximum*, *P*, and *Glycine javanica*, *G*, in the presence and absence of *Rhizobium*: (a) replacement diagrams; (b) ratio diagrams; (c) relative yield totals. (From de Wit *et al.* 1966.)

4.5 The ecological niche

It is necessary to digress slightly at this point. The term 'ecological niche' has been in the ecological vocabulary for over half a century, but for more than thirty of those years its meaning was rather vague (see Vandermeer 1972 for a historical review). Here, we shall concern ourselves only with its current, generally accepted meaning, originating with Hutchinson (1957).

If we consider a single environmental parameter (e.g. temperature), then a species will only be able to survive and reproduce within certain temperature limits. This range of temperature is the species' ecological niche *in one dimension* (Fig. 4.14a). If we *also* consider the range of humidities in which the species can survive and reproduce, then the niche becomes *two*-dimensional and can be visualized as an area (Fig. 4.14b); and if a third dimension is added (food particle size in Fig. 4.14c) then the niche becomes a volume. Yet it is clear that there are many biotic and abiotic parameters affecting a species: the number of niche-dimensions (n) greatly exceeds three. We cannot visualize such a situation; but we can, by analogy with the three-dimensional model, consider a species' ecological niche to be an 'n-dimensional hypervolume' within which a positive contribution to future generations can be made. And this, in essence, is Hutchinson's conception of the niche. There are, however, several additional, important points which must be understood.

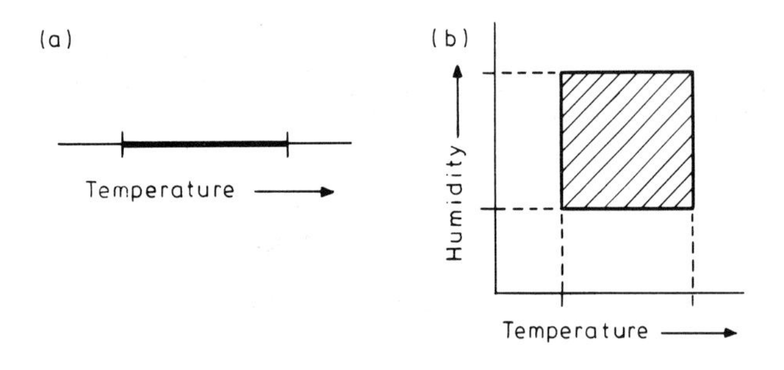

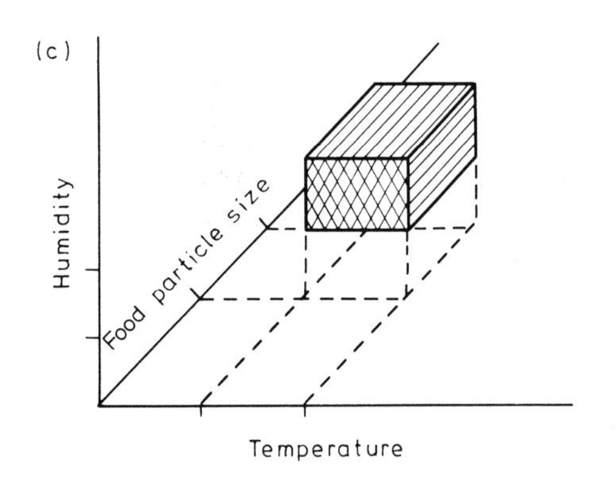

Fig. 4.14. Building up the n-dimensional hypervolume: ecological niches. (a) in one dimension (temperature); (b) in two dimensions (temperature and humidity); (c) in three dimensions (temperature, humidity and food particle size).

(1) The ecological niche of a species has been defined from the species' point of view. This allows it to be contrasted with the term 'habitat', which is an objective *description*—in n-dimensions if need be—of the environment itself. Thus, although both niche and habitat are defined in terms of environmental parameters, the term niche characterizes a species, while the term habitat characterizes an environment within which many species may live.

(2) Hutchinson gave special consideration to one particular type of environmental parameter: interspecific competitors. The connections between interspecific competition and the ecological niche should become much clearer below, but we can outline Hutchinson's ideas at this stage. He called the niche of a species in the absence of competitors from other species its *fundamental* niche, i.e. the niche which it could *potentially* occupy. In the presence of competitors the species is restricted to a *realized* niche, the precise nature of which is determined by which competing species are present. In other words, Hutchinson wished to stress that interspecific competition reduced contributions to future generations, and that as a result of interspecific competition, contributions in certain parts of a species' fundamental niche might be reduced to zero. These parts of the fundamental niche are absent from the realized niche.

(3) A species' niche might also be restricted *in practice* by the habitat: parts of a species' niche—fundamental or realized—which are simply not present at a particular location (in space and time) become, temporarily, irrelevant. This is particularly pertinent to laboratory investigations, in which the simplicity of the habitat

confines species to those parts of their niche which happen to be provided by the experimenter.

(4) Finally, it must be stressed that the Hutchinsonian niche is not supposed to be a literal description of a species' relationship with its environment. Like other models, it is designed to help us think, in useful terms, about an immensely complex interaction. For instance, the perpendicularity of the volume in Fig. 4.14c is a result of the assumption that the three parameters are independent. Interaction of variables—almost certainly the general rule—would lead to a less regularly shaped niche; but in practice a perpendicular visualization may be just as useful. Similarly, the solid lines around the niche in Fig. 4.14 ignore the existence of variability within a species. Each individual has a niche, and a species' niche is, in effect, the superimposition of as many niches as there are individuals: a species' niche should have very blurred edges. Once again, however, such sophistication is rarely necessary in practice.

4.6 The Competitive Exclusion Principle

We can now return to our main theme and consider two laboratory investigations of interspecific competition. The first is the work of Park (1954) on the flour beetles *Tribolium confusum* and *T. castaneum*. In a series of simple, sterilized cultures, Park held most environmental variables constant, but varied a single, albeit complex parameter: climate. In all conditions, both species were able to survive in monospecific cultures: the fundamental niches of both species spanned the whole climatic range (Fig. 4.15a). In mixed-species cultures,

Fig. 4.15. (a) and (b) indicate, respectively, the fundamental and realized niches of the flour beetles *Tribolium confusum* and *T. castaneum* in relation to climate; (based on the data of Park 1954). (c) and (d) indicate, respectively, the fundamental and realized niches of the flour beetles *Oryzaephilus surinamensis* and *Tribolium confusum* in relation to food, while (e) indicates their realized niches when another niche dimension, space, is added (based on the data of Crombie 1947.) For further discussion, see text.

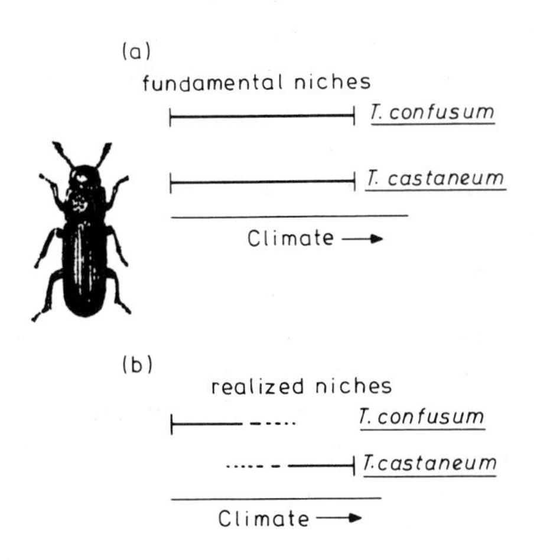

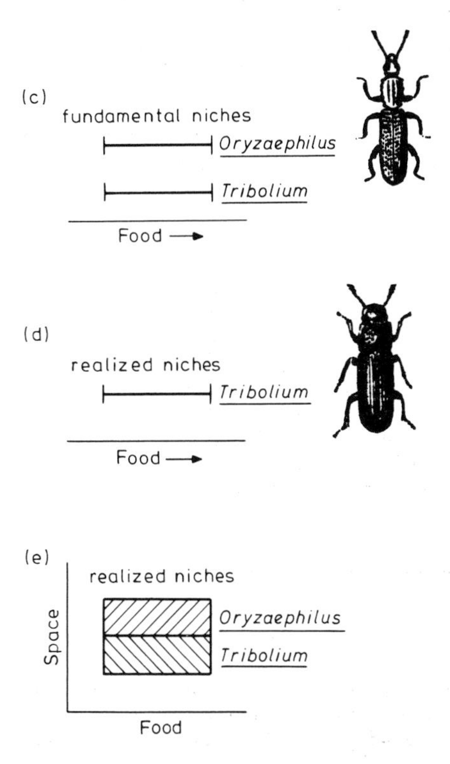

however, the results were as follows:

Climate	Percentage wins	
	T. confusum	*T. castaneum*
hot–moist	0	100
temperate–moist	14	86
cold–moist	71	29
hot–dry	90	10
temperate–dry	87	13
cold–dry	100	0

In other words, it appears that the climatic extremes are represented in the realized niche of only one of the two species. In the middle of the range there is also, invariably, elimination of one species by the other; but the precise outcome is probable rather than definite. The situation can therefore be visualized as in Fig. 4.15b, with overlapping realized niches in a single dimension.

The second experiment (Crombie 1947) also concerns two species of flour beetle: *Tribolium confusum* and *Oryzaephilus surinamensis*. Both species can maintain monospecific populations under a variety of environmental conditions in laboratory cultures of ordinary flour (Fig. 4.15c). In mixed cultures of this type, however, *T. confusum* always eliminates *O. surinamensis*; only *T. confusum* has a realized niche (Fig. 4.15d). In such circumstances, *T. confusum* has a higher rate of reproduction and survival, and is also more effective in its destruction of pre-adult individuals. (Both species exhibit this reciprocal predation making it a '− −' and therefore competitive interaction.) However, if an *extra dimension* is added to the environment, namely 'space' (in the form of small glass tubes which are available to *O. surinamensis* but too small to be accessible to *T. confusum*), then *O. surinamensis* gains added protection against predation and there is stable coexistence. Both species now have realized niches (Fig. 4.15e).

The standard interpretation of experiments like these has been elevated to the status of a Principle: the so-called 'Competitive Exclusion Principle'. This merely recasts, in terms of niches, what has already been hinted at in granivorous ants and certain plants: if there is no differentiation between the realized niches of two competing species, *or if such differentiation is precluded by the limitations of the habitat,* then one species will eliminate or exclude the other. (As the experiments described above show, such lack of differentiation is often expressed as the total non-existence of the realized niche of one of the species.) Conversely, when differentiation of realized niches is allowed by the habitat, coexistence of competitors is possible.

4.7 Competitive exclusion in the field

Laboratory habitats tend to differ from field habitats in having fewer dimensions, and narrower ranges of those dimensions that they do have. It is likely, therefore, that, in the laboratory, habitat will frequently preclude niche-differentiation, forcing potential coexistors to compete in a way that leads to the elimination of one of them. For this reason, field evidence of competitive exclusion is particularly valuable.

Connell (1961) produced such evidence, working in Scotland with two species of barnacle: *Chthamalus stellatus* and *Balanus balanoides*. Adult *Chthamalus* generally occur in an intertidal zone which is above that of adult *Balanus*. Yet young *Chthamalus* do settle in the *Balanus* zone, so that their subsequent disappearance suggests either that *Balanus* individuals exclude them, or that they are simply unable to live there. Connell sought to distinguish between these alternatives by monitoring the survival of young *Chthamalus* in the *Balanus* zone, taking successive censuses of mapped individuals over the period of a year. Most important of all, he ensured at some of his sites that the *Chthamalus* individuals were kept free from contact with *Balanus*. In other words, he carried out a 'removal experiment' allowing him to compare the responses of *Chthamalus* in the presence or absence of *Balanus*. In contrast with the normal pattern, *Chthamalus* in the absence of *Balanus* survived very well (Fig. 4.16). Thus, it seemed that competition from *Balanus,* rather than increased submergence time, was the usual cause of *Chthamalus* mortality. This was confirmed by direct observation. *Balanus* smothered, undercut or crushed *Chthamalus,* and the greatest *Chthamalus* mortality occurred during the seasons of most rapid *Balanus* growth. Moreover, the few *Chthamalus* individuals that survived a year of

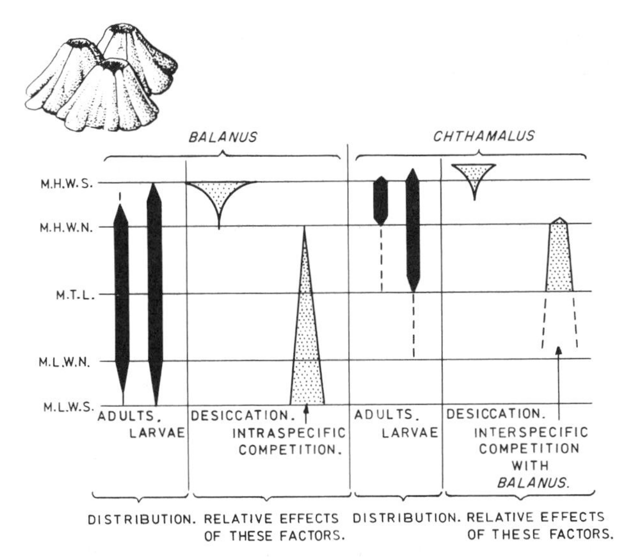

Fig. 4.16. The intertidal distribution of adults and newly settled larvae of *Balanus balanoides* and *Chthamalus stellatus*, with a diagrammatic representation of the relative effects of desiccation and competition. Zones are indicated to the left; from M.H.W.S. (mean high water–spring) down to M.L.W.S. (mean low water–spring). (After Connell 1961.)

Balanus crowding were much smaller than uncrowded ones, showing, since smaller barnacles produce fewer offspring, that interspecific competition was also reducing fecundity. It is clear, then, that the *fundamental* niches of both species extend down into the lower levels of the 'barnacle belt'; but interspecific competition from *Balanus* excludes *Chthamalus* from these levels, restricting its realized niche to the upper zones in which it can survive by virtue of its comparatively high resistance to desiccation. It is also clear that competition is markedly 'one-sided' in the lower zones. Yet *Balanus* must expend some of its energy and resources in the process of smothering, undercutting and crushing invading *Chthamalus,* and the two species do undoubtedly *compete* for the same space. This is, therefore, an example of interspecific competition bearing a superficial resemblance to amensalism.

Another example of competitive exclusion in the field is provided by the work of Inouye (1978) on two species of bumble bee: *Bombus appositus* and *B. flavifrons.* In the Colorado Rocky Mountains, *B. appositus* forages primarily from larkspur, *Delphinium barbeyi* and *B. flavifrons* from monkshood, *Aconitum columbianum.* When Inouye temporarily removed one or other of the bee species, however, the remaining species quickly increased its utilization of its less-preferred flower. Moreover, it visited more of these flowers than usual during each stay in a patch (such increased stay-times being a recognized indication that the forager is being more successful in the patch). Thus, the fundamental niches of both bee species clearly include both species of flower; and, since it is the presence of one bee species that restricts the realized niche of the other, reciprocal competitive exclusion is strongly suggested. The probable mechanism is as follows: *B. appositus* (longer proboscis) is better adapted to forage from larkspur (longer corolla tube) than *B. flavifrons,* but when either bee species is alone the difference is sufficiently slight for both flowers to be valuable sources of nectar. When both bee species are present, however, *B appositus* depletes the nectar in larkspur flowers to a level which makes them unattractive to *B. flavifrons* (with its shorter proboscis). As a consequence, *B. flavifrons* concentrates on monkshood, depleting the nectar *there* to a level which makes *them* significantly less attractive (in cost-benefit terms) to *B. appositus.* The two bee species are, therefore, attracted primarily to different flowers, and they forage accordingly. This results in reciprocal competitive exclusion.

The *Balanus–Chthamalus* interaction and Inouye's *Bombus* study are excellent illustrations of the two contrasting types of interspecific competition (Park 1954). The aggressive competition by which *Balanus* excludes *Chthamalus* in their joint pursuit of limited space is termed *interference competition*; whereas the reciprocal exclusion of the two *Bombus* species, resulting from the depletion of a resource by one species to a level which makes it essentially valueless to the other species, is termed *exploitation competition.* Thus, with interference there is no consumption of a limited resource, whereas with exploitation there invariably is. (Note that the interference/exploitation dichotomy is somewhat similar to that between scramble and contest (Section 2.4), in that contest always involves interference, while scramble usually involves only exploitation. Note, however, that interference and exploitation have none of the extreme threshold or all-or-none characteristics associated with scramble and contest.)

These two field studies are also particularly good examples of competitive exclusion because they are the results of experimental manipulations; the exclusion of *Glycine* by *Panicum* in the absence of *Rhizobium* (Section 4.4.3) was equally persuasive. But evidence is much more commonly circumstantial. Remember, for instance, that amongst the larger species of Davidson's granivorous ants there was never coexistence between species that shared both a size and a foraging strategy, i.e. species with apparently very similar realized niches. Providing the number of co-occurrences expected by mere coincidence is much greater than the number observed, reciprocal competitive exclusion is a very *plausible* explanation in such cases; but it can never be more than a plausible explanation when reliance is placed solely on observational data.

4.8 Competitive release

A class of evidence lying between simple observation and experimental manipulation is represented by observations of what is known as 'competitive release'. Some of the species in the above examples of exclusion show this phenomenon in the special context of an experimental manipulation (expanding their 'range' when a competitor is removed); but the term competitive release is more commonly applied to the results of 'natural experiments', as in the following example.

The New Guinea archipelago comprises one large, several moderately sized and very many small islands. Species-distribution on islands has been extensively studied and analysed (see, for instance, MacArthur 1972); but for our present purposes we need note only that in the New Guinea region, as elsewhere, small islands tend to lack species which are present on large islands and on the mainland. One example of this concerns ground doves (Diamond 1975). As Fig. 4.17 illustrates, there are three similar species of ground dove on New Guinea itself, and moving progressively inland from the coast one encounters them in sequence: *Chalcophaps indica* in the coastal scrub, *C. stephani* in the light or second-growth forest, and *Gallicolumba rufigula* in the rainforest. On the island of Bagabag, however, where *G. rufigula* is absent, *C. stephani* expands its range inland into rainforest; and on Karkar, Tolokiwa, New Britain, and numerous other islands where *C. indica* is also absent, *C. stephani* expands coastwards to occupy the whole habitat gradient. On Espiritu Santo, on the other hand, *C. indica* is the only species present, and it occupies all three habitats.

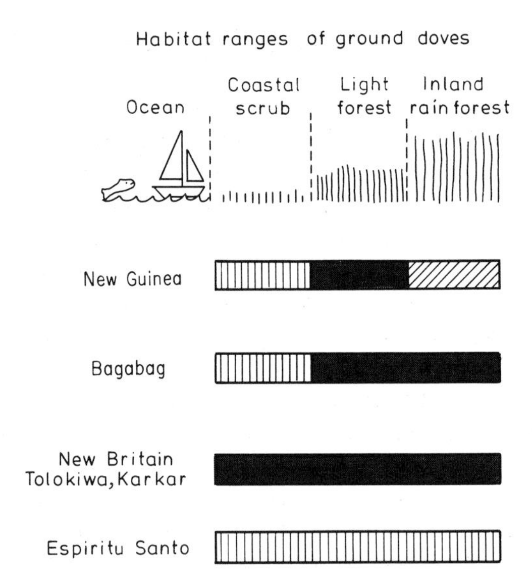

Fig. 4.17. Habitats occupied by three species of ground doves on various islands: *Chalcophaps indica* (vertical bars), *Chalcophaps stephani* (solid shading), *Gallicolumba rufigula* (diagonal bars). (After Diamond 1975.)

This, as Diamond (1975) remarks, is a particularly neat example, but there are many other natural experiments with similar results. Direct interspecific competition has not been positively established, but since *C. stephani* and *C. indica* only occupy rainforest in the absence of *G. rufigula, C. stephani* only occupies coastal scrub in the absence of *C. indica,* and *C. indica* only occupies light forest in the absence of *C. stephani,* the conclusion is more or less unavoidable that there is competitive exclusion on New Guinea and competitive release elsewhere.

4.9 Coexistence: resource partitioning

We have seen, from both direct and circumstantial evidence, that competitive exclusion can influence the range and distribution of species. We return now to what is, in a sense, the reverse question: When do similar or competing species coexist?

Vance (1972a) studied competition and the mechanism of coexistence in three sympatric species of intertidal hermit crabs from the San Juan Islands of Washington State: *Pagurus hirsutiusculus, P. granosimanus* and *P. beringanus*. The first priority was to confirm that competition was occurring. These hermit crabs are generalized omnivorous feeders and food is apparently abundant throughout the year: *a priori*, the most likely limiting resource was the empty gastropod shells which the crabs inhabit. Vance chose an experimental and a control site, and to the former he added 1000 marked, appropriately sized shells once per month for one year. He concentrated on *P. hirsutiusculus* and estimated densities at both sites at the beginning and end of the year, noting separately (at the experimental site) the numbers in marked and unmarked shells. Despite the fact that many of the empty shells were removed by strong currents, there was a significant increase in total numbers of crabs at the experimental site. However, there was no such increase at the control site, nor in the unmarked shells at the experimental site. Empty shells are obviously a limited resource for *P. hirsutiusculus*. To establish the generality of shell limitation, Vance gathered data on the size distributions of the three species of crabs, of the shells occupied by them, and of unoccupied shells; and he compared these data with the preferred shell sizes of the crabs, determined in preference experiments. Except for small size-classes, empty shells were rare, and all but small crabs of the three species occupied shells which were smaller than their preferred size. Thus, it appears that empty shells constitute for these three species of hermit crab a common, necessary and limiting resource for which they compete.

Vance went on to examine the mechanism of coexistence of these three species. He found, first of all, that there was partitioning of the resource: *P. hirsutiusculus* prefer short-spired light shells, whereas both *P. beringanus* and *P. granosimanus* prefer relatively taller and heavier shells. He also found that there was partitioning of the habitat: *P. hirsutiusculus* predominates in the upper intertidal and amongst *Hedophyllum sessile*, a brown alga living on horizontal rock faces in the lower intertidal; *P. beringanus* predominates in tide-pools in the lower intertidal; and *P. granosimanus* predominates under large loose stones and in shallow tide-pools of the mid-intertidal. Thus, there is differentiation amongst the *realized niches* of the three species, allowing them to compete and yet coexist.

The mode of competition between these species appears, essentially, to be interference. Shell-occupancy is determined by intra- and interspecific fighting: crabs in less-preferred shells attempting to displace crabs in more-preferred shells, the loser taking the less-preferred shell (Vance 1972b). We have seen that the outcome of such interspecific encounters is determined to some extent by resource partitioning on the basis of shell shape and weight. However, the basic resource—empty shells—is also divisible into 'shells-amongst-brown-algae', 'shells-in-shallow-tide-pools' and so on; and each species is apparently specialized in winning fights in a particular area. Thus, what we have called 'habitat partitioning' could also be considered as resource partitioning. Resource partitioning is, therefore, the means by which these three competitors coexist. As Table 4.5 shows, moreover, both types of resource partitioning are necessary for coexistence. *P. beringanus* and *P. granosimanus* are well separated by habitat, but *P. hirsutiusculus* is only truly separated from the other two in shells within its own preference range.

We have obviously met other, similar examples already, though in a slightly different context. *Bombus appositus* and *B. flavifrons* partition the nectar resource on the basis of corolla length: they are adapted to do so by differences in proboscis length. *Balanus balanoides* and *Chthamalus stellatus* partition the space resource on the basis of intertidal zone: *Balanus* is adapted to physically oust its competitor from the lower zones and *Chthamalus* to resist desiccation in the higher zone. *Glycine* and *Panicum* partition the 'total nitrogen' resource: *Glycine* is adapted to utilize free nitrogen by its intimate association with *Rhizobium*. Finally, granivorous ants partition the seed resource on the basis of the size, density and micro-distribution of the seeds: they are adapted to do so by differences in their own size and in foraging strategy. In all these cases the basic pattern is the same: competing species appear to coexist as a result of resource partitioning, or (in other words) by virtue of the differentiation of their realized niches.

Table 4.5 Hermit crabs occupying *Littorina sitkana* shells (which fall in the *Pagurus hirsutiusculus* preference range) and *Searlesia dira* shells (which fall in the *P. beringanus–P. granosimanus* preference range) collected from various physical habitats at a single site. (After Vance 1972a.)

Habitat	*P. hirsutiusculus*	*P. beringanus*	*P. granosimanus*
	Shell species: *Littorina sitkana*		
Hedophyllum beds	20	0	0
Deep tide-pools, lower intertidal	10	16	2
Cobble bed, mid-intertidal	6	0	32
	Shell species: *Searlesia dira*		
Deep tide-pool, lower intertidal	0	18	1
Shallow tide-pools, mid-intertidal	0	0	26

Moreover, since the same examples have served to illustrate both phenomena, it is clear that resource partitioning and competitive exclusion are often, though not always, alternative aspects of the same process. Competitive exclusion between species from portions of their fundamental niches leads to differentiation of their realized niches.

In fact, there are two further patterns which have emerged from the examples we have considered. The species pairs—bees, barnacles and plants—partitioned their resource along a single dimension: corolla length, intertidal zone and total nitrogen, respectively. The three species of hermit crab, on the other hand, required at least two dimensions: shell shape and weight, and shell location; while the guild of granivorous ants partitioned resources on the basis of seed size and seed density/distribution, and even these two dimensions were apparently not enough to account for the coexistence of the smaller species. There appears, in other words, to be a correlation between the number of species in a guild and the number of niche dimensions involved in the partitioning of the resource. Schoener (1974) reviewed the literature on resource partitioning, and found that this correlation was, indeed, statistically significant. Interestingly, this parallels the results of theoretical investigations by MacArthur (1965) and Levins (1968).

In addition, Schoener (1974), like several others, pointed to the *niche complementarity* frequently involved in resource partitioning. As illustrated by the examples of hermit crabs and, more especially, Davidson's ants, species which are not differentiated along one niche dimension tend to be separated along another. Realized niches are, therefore, fairly evenly distributed in multidimensional space, and species compete simultaneously with several species in several dimensions (called *diffuse competition* by MacArthur 1972).

Finally, however, it should be clear that we have avoided the most important aspect of competitive exclusion. We have seen that when there is no niche differentiation, competitive exclusion occurs; and we have seen that when competitors coexist there is resource partitioning. But we have ignored a much more profound question: *How much* niche differentiation is necessary for the coexistence of competitors? We return to this question in Section 4.15.

4.10 Character displacement

A particular type of resource partitioning, allowing competitors to coexist, is known as character displacement. It involves the modification of the morphological form of a species as a result of the presence of interspecific competitors. Fenchel (1975) investigated the coexistence of hydrobiid mud snails in Limfjord, Denmark, and paid particular attention to two species: *Hydrobia ventrosa* and *H. ulvae.* These deposit feeders seem to ingest their substrate indiscriminately and utilize the attached micro-organisms, and Fenchel found that for both species there is a single linear relationship between shell length and food particle size. Some of Fenchel's results are illustrated in Fig. 4.18. It is clear from this figure that when the two species live alone (which they do in a range of habitats), their sizes are more or less identical, as are the sizes of their food particles. When the two species coexist, however, there is character displacement. *H. ulvae* is larger, *H. ventrosa* is smaller, and the sizes of their food particles are similarly modified. The evidence for interspecific competition is not direct, but there is no simple alternative to the suggestion that the character displacement allows the partitioning of a potentially limiting resource, and thus allows the coexistence of two competitors.

A similar situation, of course, was that described previously for the ant, *Veromessor pergandei* (Davidson 1978), which shows character (size) displacement in response to its competitive milieu: the number and nature of its interspecific competitors. Thus, we can note that, in our newly defined terms, Davidson's ants show competitive exclusion, coexistence through differentiation of realized niches (resource partitioning), and character displacement.

Fig. 4.18. Coexistence through character displacement. (a) Average lengths (plus standard deviations) of *Hydrobia ulvae* (open circles) and *H. ventrosa* (filled circles) at a variety of sites at which they coexist or live alone. (b) Distributions of food particle size of the same species at typical sites at which they coexist or live alone. (After Fenchel 1975.)

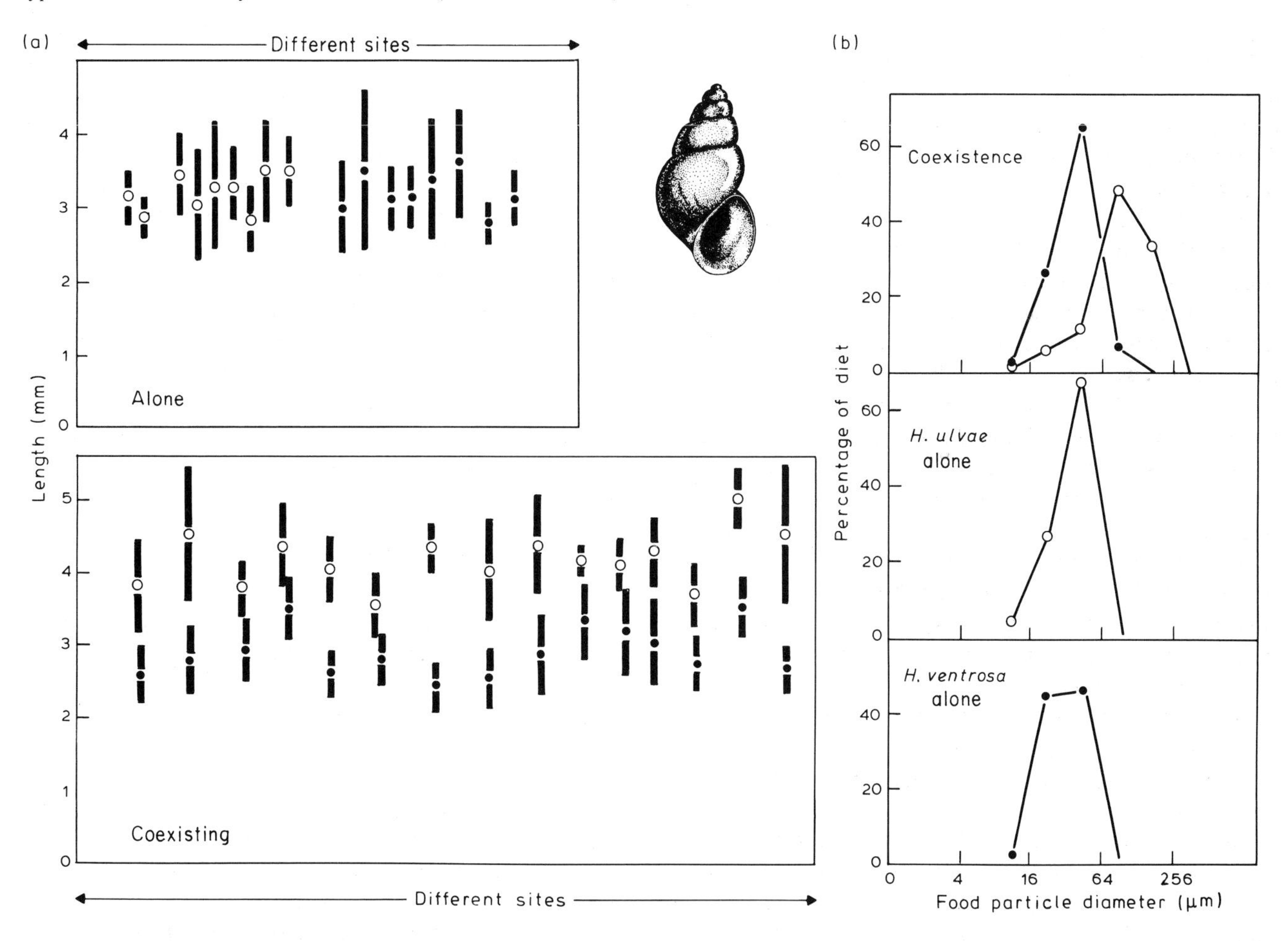

4.11 Competition: its avoidance or its non-existence?

We have seen that species can coexist when, as competitors, they partition resources between them; and also that such partitioning may be achieved, in special cases, by character displacement. But species may also coexist simply because they do not compete.

We have tried, until now, to include only those examples in which competition has been positively established. Suppose, however, that Inouye, instead of carrying out his experimental manipulations, had observed simply that two different species of bee forage from two different species of flower, and that the bee species with the longer proboscis foraged from the flower species with the longer corolla tube. The explanation that has been suggested, involving competition, competitive exclusion and resource partitioning, would certainly have been plausible; but there are two alternative explanations. It might have been suggested that each bee species has become so well adapted to its own flower that there was no (even potentially) shared resource, and therefore no interspecific competition. This explanation implies the existence of interspecific competition in the past, but proposes, in addition, that selection for resource partitioning has been so strong and so longstanding that the partitioning is now complete and irreversible. It might also have been suggested, however, that the difference in proboscis length between *B. appositus* and *B. flavifrons* merely reflects the fact that they *are* two different species, and has nothing to do with competition. There would, therefore, have been a real problem of interpretation if Inouye had relied solely on observational data; and to illustrate this problem we can consider an example of the sort of information that is normally described as 'field evidence of interspecific competition'.

Lack (1971) described the coexistence of five species of tit in English broadleaved woodlands: the blue tit (*Parus caeruleus*), the great tit (*P. major*), the marsh tit (*P. palustris*), the willow tit (*P. montanus*) and the coal tit (*P. ater*). Four of these congeneric species weigh between 9.3 g and 11.4 g on average (great tit 20.0 g); all have short beaks and hunt for food chiefly on leaves and twigs, but at times on the ground; all eat insects throughout the year, and also seeds in winter; and all nest in holes, normally in trees. Nevertheless, in Marley Wood, Oxford, all five species breed and the blue, great and marsh tits are common.

All five species feed their young on leaf-eating caterpillars, and all except the willow tit feed on beechmast in the winters when it is plentiful; but both of these foods are temporarily so abundant that competition for them is most unlikely. The small blue tit feeds mainly on oak trees throughout the year, concentrating on the smaller twigs and leaves of the canopy to which it is suited by its agility. It also strips bark to feed on the insects underneath, and generally takes insects less than 2 mm in length. It eats hardly any seeds, except those of birch which it takes from the tree itself. The heavy great tit, by contrast, feeds mainly on the ground, especially in winter. Most of the insects it takes exceed 6 mm in length; it eats more acorns, sweet chestnut and wood sorrel seeds than the other species; and it is the only species to take hazel nuts. The marsh tit has a feeding station intermediate between the other two common species: in the shrub layer, in large trees on twigs and branches below 20 feet, or in herbage. It is also intermediate in size between the other two species, and generally takes insects between 3 and 4 mm in length. In addition it takes the fruits and seeds of burdock, spindle, honeysuckle, violet and wood sorrel. The coal tit is another small species feeding on oak, but also, later in the winter, on ash. It generally takes insects which are less than 2 mm long, and which are indeed shorter, on average, than those taken by the blue tit. Moreover, unlike the blue tit, the coal tit feeds mainly from branches rather than in the canopy. Finally, the willow tit is most like the marsh tit, feeding on birch and, to a lesser extent, elder, and in herbage. Unlike the marsh tit, however, the willow tit avoids oak and takes very few seeds.

As Lack concluded, the species are separated from each other at most times of the year by their feeding station, the size of their insect prey and the hardness of the seeds they take; and this separation is associated with differences in overall size, and in the size and shape of the beaks. Yet, as we have seen, there are three possible interpretations of this situation. The first two ('current competition' and 'competition in the past') are based on the assumption that the differences reflect the partitioning by the tits of a potentially limited resource; but the third makes no such assumption. It states simply that the five species, in the course of their evolution, have adapted to their environment in different ways; but in ways that have nothing to do with interspecific competition. And on the basis of the evidence presented, *it is impossible to reject this interpretation*. It has not been shown that the birds would expand their niches in the absence of the other species, and it has not even been shown that food is a limited resource. There is, therefore, no direct evidence of competition, and no overriding reason for involving it in our interpretation.

Nevertheless, the possibility of the first two interpretations does remain. We have seen in several examples—ants, barnacles, bumble bees, etc.—that competing species can coexist by resource partitioning, and can retain the ability to expand their range in the absence of their competitor. Thus, it may be the case that this is also the correct interpretation of the tits' ecology, but that the appropriate experimental manipulations have simply not been carried out.

Moreover, we have also seen—e.g. in the examples of character displacement—that species can evolve morphological adaptations which allow them (largely if not totally) to avoid competition. This (second) interpretation, like the third interpretation (above), denies the existence of current interspecific competition, but unlike the third interpretation it invokes interspecific competition as the evolutionary driving-force behind the differences currently observed.

Indeed, the first and second interpretations are based on alternative outcomes of a shared evolutionary process, which undoubtedly does pertain in some cases. The process occurs as follows. Natural selection favours the survival and reproduction of those individuals with the greatest fitness, but interspecific competition reduces fitness. Individuals that avoid interspecific competition will therefore evolve, and interspecific competition is avoided by resource partitioning. Evolution of a realized niche which is too small, however, will increase *intra*-specific competition, and this, too, reduces fitness. We can, therefore, expect each species to evolve towards a form in which inter- and intraspecific competition are optimally offset. Sometimes this will evolve relatively flexible resource partitioning (first interpretation); sometimes the partitioning will be inflexible (second interpretation). Sometimes there will be a lessening of interspecific competition; sometimes its total avoidance. Yet in either case, and with either interpretation, interspecific competition will be of paramount importance.

On the available evidence, however, it is impossible to determine whether these tit data indicate interspecific competition (first interpretation), its evolutionary avoidance (second interpretation), or its total non-existence, now *and* in the past (third interpretation); and this would be true of almost all examples of apparent interspecific competition in the field. There are undoubtedly cases of current resource partitioning amongst competitors; and there are undoubtedly cases in which species' ecologies have been moulded by interspecific competition in the past. But differences between species are *not*, in themselves, indications of the ways in which those species coexist; and interspecific competition *cannot* be studied by the mere documentation of these interspecific differences.

Note, finally, that while the first interpretation is based on a '− −' interaction, the second interpretation assumes that the interaction is essentially '00'. In other words, it is assumed that evolution, acting to avoid 'minuses', converts them to 'zeros'; and this is probably also the basis of most cases of amensalism (−0). The plant species that produces a toxic metabolite causing growth-reduction in a second species presumably does so as an *evolutionary* response to the harmful, competitive effects that the second species had on its growth in the past. At that time the interaction would have been competitive: '− −'. Now, however, evolution has led to the production of toxin by the aggressive species which occurs *whether or not* potential competitors are present. The aggressive species is, therefore, unaffected by these other species, and the interaction is amensal. Evolution has, in this case, converted only one of the minuses of a '− −' interaction into a zero. In other cases (as in the second interpretation, above) it converts both.

4.12 Competition and coexistence in plants

The theory of the niche and the Competitive Exclusion Principle have their origins firmly rooted in zoological study, and it is intuitively less easy to see how niche-differentiation can occur in autotrophic plants, when all have essentially the same basic growth requirements (light, water and nutrients). It might be imagined that plants have evolved specializations to capture energy for photosynthesis from different wavelengths of light, or that nutrients might be utilized in unique and separate ways; but comparative physiological studies show that this is not so. Plant growth requirements ('food') are not usually discrete packages that can be simply partitioned amongst competing species. (An important exception to this, however, is nitrogen utilization. The legumes (as we have seen, Section 4.4.3), and some other genera such as *Alnus*, do not place total reliance on fixed nitrogen in the soil for growth. Instead, by virtue of their symbiotic relationship with nitrogen-fixing micro-organisms, they utilize free nitrogen from the air.) Moreover, the very nature of the effect of limiting resources on plant growth is complex. Limitation in water supply resulting from intense root competition, for instance, will limit leaf growth; but this may contribute in turn to a reduction in the growth of new roots. Thus, shortage of one limiting resource (water) affects the competitive struggle to obtain both light and water itself. Disentangling the web of cause and effect experimentally has proved very difficult (reviewed by Harper 1977); and it is these difficulties that led de Wit (1960) to side-step the problems by proposing that plant species may compete for the 'same space' or for 'different space' (where 'space' is defined as an all-encompassing term subsuming all limited resources in common demand by competitors). He argued that plant species will only coexist when they are competing for space which is to some extent different.

In practice, most attempts to explain the coexistence of plant species have rested largely on a demonstration of the fact that potential competitors differ in ways which might reduce competition. Attention has focused particularly on differences in life-form, differences in the timing of various stages of growth—particularly germination and flowering—and differences in preferred levels of abiotic factors (see Grubb 1977 and Werner 1979 for reviews). However, we have seen from the tits in Marley Wood that such differences, taken alone, are impossible to interpret with confidence. The conclusions that can be drawn from such data are severely limited.

Perhaps the most that can be obtained from the interpretation of these differences is illustrated by the work of Werner and Platt (1976). They studied six species of golden rod (*Solidago*) that commonly occur together both in old hayfields undergoing successional change, and in mature, stable prairie communities in North America. In both habitats, but particularly in the prairies, they were able to relate the frequency of occurrence of each species to the availability of soil moisture (Fig. 4.19): different species appear to 'prefer' different moisture levels. There is, of course, no direct evidence that interspecific competition is the driving force behind this arrangement; but the circumstantial evidence is strengthened in this case by the fact that the degree of separation between species is much greater in the mature prairies than in the young hayfield successional habitats. The position taken up by a species along this environmental gradient is apparently the one at which it

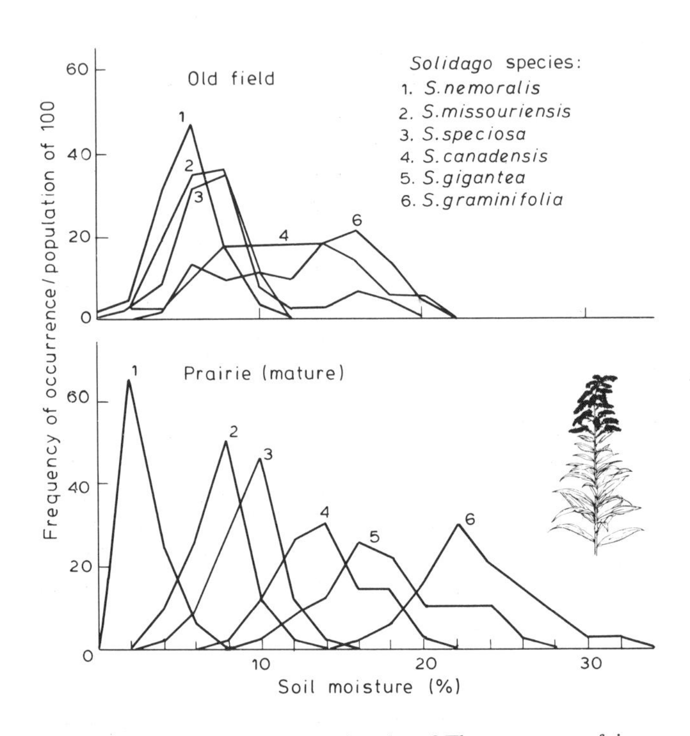

Fig. 4.19. Coexistence of competing plants? The occurrence of six species of *Solidago* in relation to available soil moisture in a hayfield undergoing succession, and in a mature prairie in North America. Soil moisture percentages were determined in summer. (From Werner & Platt 1976.)

has been most successful in competition with the others: interspecific competition, with time, seems to *realize* the niche of each species.

Interspecific competition can also be implicated (this time as the driving-force behind a *temporal* separation in resource-utilization) in the coexistence of barley (*Hordeum sativum*) and the weed, white persicaria (*Polygonum persicaria*) (Aspinall 1960; Aspinall & Millthorpe 1959). Barley, during its period of active growth, very strongly suppresses white persicaria through both root and shoot competition, but the weed is not eliminated. This is partly due to its ability to persist in a suppressed state; but it also results from the persicaria only initiating its own very rapid development *after* the barley has begun to divert its photosynthate toward the production of seed, and away from those vegetative organs directly active in competition.

It is instructive to examine another example of temporal heterogeneity in resource utilization to discover the sorts of differences in growth-pattern which can lead to coexistence. Khan *et al.* (1975) grew two varieties of *Linum usitatissimum*, flax and linseed, together in a replacement series, and also separately. In very broad terms, both varieties were similar in dry matter accumulation and in the development of leaf area over the first nine weeks (Fig. 4.20a); and after 18 weeks the biomass totals of both flax and linseed were the same. Comparison of the seed output by the mixture and two monocultures in the replacement diagram (Fig. 4.20b) shows that both varieties suffer interspecific competition, but that the yield in 50/50 mixture is the largest of all; and the relative yield total of the mixture is greater than one (Fig. 4.20c). In other words, the two varieties are not competing for precisely the same 'space' in de Wit's sense: there is niche-differentiation. The formal growth analysis shows that flax matures earlier than linseed and dies back to leave predominantly stem and shrivelled leaves. It is at precisely

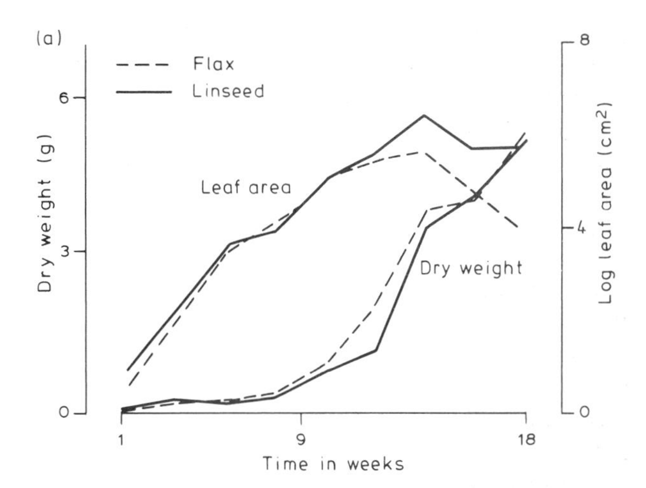

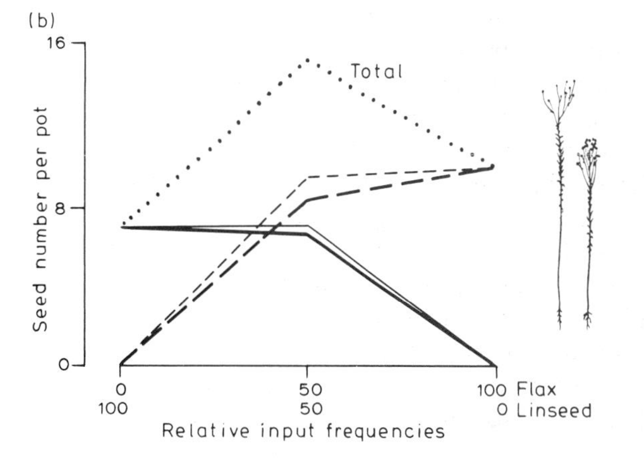

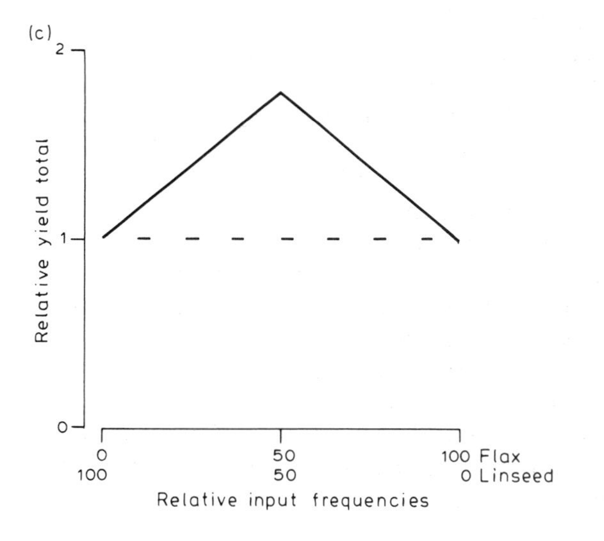

Fig. 4.20. Coexistence in two varieties of *Linum usitatissimum*, flax and linseed. (a) Dry matter accumulation and leaf area development; (b) a replacement diagram (thin lines are yield expectations in monoculture; thick lines are yield expectations when the varieties are grown together); (c) relative yield totals. (From Kahn 1963, after Khan *et al.* 1975.)

this time that linseed initiates seed filling and maturity, and thus benefits by released 'space' in the canopy due to the senescence of flax leaves. The yield response in mixture exceeds both of the monoculture yields, because both varieties escape the intensity of inter-varietal competition for light that they would have experienced in monoculture. It is clear that even slight temporal separation in the utilization of a resource *can* represent effective niche-differentiation between co-existing competitors.

In a few plant examples, however, it is possible to do more than merely implicate interspecific competition as an important interaction between coexisting species. Sharitz & McCormick (1973), for instance, studied the population dynamics of pairs of annual plant species (*Sedum smallii* and either *Minuartia uniflora* or *M. glabra*) which dominate the vegetation growing on granite outcrops in the south-eastern United States. As Fig. 4.21 shows, there is very strict zonation of the adults of the two species associated with the soil depth around the outcrops; and soil depth itself is strongly correlated with soil moisture. The experimental results in Fig. 4.21b, c, however, indicate that this zonation is not simply a reflection of the tolerance ranges of the species. In fact, their fundamental niches cover the same range of experimental conditions. Nevertheless, while *Sedum* is clearly more capable than *Minuartia* of tolerating the lack of moisture at the low end of the range, *Minuartia* is obviously much less affected than *Sedum* by interspecific competition at the high end. It is apparent, in other words, that interspecific competition in nature restricts these species to realized niches (in practice, 'zones') that are significantly smaller than their fundamental niches. Further evidence of this is provided by the very incomplete zonation at the seedling stage, prior to any substantial competitive interaction (Fig. 4.21a). The parallel with the barnacles in Section 4.7 is quite striking.

Another, very enlightening example of coexistence amongst competing plants is provided by the work of Tilman (1976) on two planktonic algal species, *Asterionella formosa* and *Cyclotella meneghiniana*. Both species are potentially limited in their growth by the concentrations of silicate and phosphate in their aqueous environment; but in practice, the growth-rate of each species is determined by the concentration of whichever nutrient leads to a lower growth-rate. Tilman established experimentally that *A. formosa* was limited by silicate when the silicate-to-phosphate ratio was less than 97, but was limited by phosphate when the ratio exceeded 97. The corresponding boundary ratio for *C. meneghiniana* was 5.6. Tilman then ran a series of competition experiments, the outcomes of which are shown in Fig. 4.22. Both species were able to live in monoculture under all experimental conditions. In competition, however, *A. formosa* excluded *C. meneghiniana* at ratios exceeding 97, *C. meneghiniana* excluded *A. formosa* at ratios less than 5.6, and the species coexisted at intermediate ratios. In other words, when the species were limited by different resources (*A. formosa* by silicate, *C. meneghiniana* by phosphate) they coexisted; but when they were limited by the same resource, the species which was least severely limited was competitively dominant.

In at least some cases, therefore, plant species which are potential competitors for a limited resource can coexist by virtue of a differentiation of their realized niches. Yet it must be recognized that there are very few instances in which this has been positively established (see Werner 1979). It is very easy, in plants, to demonstrate the reductions in fitness which can result from interspecific competition; but it has proved very difficult so far to demonstrate the mechanisms which allow potential competitors to coexist. There are two reasons for this. Grubb (1977) has suggested that this difficulty may stem from an overemphasis on adult plants. Irrespective of any niche differentiation, it is almost impossible for a seedling of one species to out-compete an established adult of another species. The most important competition between plants, therefore, may be *pre-emptive competition* for the *regeneration niche*, i.e. competition amongst seedlings to become established in a part of the environment which has recently become vacant. It seems certain that in future, the study of pre-emptive competition will teach us a great deal about the coexistence in nature of competing plants. A very similar phenomenon, restricted to single species and referred to as 'space capture', is discussed in Section 7.10.

The second reason is a neglect of the fact that

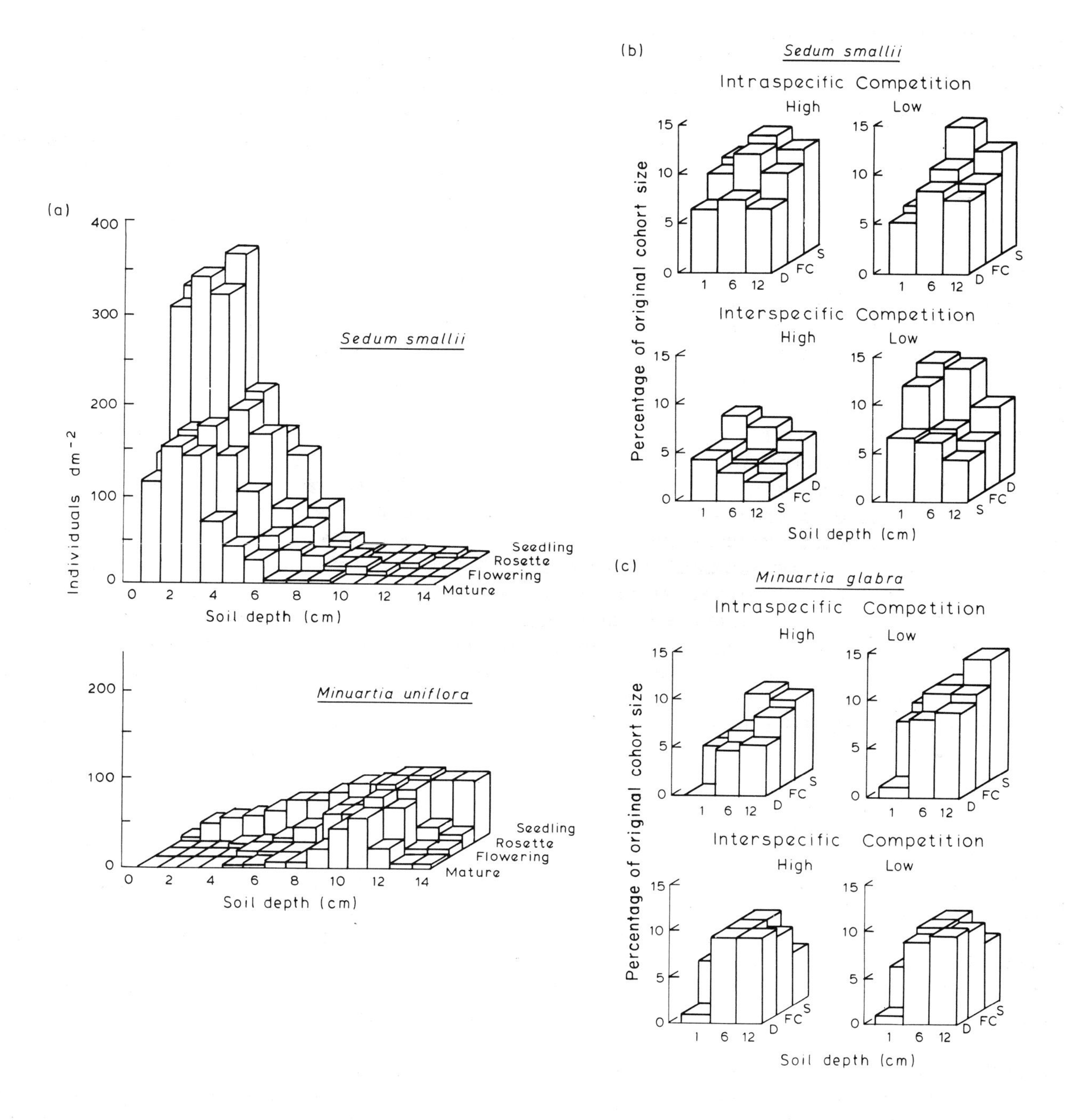

Fig. 4.21. (a) The zonation of individuals, according to soil depth, of two annual plants *Sedum smallii* and *Minuartia uniflora* at four stages of the life cycle. (b) and (c) The consequences of competitive interaction between *Sedum smallii* and *Minuartia glabra*. For each species the final density at plant maturity (as a percentage of the initial seed sown) is shown when grown alone and with the other species. The experiment was conducted at three soil depths and three relative moisture levels: S, saturation; FC, field capacity; D, 1/3 field capacity. (From Sharitz & McCormick 1973.)

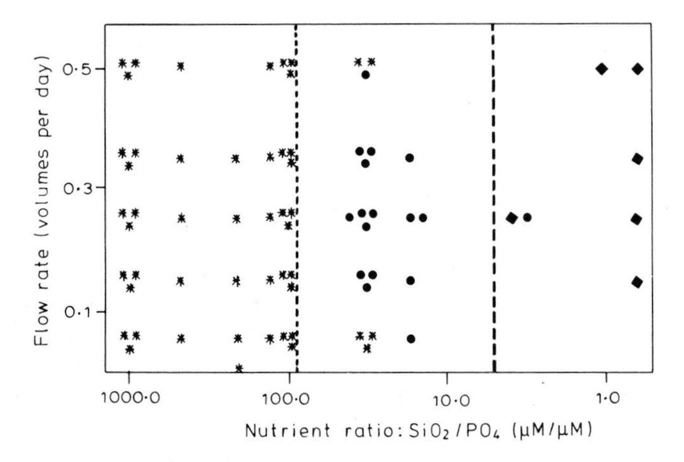

Fig. 4.22. Competition between two algae, *Asterionella formosa* and *Cyclotella meneghiniana*. Symbols indicate the winners in competition experiments in aqueous environments of varying silicate: phosphate concentration. (✱) *A. formosa* wins; (■) *C. meneghiniana* wins; (●) stable coexistence of both species. (From Tilman 1976.)

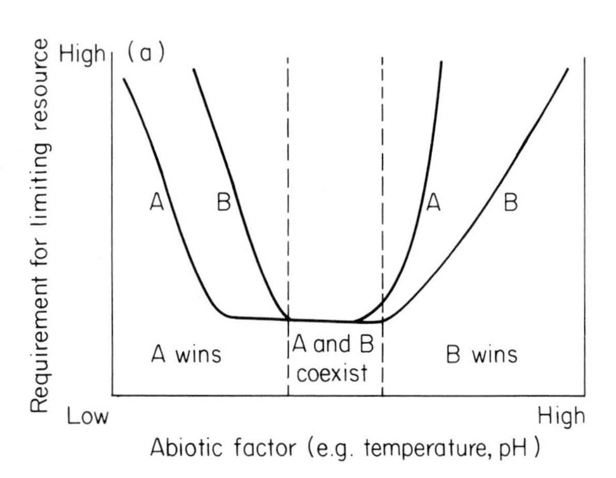

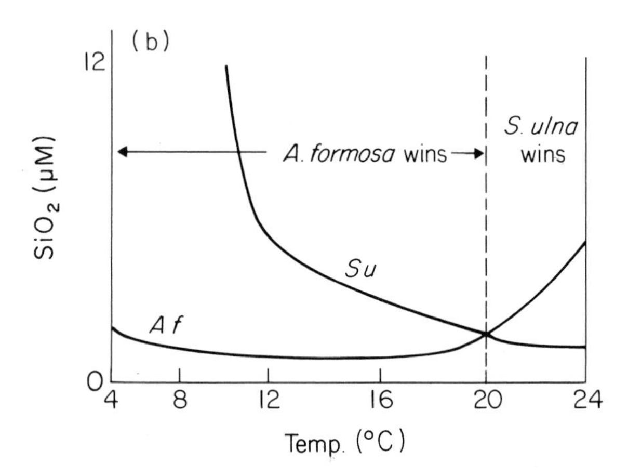

Fig. 4.23. (a) Competition for one limiting resource in relation to an abiotic factor. (b) The outcome of competition for silicate for two species of planktonic algae over a wide temperature range. *A.f.* = *Asterionella formosa*, *S. u.* = *Synedra ulna*. (Modified from Titman 1982.)

environmental parameters indirectly interact to determine the intensity of competition. Utilization of resources in limited supply may be determined by the species' response to other (non-limiting) environmental factors which govern the competitive outcome. Fig. 4.23a portrays the requirements of two hypothetical species for a limiting resource, determined by an abiotic factor such as temperature or pH. At the lower part of the range, species A, by virtue of a lower requirement for the limiting resource will tend to outcompete species B. The converse will be true in the upper part of the range. In mid range, all other factors being equal, growth is limited to the same extent and coexistence occurs. Partial support for this view comes from a further experiment using the planktonic algal species *A. formosa* and *Synedra ulna* (Fig. 4.23b). Over much of the temperature range (4–20°C), *A. formosa* has the lower requirement for SiO_2 and outcompetes *S. ulna*, but the reverse is true above 20°C. Although there is no temperature range in which both species have the same demand for silicate, we can envisage coexistence if there are fluctuations in temperature over the range in which alternative competitive displacement can occur. Indeed, diurnal temperature fluctuations between 16 and 24°C are very likely in the surface waters of the lakes in which these species live. The only proviso that we must add in final explanation of this mechanism is that it clearly depends on the time scale over which population growth responses can occur.

4.13 A logistic model of two-species competition

Having examined what is known about interspecific competition, it will be valuable to turn (as we did with single-species populations) to some simple models, to see whether they can improve our understanding of the interaction.

The conventional starting-point for such models is the differential *logistic equation* (following Lotka 1925, Volterra 1926). Obviously, therefore, in conforming to this convention, we will be incorporating into our model all of the logistic's shortcomings. Nevertheless, as will become apparent below, a useful model can be constructed.

The logistic equation:

$$\frac{dN}{dt} = rN\left(\frac{K-N}{K}\right)$$

contains, within the brackets, a term which is responsible for the incorporation of intraspecific competition. We can proceed by replacing this term with one which incorporates not only intra- but also interspecific competition. We will denote the numbers of our original species by N_1 (carrying capacity, K_1; intrinsic rate of increase, r_1), and those of a second species by N_2.

Suppose that, together, 10 individuals of species 2 have the same competitive, inhibitory effect on species 1 as does a single species 1 individual. The *total* competitive effect on species 1 (inter- and intraspecific) will then be equivalent to $(N_1+\{N_2/10\})$ species 1 individuals. We call the constant—1/10 in the present case—a *coefficient of competition*, and denote it by α_{12} since it measures the competitive effect *on* species 1 *of* species 2. In other words, multiplying N_2 by α_{12} converts it to a number of 'N_1-equivalents'. (Note that $\alpha_{12} < 1$ means that species 2 has less inhibitory effect on species 1 than species 1 has on itself, while $\alpha_{12} > 1$ means that species 2 has a greater inhibitory effect than species 1 has on itself.) We now simply need to replace N_1 in the bracket of our logistic equation with a term which signifies: 'N_1 plus N_1-equivalents', i.e.

$$\frac{dN_1}{dt} = r_1N_1\left(\frac{K_1-[N_1+\alpha_{12}N_2]}{K_1}\right)$$

or

$$\frac{dN_1}{dt} = r_1N_1\left(\frac{K_1-N_1-\alpha_{12}N_2}{K_1}\right).$$

We can, of course, write a similar equation for species 2:

$$\frac{dN_2}{dt} = r_2N_2\left(\frac{K_2-N_2-\alpha_{21}N_1}{K_2}\right).$$

This is our basic (Lotka–Volterra) model.

To describe the properties of this model, we must ask the following questions: When (under what circumstances) does species 1 increase in numbers? When does it decrease? And when does species 2 increase and decrease? In order to answer, we construct what are, in essence, the equivalents of maps. Thus, while in maps there are areas of land and areas of sea, with a coastline (neither land nor sea) dividing them; in our case we will have areas of N_1 (or N_2) increase and areas of N_1 (or N_2) decrease, with a *zero isocline* (neither increase nor decrease) dividing them. Moreover, if we begin by drawing a zero isocline (coastline), we will know that there is increase (land) on one side of it and decrease (sea) on the other. As Fig. 4.24 shows, the axes of our 'map' will be N_1 and N_2: the bottom left-hand corners are areas where there are low numbers of species 1 and 2, and the top right-hand corners areas where there are high numbers of species 1 and 2.

In order to draw the N_1-isocline we will use the fact that *on* it $dN_1/dt = 0$, i.e.

$$r_1N_1\left(\frac{K_1-N_1-\alpha_{12}N_2}{K_1}\right) = 0.$$

This is true for two trivial cases (when r_1 or N_1 are zero), but also for an important case:

$$K_1-N_1-\alpha_{12}N_2 = 0$$

or

$$N_1 = K_1-\alpha_{12}N_2.$$

Indeed, the straight line represented by this equation is our isocline, and since it *is* a straight line we can draw it by finding two points on it and joining them. Thus:

when $N_1 = 0, N_2 = \dfrac{K_1}{\alpha_{12}}$ (point A, Fig. 4.24a)

and when $N_2 = 0, N_1 = K_1$ (point B, Fig. 4.24a).

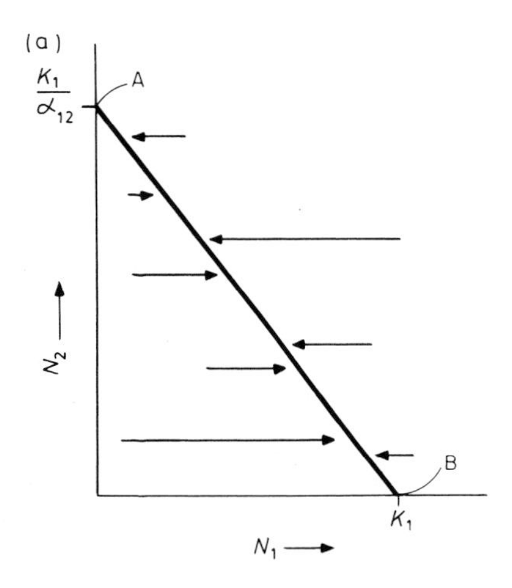

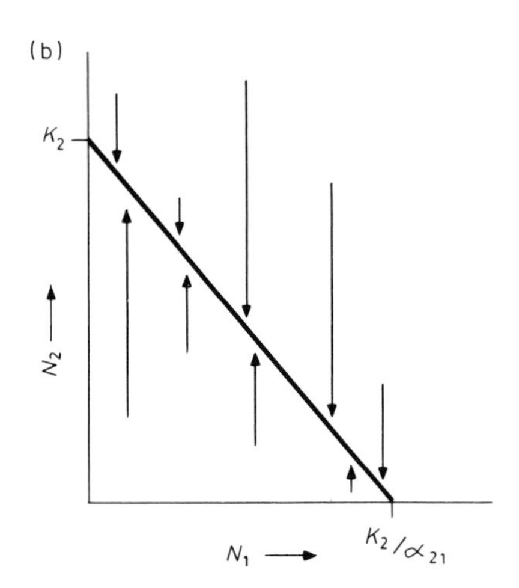

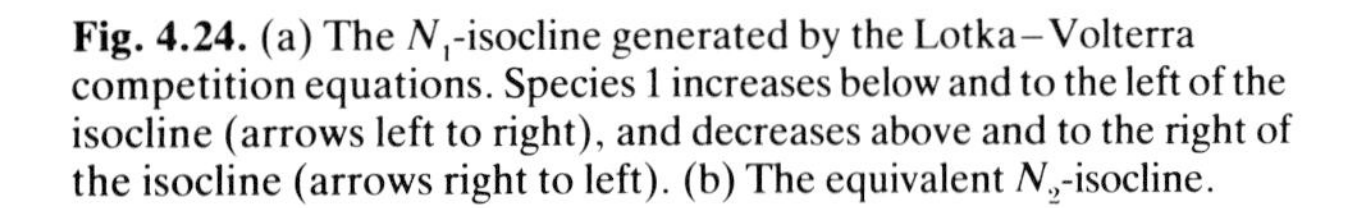

Fig. 4.24. (a) The N_1-isocline generated by the Lotka–Volterra competition equations. Species 1 increases below and to the left of the isocline (arrows left to right), and decreases above and to the right of the isocline (arrows right to left). (b) The equivalent N_2-isocline.

The line in Fig. 4.24a is, therefore, the N_1-isocline. Below and to the left of it, numbers are low, competition is comparatively weak and species 1 increases in abundance (arrows from left to right, N_1 on the horizontal axis); above and to the right of it, numbers are high, competition is comparatively strong and species 1 decreases in abundance (arrows from right to left).

Based on an equivalent derivation, Fig. 4.24b has areas of species 2 increase and decrease separated by the (straight) N_2-isocline; arrows, like the N_2-axis, are vertical.

All that is required now is to put the N_1- and N_2-isoclines together on a single figure. In so doing, it should be noted that the arrows in Fig. 4.24 are actually vectors—with a strength as well as a direction—and that, to determine the behaviour of a joint N_1–N_2 population, the normal rules of vector addition should be applied (see Fig. 4.25). It is clear from Fig. 4.26 that there are four different ways in which the two isoclines can be arranged. In Fig. 4.26a, b, one isocline lies entirely beyond the other, and the vectors indicate that, as a consequence, the species with the inner isocline becomes extinct, while the other species attains its own carrying capacity. Such situations can be defined by the intercepts of the isoclines. In Fig. 4.26a, for instance:

$$\frac{K_1}{\alpha_{12}} > K_2 \quad \text{and } K_1 > \frac{K_2}{\alpha_{21}}$$

i.e.

$$K_1 > K_2\alpha_{12} \text{ and } K_2 < K_1\alpha_{21}.$$

Species 1 exerts more effect on itself than species 2 exerts on it, but also exerts more effect on species 2 than species 2 does on itself. In other words, species 1 is a

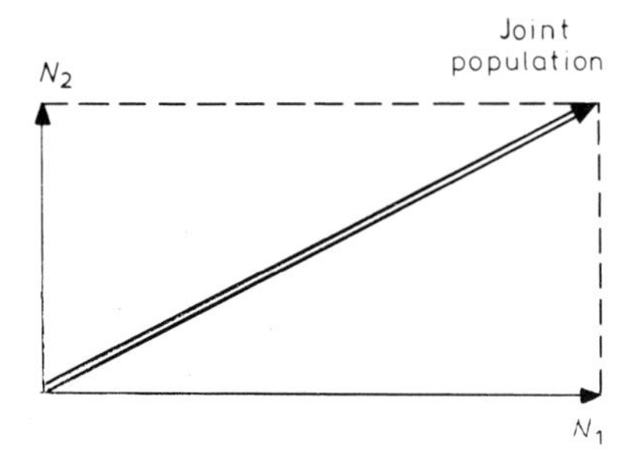

Fig. 4.25. Vector addition. When species 1 and 2 increase in the manner indicated by the N_1 and N_2 arrows (vectors), the joint population increase is given by the vector along the diagonal of the rectangle, generated as shown by the N_1 and N_2 vectors.

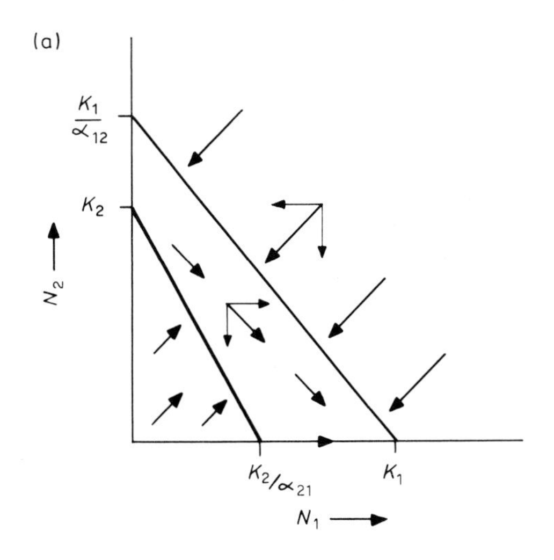

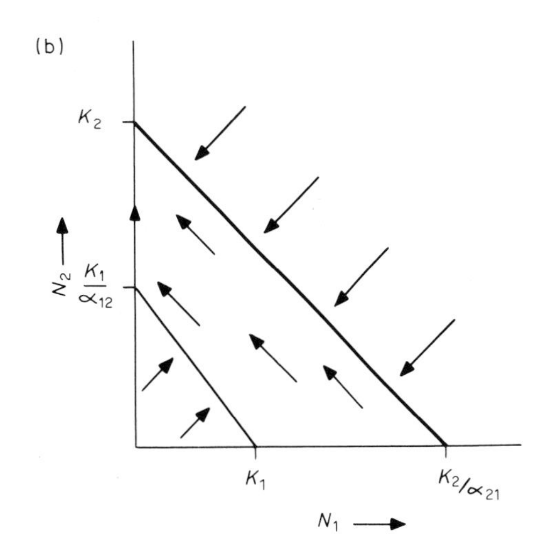

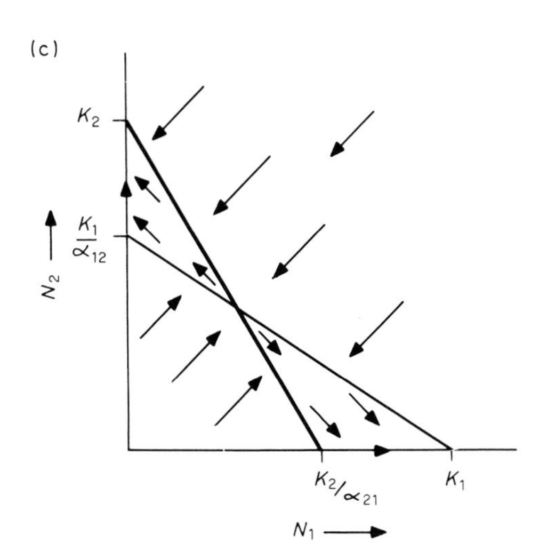

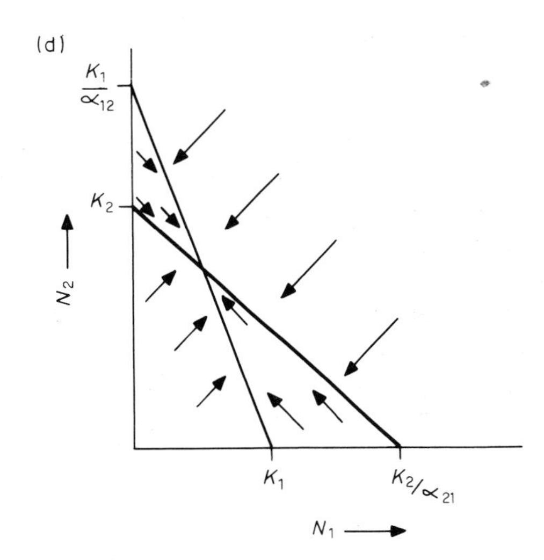

Fig. 4.26. The outcomes of competition generated by the Lotka–Volterra competition equations for the four possible arrangements of the N_1 and N_2 isoclines. Vectors, generally, refer to joint populations and are derived as indicated in (a). For further discussion, see text.

strong interspecific competitor, species 2 is a weak interspecific competitor, and species 1 drives species 2 to extinction. The situation is reversed in Fig. 4.26b.

In Fig. 4.26c:

$$K_2 > \frac{K_1}{\alpha_{12}} \quad \text{and } K_1 > \frac{K_2}{\alpha_{21}}$$

i.e.

$$K_1 < K_2\alpha_{12} \text{ and } K_2 < K_1\alpha_{21}.$$

Interspecific effects are more important than intraspecific effects: both species are strong interspecific competitors. There are two stable points ($N_1 = K_1$, $N_2 = 0$ and $N_2 = K_2$, $N_1 = 0$) and an unstable equilibrium combination of N_1 and N_2. In other words, one species always drives the other to extinction, but the precise outcome depends on the initial densities.

Finally, in Fig. 4.26d:

$$\frac{K_1}{\alpha_{12}} > K_2 \quad \text{and} \quad \frac{K_2}{\alpha_{21}} > K_1$$

i.e.

$$K_1 > K_2\alpha_{12} \text{ and } K_2 > K_1\alpha_{21}.$$

Intraspecific effects are now more important than interspecific effects, both species are weak interspecific competitors, and there is stable coexistence at a particular, equilibrium combination of N_1 and N_2.

4.13.1 The model's utility

We can now proceed to examine this simple model's utility. Clearly, it can produce the full range of outcomes of interspecific competition: stable coexistence, predictable exclusion of one species by another, and exclusion between the two species with an indeterminate outcome. It should be recognized, however, that these are the only conceivable outcomes; to be useful, the model must produce the right outcome at the right time.

The model indicates (Fig. 4.26a, b) that one species will outcompete and exclude a second if the first species is a stronger competitor on the second than the second is on itself, and the effect is not reciprocated. We saw in Section 4.7 that this was, indeed, the case for each bumble bee on its own flower (exploitation competition), and for *Balanus* excluding *Chthamalus* (interference competition). In other words, Fig. 4.26a, b successfully describes situations in which the second species lacks a realized niche in competition with the first.

In Fig. 4.26c, both species are stronger competitors on the other species than they are on themselves: there is reciprocal interference competition. This will occur when each species produces a substance that is toxic to the other species but harmless to itself, or when there is reciprocal predation. In fact, this latter situation is the mechanism by which Park's flour beetles compete (Section 4.6), and it is satisfying, therefore, to see that the model's predictions are borne out by Park's data: there is competitive exclusion, but the precise outcome is indeterminate. Whichever species starts with (or, at some point, attains) a more favourable density will 'outpredate' (or outpoison) the other.

Finally, Fig. 4.26d indicates, quite reasonably, that stable coexistence is only possible when, for both species, intraspecific competition is more inhibitory than interspecific competition, i.e. when there is niche differentiation. We have seen repeatedly that this is the case. However, the model avoids the more profound question of *how much* niche differentiation is necessary for stable coexistence.

It is clear, then, that in broad terms the model is successful in spite of its limitations. Indeed, it is important to realize that this model (given 'experimental teeth' by the laboratory work of Gause 1934) was the original inspiration of the Competitive Exclusion Principle; investigation of the relevance of the Principle, and its applicability in the real world, came later. Yet, the most important aspect of the model is that it makes exact, *quantitative* predictions about coexistence, based on the numerical values of the Ks and αs. Ultimately the model's utility must be tested in these quantitative terms.

4.13.2 A test of the model: fruit fly competition

There have been several studies in which experimental results have tended to support the model's predictions. It will be more instructive, however, to consider an example in which results and predictions disagree, because this will illustrate a point of general importance: models are of their greatest utility when their predictions are *not* supported by real data, *as long as the reason for the discrepancy can subsequently be discovered.* Confirmation of a model's predictions represents consolidation; refutation with subsequent explanation represents progress.

Ayala *et al.* (1973) reviewed Ayala's own findings on laboratory competition between pairs of *Drosophila* species. Their results are illustrated in Fig. 4.27. Ayala's basic procedure was to maintain either one- or two-species populations in culture bottles using a 'serial transfer' technique—transferring adults to new food at regular intervals—and to monitor the populations for several generations until an approximate equilibrium was reached. This allowed him to estimate the carrying capacities (K_1 and K_2) in single-species populations, and the numbers for stable coexistence $\bar{N}_1$ and $\bar{N}_2$) in two-species populations. Fig. 4.26d shows that our simple model predicts that stable coexistence should only occur at a point above and to the right of the line joining K_1 and K_2. It is clear from Fig. 4.27, however, that in seven out of eight cases Ayala found coexistence below and to the left of this line. Our simple model is, therefore, unable to account for Ayala's results.

The simplest explanation for such a discrepancy is illustrated in Fig. 4.28. If the isoclines are concave, rather than straight, the values of K_1, K_2, $\bar{N}_1$ and $\bar{N}_2$ immediately become compatible. In fact, Ayala *et al.* (1973) were able to support this solution, in a particular species-pair, by following a number of populations for a single generation and obtaining an actual series of vectors. These, too, are shown in Fig. 4.28, in which we can see that the curved isoclines are, indeed, satisfactory lines of demarcation between areas of increase and decrease.

Ayala *et al.* took their analysis a stage further by considering a range of possible alternative modes which would generate the appropriately curved isoclines. They judged these on a number of criteria, including simplicity, biological relevance of the parameters and quality of fit to the data, and found that one equation was particularly satisfactory. This equation:

$$\frac{dN_i}{dt} = r_i N_i \left(\frac{K_i^{\theta_i} - N_i^{\theta_i}}{K_i^{\theta_i}} - \frac{\alpha_{ij} N_j}{K_i} \right)$$

is a modification of the original, simple model, but it contains, in addition, a parameter, θ_i, which modifies the underlying logistic equation, such that the function relating growth-rate to density need no longer be a straight line symmetrical about $K/2$. Thus, the work of Ayala *et al.* confirms our original misgivings regarding

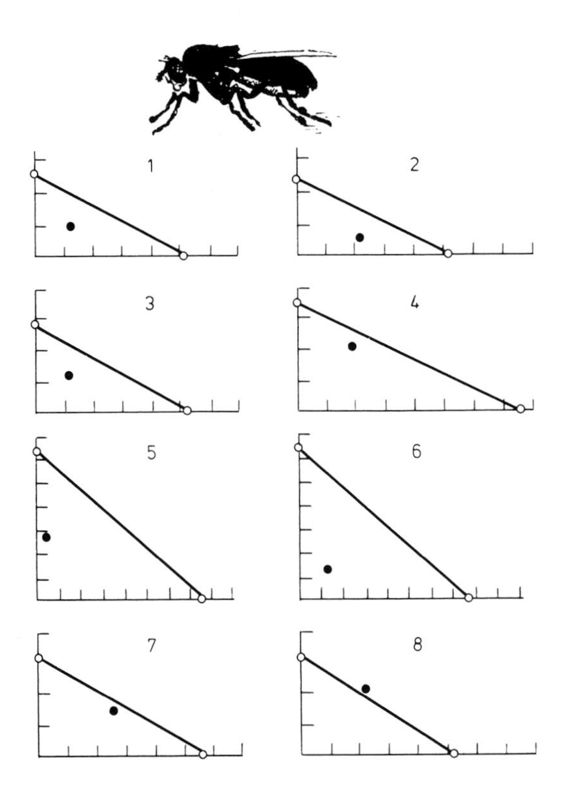

Fig. 4.27. Carrying capacities (open circles) and stable two-species equilibrium points (filled circles) for eight combinations of two species of *Drosophila*. Each division along the co-ordinates corresponds to 250 flies. In all but the last case, *the point of stable two-species equilibrium falls below the straight line joining the carrying capacities.* (After Ayala *et al.* 1973.)

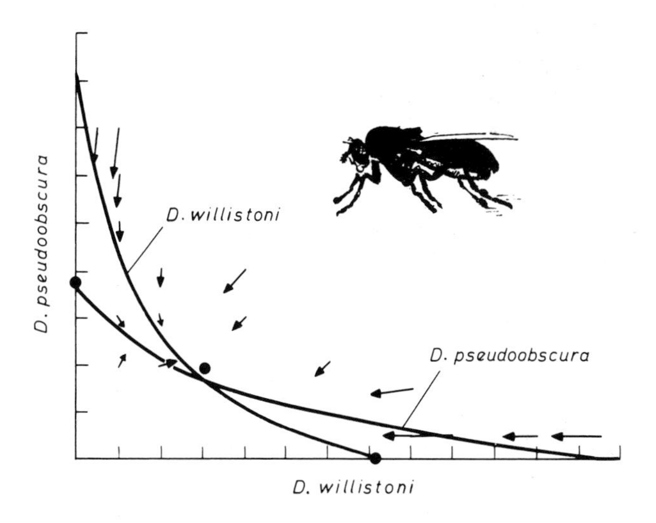

Fig. 4.28. Isoclines for two species of *Drosophila* fitted by visual inspection of the vectors, which were derived empirically and are reduced to one-third of their actual length for clarity. Each division along the co-ordinates corresponds to 200 flies. Filled circles indicate the carrying capacities and the point of stable two-species equilibrium, which were also derived empirically. (After Ayala *et al.* 1973.)

the incorporation of the logistic equation into our model. It also points to an alternative model which, in quantitative terms, is certain to be more generally applicable that the original. Note, however, that the more qualitative conclusions concerning exclusion and coexistence remain unaffected.

4.14 The analysis of competition in plants

The replacement and ratio diagrams developed by de Wit and his colleagues provide, as we have seen, an elegant graphical means of assessing the intensity and consequences of interspecific competition at particular plant densities. This methodology has not, up to now, included a formal definition of the competitive effect of one species on another, in the sense of the competition coefficients of the Lotka–Volterra equations. Moreover, it would certainly be wrong to attempt to directly compare the de Wit model with that of Lotka–Volterra, since the latter includes the effects of both density and frequency whilst the former relates only to frequency. Nor is it correct to assume that de Wit's model is the only model available for the measurement and analysis of competitive effects amongst plants. There exist in the literature a variety of algebraic and statistical models for measurement of the aggressiveness of one species towards another. De Wit's approach, however, does have distinct theoretical advantages over many of these alternatives (see review by Trenbath 1978).

The experimental evidence of de Wit and his coworkers, gained over many years, convincingly demonstrates that in experimentally controlled situations, in which two species will be competitively, mutually exclusive, it is incorrect to automatically assume that an increase in absolute yield of one species is matched by a corresponding decrease in absolute yield of the other. Such a situation will only occur where their absolute yields in monoculture are equal (de Wit & van den Bergh 1965). The importance and consequence of a differential between pure-stand yields on the form of the replacement diagrams and the outcome of competition are illustrated in Fig. 4.29.

Let us now look in more detail at the mathematics underlying de Wit's analysis. Consider a unit area in which two species, I and J, are grown in a proportional mixture always at the same overall density. Let us assume that the yield of each species in mixture is proportional to its yield in monoculture (100% of the total density), the proportionality reflecting, in some way, the area available to each species; i.e. yield of I in mixture $(I+J)$ is proportional to yield of I in monoculture

or
$$Y_{I(mixt.)} = S \times Y_{I(monoc.)} \qquad (4.1)$$

where S is a value that reflects the space available to I, a space restricted by the presence of species J. S must be expressed in a manner that reflects the proportions of the species in the mixture, and it seems reasonable to attempt to equate space available to sown-seed input or starting frequency:

$$\frac{\text{Space}_I}{\text{Space}_J} \propto \frac{\text{Sown}_I}{\text{Sown}_J}$$

Direct replacement of the proportionality sign by an equals would imply that the space occupied by a species was always in direct proportion to its sowing frequency; but in competitive situations, where one species gains at the expense of the other, this is not so. We must allow for this by introducing terms, s_I and s_J, that determine the amount of space acquired by each species:

$$\frac{\text{Space}_I}{\text{Space}_J} = \frac{s_I\,\text{Sown}_I}{s_J\,\text{Sown}_J}$$

Furthermore, to satisfy the condition that the total space potentially available to both species is constant, we must express the sowing or starting frequencies as relative frequencies.

Thus,
$$\frac{\text{Space}_I}{\text{Space}_J} = \frac{s_I\,Z_I}{s_J\,Z_J} \qquad (4.2)$$

where Z_I and Z_J are relative frequencies (adding up to unity).

The 's' terms in equation 4.2 are called crowding coefficients and may be replaced by a single term C_{IJ} $(= s_I/s_J)$ which is the relative crowding coefficient of species I to J. Alternatively, if we wish to focus attention on species J, then equation 4.2 may be inverted and the relative crowding coefficient of J to I, C_{JI} $(= s_J/s_I)$ considered. As long as 'space$_I$ plus space$_J$' is a constant, and the two species are competing for exactly the same space, C_{IJ} will equal C_{JI}^{-1}. Thus, *under these conditions*, a relative crowding coefficient, a dimensionless constant, is an agglomerative index of the extent to which one species competes with the other for the *same* growth resources.

It follows from equation 4.2 that the proportion of space occupied (loosely equivalent to resources gained) by I in a mixture is:

$$\frac{s_I Z_I}{s_I Z_I + s_J Z_J} \quad \text{or} \quad \frac{C_{IJ} Z_I}{C_{IJ} Z_I + Z_J}$$

and by J is:

$$\frac{s Z_J}{s_I Z_I + s_J Z_J} \quad \text{or} \quad \frac{C_{JI} Z_J}{Z_I + C_{JI} Z_J}.$$

Substitution of these terms for space in equation 4.1 gives

$$Y_{I(mixt.)} = \frac{C_{IJ} Z_I}{C_{IJ} Z_I + Z_J} \times Y_{I(monoc.)}. \quad (4.3)$$

and

$$Y_{J(mixt.)} = \frac{C_{JI} Z_J}{Z_I + C_{JI} Z_J} \times Y_{J(monoc.)}. \quad (4.4)$$

In these equations we have a mathematical relationship between yield in mixture and proportion sown: the y and x variates of the replacement diagram. The numerical sizes of C_{IJ} and C_{JI} determine the shape of the yield response curves in the replacement series, and to calculate them without recourse to statistical methods we may utilize the following formula:

$$C_{IJ} = \frac{Y_{I(mixt)}.\, Z_J}{(Y_{I(monoc.)} - Y_{I(mixt.)}) Z_I}$$

calculating C_{JI} with the appropriate terms in a similar way (Hall 1974).

To comprehend the utility of the relative crowding coefficients, we must first realize their connection with the relationships examined in the ratio diagram (e.g. Fig. 4.11). There, the ratio of the output yields was

Fig. 4.29. De Wit diagrams: a graphical portrayal of the effect of absolute yield on competitive outcomes.

Consider two hypothetical species, I and J, grown in mixture in a replacement series. I has a monoculture yield of 100 units and a convex response curve. By contrast, J has a monoculture yield of 100 but in the first instance a concave response curve (species J: response 1 in (a)). The outcome as seen in the ratio diagram (b) is that I outcompetes J and will drive J to exinction. If however, J responds linearly (species J: response 2) a stable mixture will be attained in which I predominates.

In the second instance, J' ('J-prime') has a monoculture yield of 50. This hastens the process of extinction of J' (I and J': response 1), and effectively nullifies the stable mixture outcome (I and J': response 2).

Whilst these illustrations are entirely artificial, it is clear that a difference in monoculture yield does interact with the shape of the response curve to determine the competitive outcome in the ratio diagram.

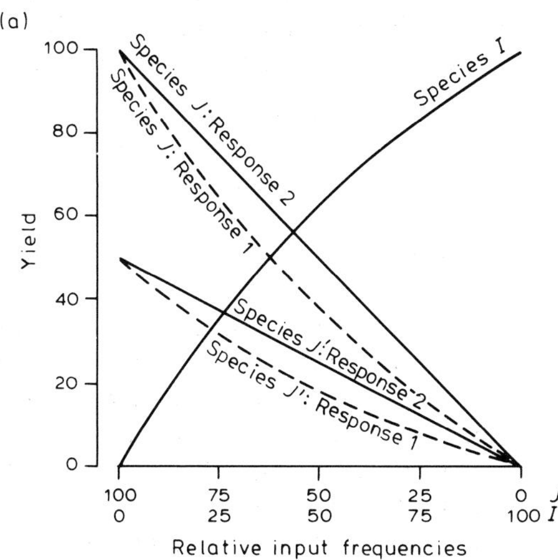

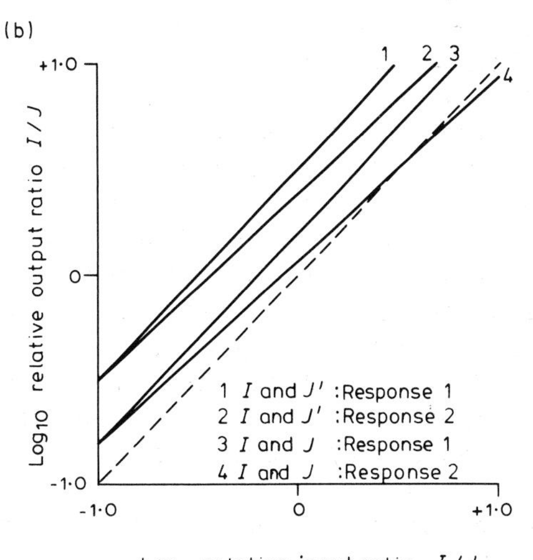

plotted against the input ratio logarithmically. Thus, we would plot log $Y_{I(mixt.)}/Y_{J(mixt.)}$ against log Z_I/Z_J, and we can see from equations 4.3 and 4.4 that the output yields, $Y_{I(mixt.)}$ and $Y_{J(mixt.)}$, are themselves functions of C_{IJ} and C_{JI}. The shape of the curves relating input and output ratios in the ratio diagram is determined by the product of C_{IJ} and C_{JI} (Hall 1974), whilst its position relative to the axes is decided (as we have already seen in Fig. 4.29) by the values of C_{IJ} and C_{JI} themselves *and* the monoculture yields of each species. To disentangle the effects of numerical superiority from competitive superiority we must consider relative output yields, as we did when considering Relative Yield Totals (Section 4.4.3). We are now in a position to assess the significance of the relationship between C_{IJ} and C_{JI}.

To start, let us consider situations where $C_{IJ} \times C_{JI} = 1$. Here, according to de Wit, both species are competing for the 'same space', and the intensity of the competitive interaction is reflected by the magnitude of the difference between the relative crowding coefficients. In the special case where this difference is zero, i.e. $C_{IJ} = C_{JI} = 1$, the growth of an individual of one species is considered to be no more affected by the presence of individuals of another species than by the presence of an equal number of its own species (intraspecific and interspecific effects are equal). This case gives us our midline, of slope unity, that *halves* the ratio diagram (Fig. 4.30). Lines of similar slope which are displaced from the midline reflect interactions in which the process of competitive exclusion is occurring.

In competitive interactions in which the product of the relative crowding coefficients is greater than 1, species may still be competing for limited resources, but such competition is not absolutely for the same space and may not lead to mutual exclusion. De Wit has described this as 'partially competing for different space'. Where this relationship between coefficients ($C_{IJ} \times C_{JI} > 1$) holds, the curve in the ratio diagram becomes 'S-shaped', the angle of intersection with the midline of unit slope increasing as the product of C_{IJ} and C_{JI} itself increases (Fig. 4.30).

Before moving on to apply this analysis of C values to some experimental results, we must note in passing an important assumption on which it rests. This is that our values for the relative crowding coefficients do not vary

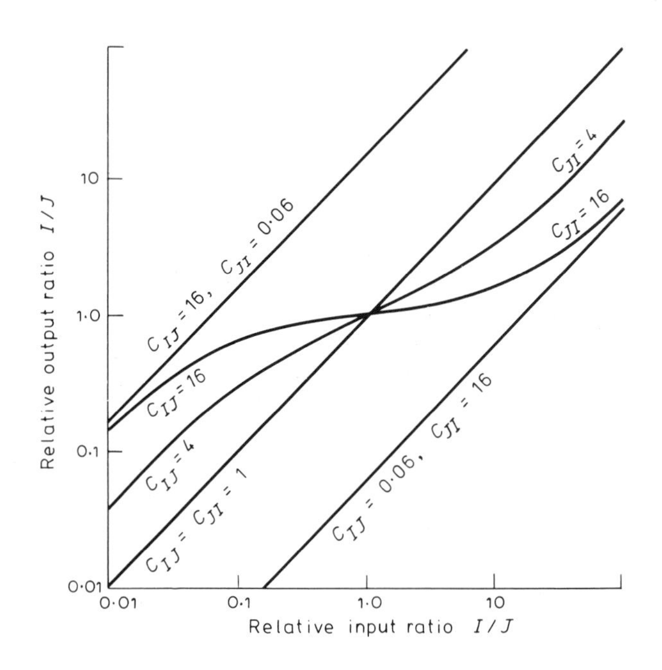

Fig. 4.30. A ratio diagram for the relative yields of two species, I and J. A range of crowding coefficients illustrates the various outcomes. See text for explanation. (After Hall 1974.)

according to changes in the proportion of the two species in the mixture; should they do so, the analysis would clearly be inappropriate.

A good example of the application of this approach is the analysis by Hall (1974) of the results of an experiment by Vallis *et al.* (1967), in which Rhodes grass (*Chloris gayana*) and a legume, Townsville stylo (*Stylosanthes humilis*), were grown in a replacement series. A prime objective of the experiment was the assessment of the intensity of competition for soil nitrogen by both species, in an association where the legume had in part an alternate source—symbiotically fixed nitrogen. Distinction between these nitrogen sources was achieved by labelling the mineralized soil nitrogen isotopically. The results after 13 weeks growth in mixture are shown (Fig. 4.31a) not only for dry matter, but also for total nitrogen (soil and fixed nitrogen) and soil nitrogen in the plant parts. Whilst we do not have the response curves of each species on their own at equivalent densities to those experienced in mixture, it is clear that the growth of legume has suffered in the presence of the grass, comparing dry matter

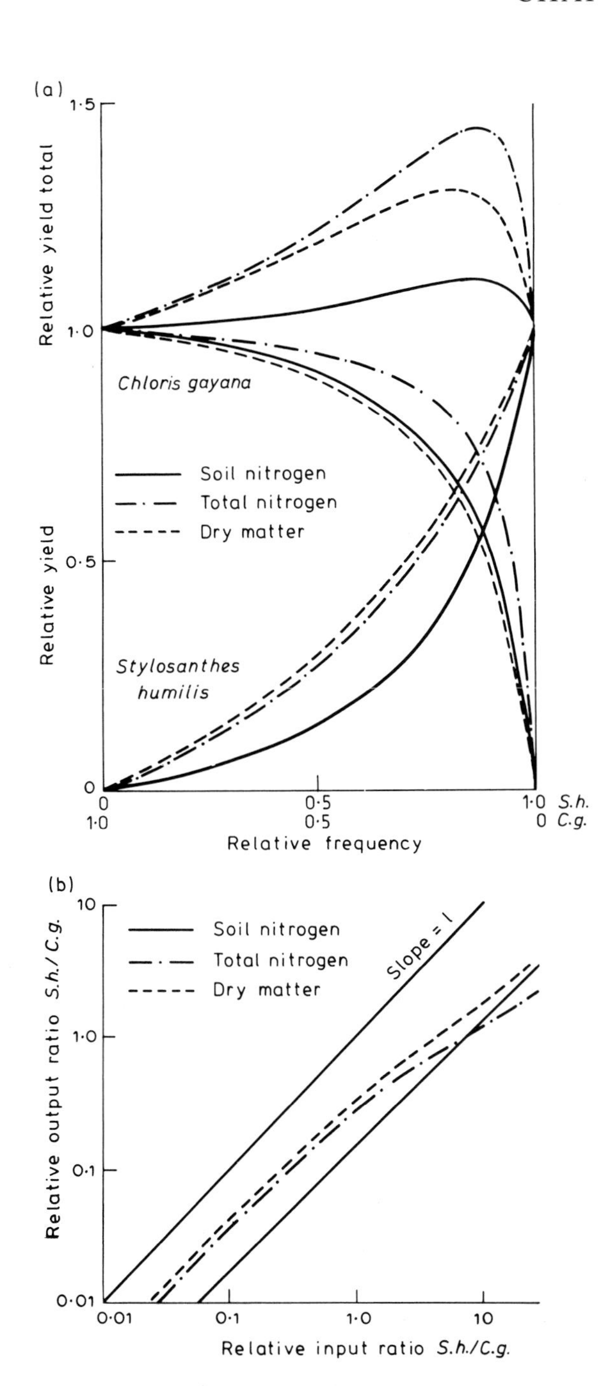

Fig. 4.31. (a) Relative yields and relative yield totals of Rhodes grass, *Chloris gayana* (*C.g.*), and the legume Townsville stylo, *Stylosanthes humilis* (*S.h.*), grown in a replacement series. (b) The ratio diagram from relative yields derived from the data in (a). (From Hall 1974.)

relative yields. Relative crowding coefficients for the three sets of data are shown in Table 4.6, from which it can be seen that the coefficient products for dry matter and total nitrogen are far greater than one, in contrast to that for soil nitrogen which is close to unity. These data suggest that whereas both species would appear to be competing for the same resource in soil nitrogen, when the total nitrogen budget is considered they are not. This is, of course, to be expected, since rhizobially-fixed nitrogen will initially only be available to the legume. Moreover, the intensity of competition for soil nitrogen in favour of Rhodes grass is clearly indicated by the magnitude of the relative crowding coefficient of 9.02 in comparison to 0.17. Plotting the relative yield data as a ratio diagram (Fig. 4.31b), we come to see the same set of conclusions. The curve for soil nitrogen is linear and displaced from the midline of unity, whilst those for dry matter and total nitrogen are S-shaped. The former indicates competition for the same resource; the latter indicates niche differentiation.

Summarizing, we see that by the use of relative crowding coefficients it is possible to examine the nature of limiting resources for which species may be in competition. In this instance, there is obvious niche overlap for soil nitrogen between the two species, and the grass, with the greater number of roots than the legume, has the advantage in exploiting the soil system. Exclusion of the legume is unlikely to occur, however, since there is no overlap for fixed nitrogen. The realized niche of each species for total nitrogen is one in which there is only partial overlap. Both Rhodes grass and Townsville stylo are only partly competing for this resource—a situation which may lead to coexistence. The data of Fig. 4.31 do not, however, allow us to discern at what proportion in mixture the two will coexist, since the yields are relative.

Table 4.6 Relative crowding coefficients of Townsville stylo and Rhodes grass measured in three ways. (After Hall 1974.)

	Townsville stylo C_{IJ}	Rhodes grass C_{JI}	$C_{IJ} \times C_{JI}$
Dry matter yield	0.42	8.42	3.54
Total nitrogen	0.37	18.33	6.78
Soil nitrogen	0.17	9.02	1.53

To do this, we would need to plot the ratio diagrams based on absolute yields.

The replacement series, as an experimental and analytical approach to the study of plant competition, has gained wide acceptance in plant ecology. Its principal value lies in the rationalization of the variety of complex interactions that may occur when two plant species associate in mixture. Interpretation of replacement diagrams, as we have seen, requires very careful attention to detail, but does allow a means of testing whether two species are in competition for the same space. Moreover, by measurement of potentially limiting resources within the plant, as well as plant yields themselves, and analysis of relative crowding coefficients, specific resources that are in common demand may be identified. This application, suggested by Hall (1974, 1978), has largely countered the criticism levelled by some workers (discussed in Donald 1978) that using the term 'space' was a convenience which evaded the need to recognize the resources for which species were in competition.

A limitation of replacement series experiments is that analysis is necessarily confined to mixtures at one overall density, which curtails the consideration of interspecific competition. If we wish to analyse interspecific interactions taking into account density-dependence in both species in the mixture, then we must reconsider the additive model developed earlier for intraspecific competition (equations 3.5 and 3.6). Hassell & Comins (1976) and Watkinson (1980) have pointed out that 'the model' may be easily extended to include the influence of a second species.

Thus, with a notation similar to that used in equation 3.5 in Section 3.2.2 the mean yield w_A of a plant of species 1 in a mixed population containing species 1 and 2, at densities N_1 and N_2 is

$$w_A = w_{m,1}[1+a_1(N_1+\gamma_{12}N_2)]^{-b} \qquad (4.5)$$

in which γ_{12} is a 'competition coefficient' or 'equivalence value' converting the density of species 2 into numerical terms equivalent to species 1. Likewise, the effect of species 2 on the mortality of 1 is

$$N_1 = N_{i,1}[1+m_1(N_{i,1}+\delta_{12}N_{i,2})]^{-1}$$

where δ_{12} is the appropriate equivalence value determining the density-dependent mortality in species 1 attributable to species 2. This pair of equations can also be written for species 2 with (of course) different equivalence coefficients.

We can appreciate the value of this model by re-examining the data of Marshall & Jain (1969) on oats (Fig. 4.9). Whereas from our earlier interpretation of the de Wit diagrams (Section 4.4.2) we could qualitatively conclude that *Avena fatua* was the superior competitor in mixture with *A. barbata* we can now quantify this. Fitting equation 4.5 to the results (Firbank & Watkinson 1985) gave the following pair of equations:

$$S_B = 188[1+0.41(N_B+1.44N_F)]^{-0.96}$$

and

$$S_F = 141[1+1.02(N_F+0.19N_B)]^{-0.72}$$

The subscripts B and F refer to *A. barbata* and *A. fatua* respectively, whilst S is the mean number of seeds produced per plant and N is the density of each species in the mixture. This pair of equations neatly encapsulates the absolute differences between the two species and their competitive interactions. On average an isolated plant of *A. fatua* produced 47 seeds (188–141) less than *A. barbata* but required more space to do so since its ecological neighbour area (1.02) is larger than for *A. barbata* (0.41). This numerical superiority in seed output is offset, however, by the fact that on a one-to-one basis *A. fatua* is much more competitive as judged by the competition coefficients. In mixture, *A. barbata* 'perceives' each *A. fatua* as equivalent to 1.44 of its own individuals. Each *A. fatua* plant on the other hand 'perceives' each *A. barbata* plant as about a fifth (0.19) of one of its own. (Note that this analysis does not assume a reciprocity in competition coefficients as in the case of the replacement series.) We can immediately see then that this approach extends our qualitative conclusions about the nature of interspecific interactions.

4.15 Niche overlap

Finally, we return to the question first posed in Section 4.9: How much niche differentiation is necessary for the coexistence of interspecific competitors? We shall see that this question is intimately related to a problem which, in evolutionary terms, confronts all species: that of offsetting interspecific competition against intraspecific competition (Section 4.11). There have been several theoretical approaches to the solution of this problem, but since, for the most part, they reach similar conclusions, we can concentrate on the one initiated by MacArthur & Levins (1967) and developed by May (1973).

Imagine three species competing for a single, unidimensional resource which is distributed continuously; food size and food at different heights in a forest canopy both conform to this description. Each species has its own niche in this single dimension within which it will (in our examples) consume food. Moreover, its consumption-rate is highest at the centre of its niche and tails off to zero at either end. Its niche can, therefore, be visualized as a resource utilization curve (Fig. 4.32). Clearly, the more adjacent species' utilization curves overlap, the more they compete. Indeed, if we make the (fairly restrictive) assumptions that the curves are 'normal' (in the statistical sense), and that the different species have similar curves, then the competition coefficient, α (applicable to both adjacent species) can be related to the standard deviation of the curves, w, and the difference between their peaks, d, by the following formula:

$$\alpha = e^{\frac{-d^2}{4w^2}}.$$

Thus α is very small when there is considerable separation of adjacent curves ($d/w >> 1$, Fig. 4.32a) and approaches unity as the curves themselves approach one another ($d/w < 1$, Fig. 4.32b).

In terms of this model, the question under consideration is: How much overlap of adjacent utilization curves is compatible with stable coexistence? Obviously, if there is very little overlap, as in Fig. 4.32a, there is very little interspecific competition and competitors can coexist. On the other hand, in such a case the species have rather narrow niches. This means, since all conspecifics are consuming very similar food, that there is intense *intra*specific competition. Moreover, the food items in those positions along the resource spectrum where the curves overlap are being almost totally ignored by the consumers. It is, therefore, likely that natural selection will favour an increased consumption of these neglected food items, an increase in niche breadth, a lessening of intraspecific competition, and thus an increase in niche overlap. The question is: How much?

MacArthur & Levins (1967) and May (1973) answered this question by what was, in essence, an extension of the search for stable coexistence pursued in Section 4.13. They assumed that the two peripheral species had similar carrying-capacities (K_1, proportional to the area under the utilization curve), and considered the coexistence between them of an intermediate species (carrying capacity K_2). Their results are illustrated in

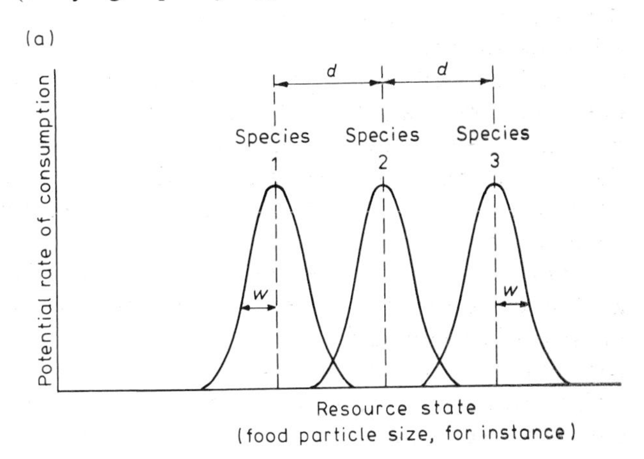

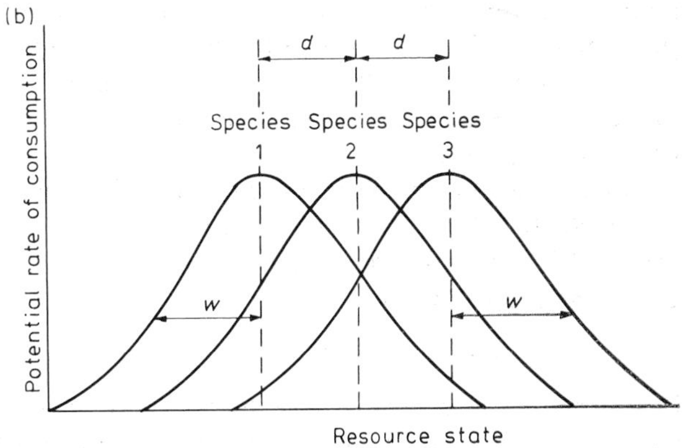

Fig. 4.32. Resource utilization curves for three species coexisting in a one-dimensional niche system; d is the distance between curve maxima, which occur at the centre of the curves; w is the standard deviation of the curves. (a) Narrow niches with little overlap ($d > w$). (b) Broad niches with considerable overlap ($d < w$).

Fig. 4.33, which indicates the values of K_1/K_2 that are compatible with stable coexistence for various values of d/w. At low values of d/w (high α) the conditions for coexistence are extremely restrictive, but these restrictions lift rapidly as d/w approaches and exceeds unity. In other words, stable coexistence is *possible* under the restrictive conditions imposed by low values of d/w, but as May (1973) points out, because conditions are so restrictive '... it may be plausibly argued that environmental vagaries in the real world will upset such an equilibrium.'

So we have seen that at high values of d/w there is intense intraspecific competition and underexploitation of resources, and at low values of d/w the equilibrium is too fragile to be maintained in the real world. Theory, therefore, suggests that coexistence of competitors (using a unidimensional resource) will be based on niche differentiation in which d/w is approximately equal to, or slightly greater than unity. Unfortunately, the testing of this suggestion is hindered by two major problems. The first is that it applies only to situations in which there is a simple, unidimensional resource (probably quite rare), and in which the utilization curves are at least approximately the same as in the model. Competition in several dimensions, and certain alternative utilization curves (Abrams 1976) would both lead to lower values of d/w being compatible with robust, stable coexistence. The second problem is the collection of the appropriate data. These are needed in order to establish the dimensionality of competition, and the form of the utilization curve; and also to determine not only d but w as well. Thus, the field evidence persuasively used by May (1973) in support of this model's suggestions was even more persuasively criticized by Abrams (1976). In particular, there is a grave danger, when attempting to 'test' the predictions of this model, that those field examples that support it will be selected, while those that do not support it are ignored. From our own examples, for

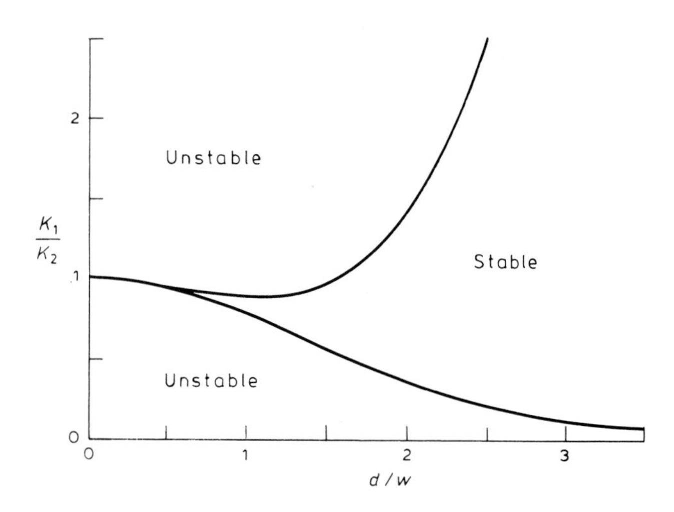

Fig. 4.33. Niche overlap and coexistence. The range of resource utilization (indicated by the carrying capacities, K_1 and K_2, where $K_1 = K_3$) which permits a three-species equilibrium community with various degrees of niche overlap (d/w). (After May 1973.)

instance, we could select the ant *Veromessor pergandei* (Fig. 4.7) and the two hydrobiid snails (Fig. 4.18) as providing empirical support for the model's prediction ($d/w \simeq 1$). Yet, while such support is gratifying, it most certainly does not prove that the model is correct. What is needed is evidence that such 'supportive' patterns occur much more frequently than would be expected by chance alone; and as yet we have insufficient data to provide this evidence.

Nevertheless, the model has shown us that there is likely to be *some* limit to the similarity of competing species; and that this limit represents a balance between, on the one hand, the evolutionary avoidance of intraspecific competition and the underexploitation of resources, and on the other hand, the evolutionary avoidance of equilibria which are too fragile to withstand the vagaries of the real world. Once again, therefore, a model, without being 'correct', has been immensely instructive.

Chapter 5 Predation

5.1 Introduction

Although this chapter is entitled 'Predation', we shall be dealing in it with a variety of interactions. In all the cases we shall be considering, however, animals will totally consume, or partly consume and also harm, other animals or plants. This broad umbrella is designed to include each of the following categories of 'predator':

(1) *True predators* that consume other animals—their *prey*—and thus gain sustenance for their own survival and reproduction.

(2) *Parasitoids*. These are insects, mainly from the order Hymenoptera, but also including many Diptera, which are free-living in their adult stage, but lay their eggs in, on or near other insects (or, more rarely, spiders or woodlice). The larval parasitoids then develop on or within their '*host*' (usually, itself, a pre-adult), initially doing little harm, but eventually almost totally consuming, and therefore killing, the host prior to pupation. Often just one parasitoid develops from each host, but in some species several individuals share a host. Nevertheless, in either case, the number of hosts attacked in one generation closely defines the number of parasitoids produced in the next. This, along with the fact that the act of 'predation' is confined to a particular phase of the life-history (adult females attacking hosts), means that parasitoids are especially suitable for study. They have provided a wealth of information relevant to predation generally.

(3) *Parasites*. These are organisms (animals or plants) that live in an obligatory, close association, usually with a single *host* individual for a large portion of their lives. They gain sustenance from their host and do their host harm, but parasite-induced host mortality is by no means the general rule, and is often rather rare.

(4) *Herbivores*. These are animals that eat plants. Some appear to act as true predators, since they totally consume other organisms for their own sustenance (seed-eaters are a particularly good example). Others are much more like parasites; aphids, for instance, live in close association with plant hosts, gain sustenance from them, and reduce their vitality. Many other herbivores, however, fall into neither category. They do not live in close association with any one plant host; but by consuming parts of plants, they are often the ultimate, if not the immediate, cause of plant mortality. Thus, they certainly do have a detrimental effect on their plant 'prey'.

These four categories are all quite distinct and special in their own way, but, as their inclusion within a single definition shows, they share important common features. For this reason there will be many statements in this chapter that apply to all categories. It would be tedious to refer individually in every such case to 'true predators and their prey, parasitoids and their hosts, parasites and their hosts, and herbivores and their food plants'. Instead, for simplicity, and where the context precludes ambiguity, the four categories will be referred to, together, as predators (or, rarely, consumers) and their prey.

We can, therefore, apply a single definition and a single, all-embracing label to these four categories. But we can also apply a single naïve expectation regarding their population dynamics. If we imagine a *single* species of predator and a *single* species of prey, then we can expect predators to increase in abundance when there are large numbers of prey. However, there should then be an increase in predation-pressure on the prey, leading to a decrease in prey abundance. This will ultimately lead to food-shortage for the predators, a decrease in predator abundance, a concomitant drop in predation-pressure, an increase in prey abundance and so on. In other words, there appears, superficially, to be an inherent tendency for predators and their prey (and parasitoids and their hosts, parasites and their hosts, and herbivores and their food plants) to undergo *coupled oscillations* in abundance, predator numbers 'tracking' those of the prey.

5.2 Patterns of abundance

Some actual examples of abundance patterns are shown in Fig. 5.1. Certainly, some do appear to show coupled oscillations (Fig. 5.1a, b), but there are also cases in which the 'prey' are kept at a constant low level (Fig. 5.1c, d), cases in which the prey exhibit periodic outbreaks (Fig. 5.1e), and cases in which the 'predators' have no noticeable, simple effect on the prey since the fluctuations in the sizes of the two populations are apparently unconnected (Fig. 5.1f, g).

It is clear, in other words, even from this limited range of examples, that *actual* predators, parasitoids, parasites, herbivores and their 'prey' all exhibit a wide variety of patterns of abundance, and we can immediately see two rather obvious reasons for this. The first is that predators and prey do not normally exist as simple, two-species systems. To understand the abundance patterns exhibited by two interacting species, these must be viewed in a realistic, multi-species context. The second reason is that our conception of even the simplest, abstracted, two-species system is itself excessively naïve. Before multi-species systems are even considered, we must abandon our expectation of universal prey–predator oscillations, and look instead, much more closely, at the ways in which predators and their prey interact in practice.

We shall examine, in turn, the individual components of the predator–prey relationship (an approach employed successfully by Hassell 1976, 1978); but the journey covering these components will be a long and fairly complicated one. Moreover, it is only *after* all the topics have been examined in detail that we shall be able to reassemble them into an integrated whole. At this stage, therefore, we provide an itinerary which can be referred to now or part-way through the journey. This is set out in Table 5.1, and in the following outline of the rest of this chapter.

In Section 5.3, the patterns of food preference shown by predators are examined, along with some possible determinants of these patterns. Then, in Section 5.4, a number of effects of time-scales and timing are considered. This is followed by an examination of the detailed effects of predators on the fitness of prey, paying special attention to the effects of herbivores on plants (Section 5.5). Then, in Section 5.6, consideration is given to the ways in which prey 'thresholds' and food quality complicate the relationship between predation-rate and the beneficial effects to the predator. Section 5.7 covers the way in which predation-rate is influenced by the availability (especially the density) of prey items; while a long Section 5.8 considers some of the consequences of environmental heterogeneity (particular attention being paid to the unequal way in which many predators distribute their harmful effects amongst individual prey). In Section 5.9, the consequences of mutual interference amongst predators are examined, and in Section 5.10, the similarities between the effects produced by the processes in Sections 5.8 and 5.9 are discussed. Then, in Section 5.11 attention is drawn to the tendency of predators to maximize their 'profits' by 'foraging optimally'; while Section 5.12 is a resumé of the preceding nine sections.

Table 5.1 Summary of the components of the predator–prey interaction which are examined in this chapter

Effect of	On	Section
evolution	predator specialization	5.3
time-scales	patterns of abundance	5.4
prey density	prey	5.5
predation-rate	prey	5.5
predation-rate	predator	5.6
prey density	predation-rate	5.7
aggregation	predation-rate	5.8
mutual predator interference	predation-rate	5.9
foraging strategies	predation-rate	5.11

Throughout these sections we will be especially concerned with the effects that these individual components have on the dynamics of the predator and prey populations; and in particular, with the regulatory, stabilizing effects of density-dependent processes, and the destabilizing effects of inversely density-dependent processes (Section 2.3 and 2.6). Then, in Section 5.13, we attempt to incorporate many of the behaviourally complex components into models of predator–prey interactions, hoping to explore further their effects on population dynamics and stability; and this allows us, in Section 5.14, to reconsider the abundance patterns of the present section in the light of Sections 5.3 to 5.13. Finally, Section 5.15 examines problems emanating from the

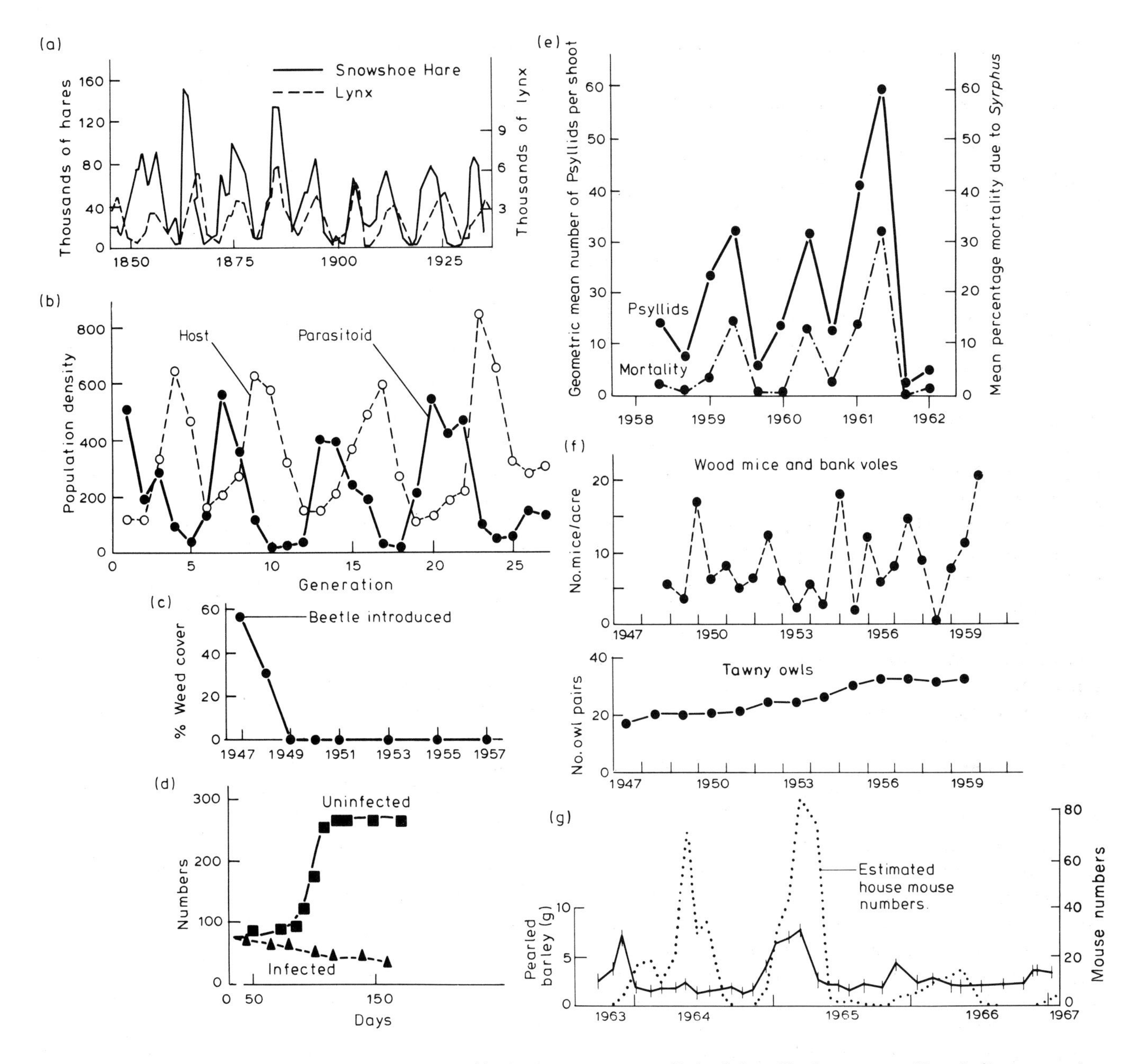

Fig. 5.1. Patterns of abundance in predator–prey systems. (a) The lynx, *Lynx canadensis*, and the snowshoe hare, *Lepus americanus*; true predator–prey (after MacLulick 1937). (b) The wasp, *Heterospilus prosopidis*, and the bean weavil, *Callosobruchus chinensis*; parasitoid–host (after Utida 1957). (c) Changes in the abundance of the Klamath weed, *Hypericum perforatum*, following the introduction of leaf-eating beetle *Chrysolina quadrigemina*; herbivore-plant (after Huffaker & Kennett 1959). (d) The effect on the beetle *Laemophloeus minutus* of being infected by the protozoan *Mattesia dispora*; parasite–host (after Finlayson 1949). (e) The psyllid *Cardiaspina albitextura* and mortality caused by *Syrphus* species; true predator–prey (after Clark 1963). (f) Tawny owls, *Strix aluco*, and wood mice and bank voles, *Apodemus sylvaticus* and *Clethrionomys glareolus*; true predator–prey (after Southern 1970). (g) The house mouse, *Mus musculus*, and barley, *Hordeum vulgare*; herbivore–plant (after Newsome 1969b).

contrived human predation involved in the process of harvesting.

5.3 Coevolution, and specialization amongst predators

Predators of all types can be classified, according to their diet width, as monophagous (feeding on a single prey type), oligophagous (few prey types) or polyphagous (many prey types), and the degree of specialization can have important effects on predator–prey dynamics. The abundance of a monophagous predator, for instance, is likely to be closely linked to the distribution and abundance of its prey; while a polyphage is very unlikely to have its abundance determined by any one of its prey types. If we are to understand the patterns of diet width amongst predators, however, we must begin by establishing two basic points. The first is that *predators choose profitable prey*. This follows, by an act of faith, from a consideration of natural selection (evolution favours those individuals with the highest fitness; fitness is increased by increasing the profitability of food-acquisition; evolution, therefore, favours predators that choose profitable prey). It is also borne out by the facts. Fig. 5.2, for instance, illustrates examples of predators actively selecting those prey items which are most profitable, i.e. prey items for which the gain (in terms of energy intake per unit time spent handling prey) is greatest.

The second point stems from the fact that all animals and plants have evolved in response to selection pressures originating, to a large extent, from the other animals and plants in their environment. The point, then, is that *predators and their prey are likely to have coevolved.* There is a continuous selection pressure on prey to avoid death (or, more generally, fitness-reduction) at the hands of their predators, and a reciprocal, continuous pressure on predators to increase *their* fitness by exploiting their prey more effectively. At perhaps its most trivial, this evolutionary arms race consists of prey that can run quickly from their predators, and predators that can run quickly after their prey, both being favoured by natural selection. However, we can see the results of analogous pressures on prey in the distasteful or poisonous chemicals in the leaves of many plants, in the spines of hedgehogs, the camouflage-coloration of many insects, and the immunological responses of hosts to parasite infection; while in predators these pressures result, for instance, in the long, stout, penetrative ovipositors of wood wasps, the hooks and suckers on the heads of tapeworms, and the silent approach and

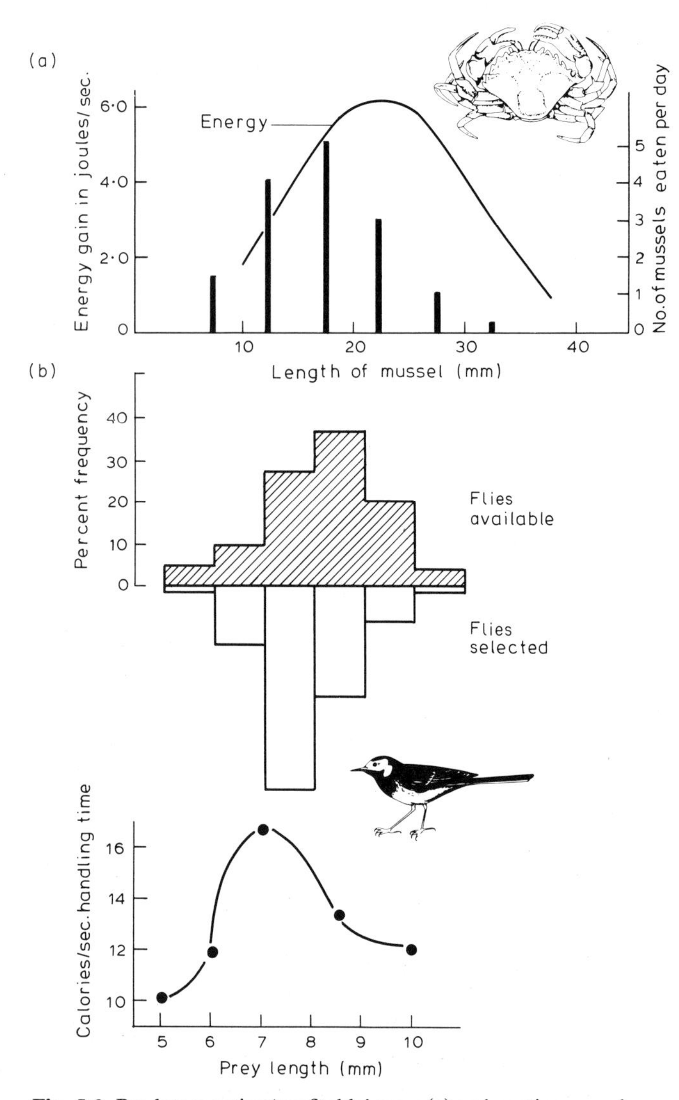

Fig. 5.2. Predators eating 'profitable' prey: (a) crabs eating mussels (Elner & Hughes 1978); (b) pied wagtails eating flies (Davies 1977). (After Krebs 1978.) 1 calorie (non-SI unit) = 4.186 joules.

sensory excellence of owls. It is clear, in short, that no natural predator–prey interaction can be properly understood unless it is realized that each protagonist has played an essential role in the evolution of the other.

It is equally clear, however, irrespective of any coevolution, that no predator can possibly be capable of consuming all types of prey; simple design constraints prevent shrews from eating owls, and humming-birds from eating seeds. Nevertheless, coevolution provides an added force in the restriction of diet width. Each prey species responds (in an evolutionary sense and on an evolutionary time-scale) in a different way to the pressures imposed by its predators, but a predator cannot 'coevolve' in a wide range of directions simultaneously. Predators, therefore, tend to specialize to a greater or lesser extent, and the better adapted a predator is for the exploitation of a particular prey species, the less likely it is to profit from the exploitation of a wide variety of prey. Parasites, in particular, because they live in such intimate association with their hosts, tend to specialize, evolutionarily, on a single host species. Their whole life-style and life-cycle is finely tuned to that of their host, and this precludes their being finely tuned to any other host species. They are, therefore, commonly monophagous. For similar reasons, the same is true of many parasitoids, although in their case, possibly because they are free-living as adults, polyphagy (or at least oligophagy) is rather more common.

With herbivores and predators, on the other hand, the situation is much less clear-cut. At one extreme there *are* examples of complete monophagy. The fruit fly, *Drosophila pachea,* for example, indigenous to the Sonoran Desert of the south-west United States, is the only species capable of consuming the rotten tissues of the Senita cactus (and its associated micro-organisms), because all other species are poisoned by an alkaloid—pilocereine—which the Senita produces (Heed *et al.* 1976). Moreover, natural selection has favoured *D. pachea's* exploitation of this exclusive niche to such an extent that the fly has an absolute requirement for certain unusual sterols which are also only produced by the Senita. Many plants produce 'secondary', protective chemicals and, probably as a consequence, many herbivorous insects are specialists (Lawton & McNeil 1979, Price 1980). However, most true predators and many herbivores feed on a variety of prey items.

A partial explanation of the range of diet widths is provided by a simple consideration of the fact that, although there is pressure from coevolution towards predator specialization, there is a counterbalancing evolutionary pressure discouraging the reliance of predators on an unpredictable or heavily exploited resource. For many parasites, the necessity for specialization clearly provides a pressure that outweighs any disadvantages stemming from resource-unpredictability; for *D. pachea,* specialization is presumably favoured by the exclusivity (and, perhaps, predictability) of the Senita, and the consequent lack of interspecific competition. To a certain extent, similar arguments can be advanced to explain diet width generally. Thus, where prey exert pressures which demand specialized morphological adaptation by the predator, there is a tendency for the predator to have a narrow range of diet; and where predators feed on an unpredictable resource, there is a tendency for them not to be specialists.

5.3.1 One explanation for the degrees of specialization

In many cases, however, some further explanation is needed, and this may be provided by the ideas of MacArthur & Pianka (1966, see also MacArthur 1972). In order to obtain food, any predator must expend time and energy, firstly in *searching* for its prey, and then in *handling* it (i.e. pursuing, subduing and consuming it). Searching will tend to be directed, to some degree, towards particular prey types; but, while searching, a predator is nevertheless likely to encounter a wide variety of food items. MacArthur and Pianka, therefore, saw diet width as being determined by the choices made by predators once they had encountered prey. Generalist predators are those that choose to pursue (and, hopefully, subdue and consume) a large proportion of the prey they encounter; specialists are those that continue searching except when they encounter prey of their specifically preferred type.

The basic conclusion MacArthur and Pianka drew from these considerations can be stated as follows. Predator choices are determined, ultimately, by the forces of natural selection, and are driven by these forces towards the maximization of profitability for the predator. Natural selection, therefore, favours a predator that chooses to pursue a particular prey item *if*, during the time it takes to handle *that* prey item, the predator cannot expect to search for *and* handle a more profitable prey item. On this basis, predators with handling-times that are generally short compared to their search-times should be catholic in their tastes, because in the short time it takes them to handle a prey item which has already been found, they can barely begin to search for another prey item. This is MacArthur and Pianka's explanation for the broad diets of many insectivorous birds that 'glean' foliage. Search is always moderately time-consuming; but handling the minute, stationary insects takes negligible time and is almost always successful. Ignoring such prey items (i.e. narrowing diet width) would, therefore, decrease overall profitability, and the birds tend to be generalists.

By contrast, many other predators have search-times that are short relative to their handling-times. In such cases, specialization will be favoured, because the predators can expect to find a more profitable food item very soon after ignoring a less profitable one. Lions, for instance, live more or less constantly in sight of their prey so that search-time is negligible; handling-time, on the other hand, and particularly pursuit-time, can be very long (and energy-consuming). Lions consequently *specialize* on those prey that can be pursued most profitably: the immature, the lame and the old. Thus, on the basis of MacArthur and Pianka's ideas we can expect 'pursuers' or 'handlers' (like lions) to have relatively narrow diets, and 'searchers' to have relatively broad ones. Along similar lines, Recher (in MacArthur 1972) found that great blue herons in the *productive* waters of Florida (in which search-times were consequently short) had a much narrower range of food-size in their diet than those inhabiting the *unproductive* lakes of the Adirondacks (where search-times were relatively long).

Clearly, we can go some way at least towards an understanding of the degrees of specialization shown by predators.

5.3.2 Food preference and predator switching

Irrespective of this range of diet width, however, polyphagy is very common; especially amongst predators and herbivores. Yet polyphagous animals are rarely indiscriminate in the various types of food they eat. Horton (1964), for example, presents the results of an accidental field experiment in which deer broke into a plantation containing four species of tree arranged at random; white pine, red pine, jack pine and white spruce. As Table 5.2 shows, the deer, with free and equal access to all four species, exhibited a fairly consistent preference for jack pine followed by white pine, with red pine being only lightly browsed and white spruce ignored. Such preference amongst polyphagous animals is, in fact, the general rule; but as Murdoch & Oaten (1975) have shown, there are two distinct forms that this preference can take.

An example of the simpler and much commoner form is shown in Fig. 5.3 (Murdoch 1969). Two types of

Table 5.2 Food preferences exhibited by deer. Percentage browsing incidence of deer on planted trees. (After Horton 1964.)

	White pine	Red pine	Jack pine	White spruce
Winter 1956–57	31	19	84	0
Winter 1958–59	9	1	48	0
Winter 1960–61	17	0	70	0

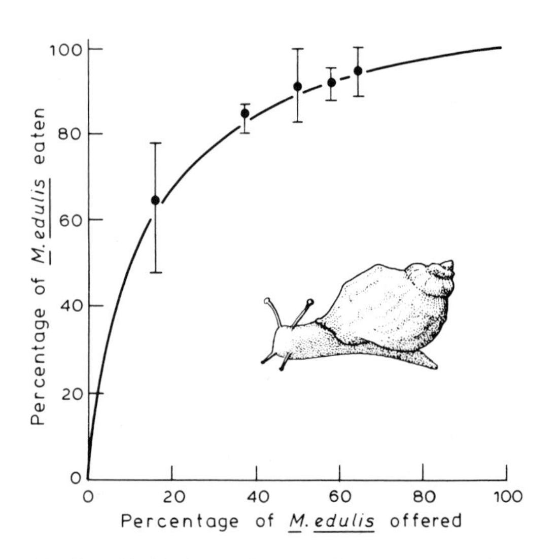

Fig. 5.3. Snails exhibiting a consistent preference amongst the mussels *Mytilus edulis* and *M. californianus* irrespective of their relative abundance (means±2 s.e.). (After Murdoch & Oaten 1975.)

predatory shore snails, *Thais* and *Acanthina*, were presented with two species of mussel (*Mytilus edulis* and *M. californianus*) as prey, at a range of prey proportions. When the proportions were equal, the snails showed a marked preference for the thinner-shelled *M. edulis*. The line in Fig. 5.3 has been drawn on the assumption that they retained this same preference at other (unequal) proportions, and this assumption is clearly justified: irrespective of availability, the snails showed the same marked preference for the less protected prey, which they could exploit most effectively.

This can be contrasted with data of a rarer sort in Fig. 5.4a, obtained from an experiment in which the predatory water bug, *Notonecta glauca*, was presented with two types of prey—the isopod, *Asellus aquaticus* and mayfly larvae (*Cloëon dipterum*)—with overall prey density held constant (Lawton *et al.* 1974). In this case the preference exhibited when the two prey types were equally available, if extrapolated to other availabilities (thin line in Fig. 5.4a), is obviously not a good indication of the overall response. Instead, *Notonecta* took a disproportionately small number of *Asellus* when they were scarce, and a disproportionately high number when they were common. This is known as *predator switching*, since it suggests that predators switch their preference to whichever prey is most common. In the case of *Notonecta*, the explanation is apparently illustrated in Fig. 5.4b: the more previous experience *Notonecta* has of *Asellus*, the more likely it is to make a successful attack. It appears, in other words, as if predator switching is based upon a learnt ability to specialize. This, essentially, is the view taken by Tinbergen (1960), who proposed that certain predators, particularly vertebrates, develop a 'specific searching image'. This enables them to search more successfully (since they effectively 'know what they're looking for'), and results in them concentrating on their 'image' prey to the relative exclusion of their non-image prey. Moreover, since the searching image develops as a result of previous experience, and since the predators (or herbivores) are most likely to experience common prey, we can expect predators to concentrate on a prey type when that type is common and switch to another prey type when it is rare.

The basis for predator switching has been discussed in more detail by Murdoch & Oaten (1975). One of their most instructive examples is the work of Murdoch, Avery and Smyth (in Murdoch & Oaten 1975) on guppies offered a choice between fruit flies and tubificid worms as prey (Fig. 5.5). Fig. 5.5a shows that there was, indeed, switching by the guppies. But more interesting than this is Fig. 5.5b, which shows that although the *population* of predators showed little preference when offered equal proportions of the two prey, the *individual* guppies showed considerable specialization. This does not contradict the idea of a search image, but does

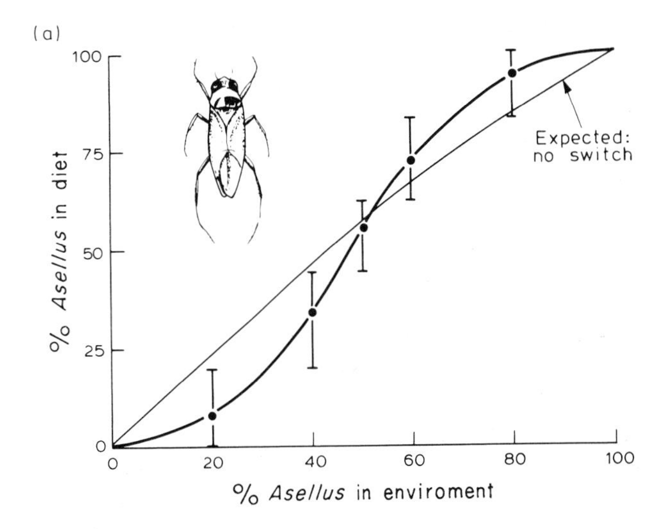

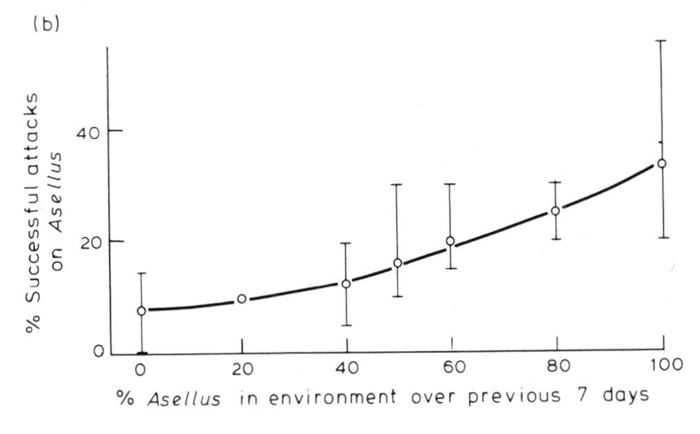

Fig. 5.4. (a) The percentage of *Asellus* in the diet of *Notonecta* as a function of their relative abundance; the straight line indicates the function expected on the basis of a consistent preference. (b) The effect of 'experience' on the success of *Notonecta* in attacking *Asellus*. (After Lawton *et al.* 1974.) (Means and total ranges are indicated.)

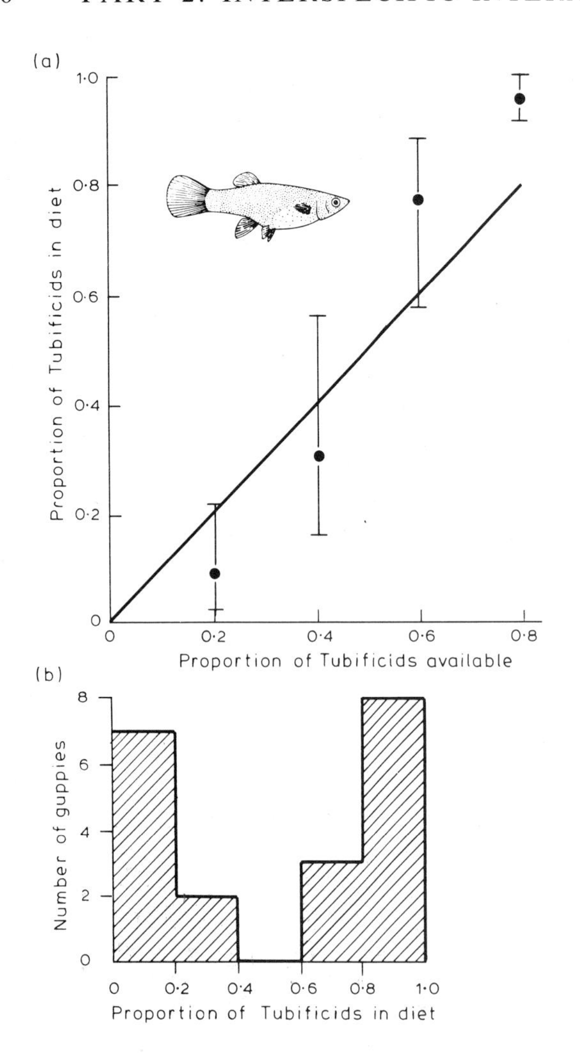

Fig. 5.5. (a) Switching in guppies fed on *Drosophila* and tubificids (with *total ranges* indicated). (b) A frequency histogram showing the number of individual guppies with particular types of diet when offered equal numbers of the two prey types. All showed a preference (< 0.4 or > 0.6) even though the population as a whole consumed approximately equal amounts of the two types. (After Murdoch & Oaten 1975.)

suggest that predator switching in a population does not result from individual predators gradually changing their preference, but from *the proportion of specialists changing*. Murdoch and Oaten provide evidence for the occurrence of this type of switching in other vertebrate predators, in some invertebrate predators, and in a herbivore: the feral pigeon feeding on peas and beans (Murton 1971). Nevertheless, it is fair to conclude that this more complex type of food preference is most common in vertebrates, where the ability to learn from experience is most highly developed. The consequences of it are discussed in Section 5.7.4.

5.4 Time and timing

In any discussion of predation there are several points that must be made regarding the effects of time and timing. The first is that levels of specialization need not involve morphological adaptation, but may, in many cases, be the result of differing degrees of 'temporal coincidence. Thus, the European rabbit flea (*Spilopsyllus cuniculi*), like very many parasites, has a life-cycle that coincides more or less exactly with that of its host: maturation of the flea can only occur on a doe in the latter part of pregnancy, and eggs are laid only on the new-born rabbit young (Mead-Briggs & Rudge 1960). The wood pigeon (*Columba palumbus*), by contrast, switches its food-preference seasonally, depending on availability (Fig. 5.6; Murton *et al.* 1964). Thus, while it may, at any one time, be fairly specialized in its feeding habits, it is temporally (as well as morphologically) a generalist; and, unlike the parasite, its population dynamics are not strongly dependent on the availability of any one type of prey.

Related to the effects of temporal coincidence are the effects of differing lengths of life cycle amongst predators and their prey. Aphids, for instance, generally pass through several generations for each time their host plant passes through a seasonal cycle (never mind a generation). They can, therefore, be expected to react quickly towards, and fairly accurately reflect, the quantity and quality of their food. The dynamics of a sycamore population, on the other hand, is unlikely to be greatly influenced by the intra-seasonal fluctuations in the abundance of its aphids. Similarly, small mammals, with a fairly high intrinsic rate of increase and a lifespan never exceeding a year, exhibit a pattern of abundance that reflects the yearly changes in environmental quality; while tawny owls in the same habitat, often living 6 years and failing to breed when food is scarce, maintain a comparative constancy of abundance irrespective of these environmental fluctuations (Southern 1970).

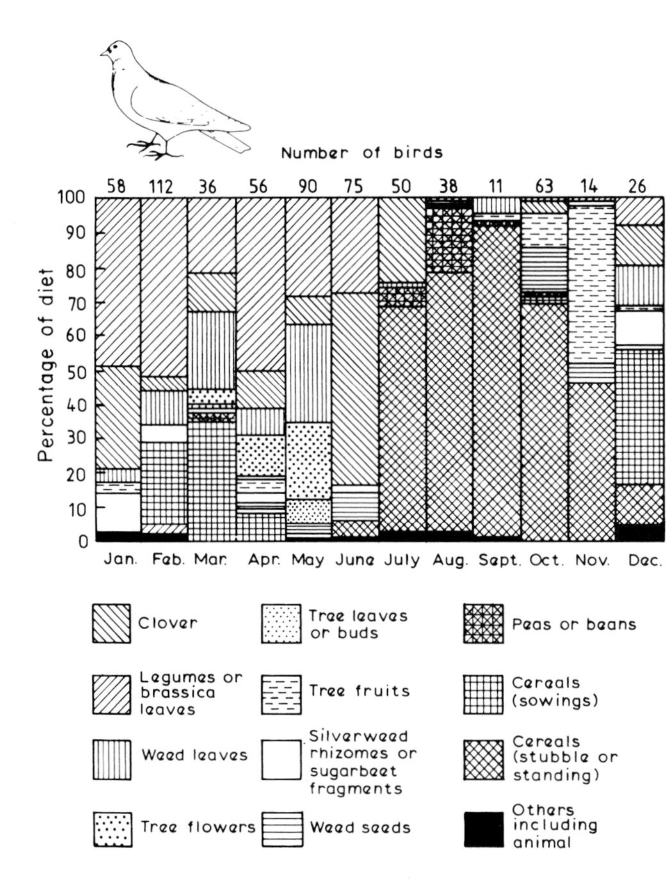

Fig. 5.6. Seasonal changes in diet. The percentage contributions of different food types to the diet of the wood pigeon *Columba palumbus* in Cambridgeshire, England. (After Murton *et al.* 1964.)

In part such 'discrepancies' are the result of making comparisons on an artificial, yearly basis, when it might be more appropriate to consider the fluctuations from generation to generation. Nevertheless, it is quite clear that generation-times, and in particular the relative generation-times of a 'predator' and its 'prey', can have important effects on predator–prey dynamics through their influence on the speeds at which species respond to changes in the environment.

Another aspect of timing relevant to predator–prey dynamics is the effect of time-lags. We have already seen (in Section 3.4.1) that time-lags tend to cause increased levels of fluctuation in dynamic systems (i.e. they tend to *destabilize* systems), and we can expect this to apply to populations of predators and their prey. Once again, populations with discrete generations will tend to oscillate in numbers more than populations breeding continuously; but there is, in a sense, an addition reinforcement of such time-lags in predator–prey systems. In Section 3.4.1 we were considering systems in which the food-resource remains constant in size (as evidenced by the constancy of the single species' carrying capacity). Now we are dealing with systems in which both consumer *and* food contribute to the effects of any time-lags. There is, consequently, a tendency for populations of predators and their prey to fluctuate out of phase with one another, and it is essentially this that leads to the 'naïve expectation' of coupled predator–prey oscillations outlined in Section 5.1. Varley (1947) coined a special term—delayed density-dependence—to describe such a relationship between predation-rate and prey density.

5.5 Effects on prey fitness

Turning to this next component, we can note that when predators and prey interact, the fitness of prey individuals is obviously affected by the predators, but it is also influenced by the prey themselves through the density-dependent process of intraspecific competition. As Chapters 1–3 made clear, this will tend to regulate the size of a single-species prey population, but in so doing it will also tend to *stabilize* the interaction between the prey and their predators. Prey populations reduced by their predators will experience a compensatory decline in the depressant effects of intraspecific competition; while those that grow large through the rarity of predators will suffer the consequences of intraspecific competition all the more intensely.

The most important effects on prey fitness in the present context, however, are attributable to predators. These can be most easily described in the case of parasitoids and their hosts: a host which is successfully attacked dies—its fitness is reduced to zero. It might appear, moreover, that the effects of true predators on their prey are equally straightforward—and they often are. But consider, by contrast, the work of Errington (1946). Errington made a long and intensive study of populations of the musk-rat (*Ondatra zibethicus*) in the north-central United States. He took censuses, recorded mortalities and movements, followed the fates of individual litters, and was particularly concerned with predation on the musk-rat by the mink (*Mustela vison*). He found that adult musk-rats that were well established in a breeding territory were largely free from mink predation; but those that were wandering without a territory, or were exposed by drought, or injured in intraspecific fights were very frequently preyed upon. Certainly, those that were killed had their fitness reduced to zero. Yet, because these were individuals that were unlikely to ever produce offspring, they had low or zero fitnesses anyway. Similar results have been obtained, in fact, for predation on other vertebrates. Those most likely to succumb are the young, the homeless, the sick and the decrepit—the very individuals whose immediate prospects of producing offspring are worst. The harmful effects on the prey population are clearly not as drastic as they might be; and while the effects of true predators on their prey are often straightforward, there are obviously cases, particularly amongst vertebrates, in which the superficial simplicity can be misleading. In particular, a predator that effectively ignores the potential contributors to the next prey generation will have very little effect on prey abundance.

Nevertheless, it is possible in both true predators and parasitoids to equate the predation-rate in a population with the prey death-rate; the difficulty is that prey death does not always lead to simple reductions in the overall vitality of the prey population. With parasites, on the other hand, it is well known that the effects on the host are often not drastic (though, of course, to conform with the definition in Section 5.1, and thus warrant inclusion in this chapter, the parasite must have some adverse effect on its host's fitness). 'Predation-rate' in parasites, therefore, cannot be equated with host death-rate. Instead, it can be taken as the rate at which host tissue and energy is diverted from hosts to parasites.

Predation-rate in parasites, then, is obviously likely to increase with increases in the mean number of parasites per host; and the effect this can have on host fitness is depicted in Fig. 5.7, illustrating the work of Lanciani (1975) on the ectoparasitic mite, *Hydryphantes tenuabilis*, and its host, the water bug *Hydrometra myrae.* It is clear from Fig. 5.7a, b that the mite affects the survival of its host, and from Fig. 5.7c, d that it affects its host's fecundity. Taken all in all, therefore, the mite affects the host's reproductive potential (Fig. 5.7e), and this effect increases with increasing intensity of infection. In other words, the greater the proportion of parasites to hosts, the greater their depressant effect will be; there is inverse density-dependence tending to *destabilize* the host–parasite interaction.

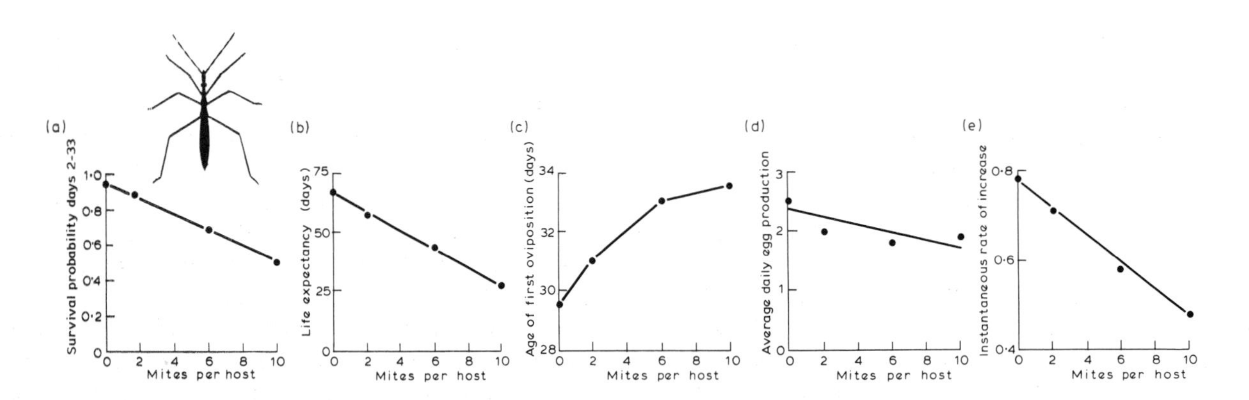

Fig. 5.7. The relationship between the rate of infection of the water bug, *Hydrometra myrae*, (host) with the mite, *Hydryphantes tenuabilis*, (parasite) and (a) host survival, (b) host life-expectancy, (c) host maturity, (d) host fecundity and (e) host rate of increase. (After Lanciani 1975.)

5.5.1 *The effects of herbivores on plant fitness*

Examination of plants occurring naturally in the field rarely reveals a perfect, pristine individual; they almost always bear witness to the ravages of herbivores: leaves are trimmed and holed, seeds and fruits are bored and cracked. Yet, while the number of herbivore–plant interactions that can be listed is potentially immense, we may initially categorize the effects on the plant as either the 'taking' of plant parts with the donor remaining alive, or the predation of all or most of the individual precipitating its death immediately or in the short-term. And between these end-points lies a spectrum of predatory events that may well result in the ultimate death of an individual. We might, for instance, envisage a plant pathogen reducing the photosynthetic area of an individual to such a degree that its ability to compete with neighbours is lowered, and its death hastened.

To seed-eating animals, the crops of seed represent a source of highly nutritious food, which is packaged in a discrete, compact way; and it would be easy to argue that each seed eaten represents a measurable reduction in fitness, since it constitutes the death of a whole individual. It is perhaps more valid, however, to consider not the fitness of the seeds themselves, but the fitness of the parent that produced them. We can see, for instance, in the related case of fruit-eating herbivores, that although a significant quantity of plant material is lost with every fruit eaten, and many seeds (i.e. individuals) are destroyed in the process, the herbivore—by acting as an essential agent of seed-dispersal—is, from the parent plant's point of view, a net *contributor* to fitness. (Similar comments apply to animal pollinators.) And while the situation is not quite so clear-cut in the case of specialist seed-eaters, there is certainly some evidence to suggest that a parent plant's fitness is not incrementally reduced each time one of its seeds is eaten.

Lawrence & Rediske (1962), for instance, monitored the fates of experimentally sown Douglas fir seeds, both in open plots and in plots supposedly screened from the attacks of vertebrate herbivores. Their results (Table 5.3) indicate that their screening was effective in that vertebrate granivores were largely excluded; but they also illustrate two further points. The first is that these

Table 5.3 Compensatory mortality when predation is prevented. The fates, in percentage terms, of Douglas fir seeds sown in open and screened plots. (After Lawrence & Rediske 1962.)

	Plots	
	Open	Screened
Pre-germination period		
Loss due to:		
fungi	19.0	20.1
insects	9.5	12.8
rodents	14.0	1.8
birds	4.1	0.9
unknown	6.8	1.8
Total loss, pre-germination	53.4	37.4
Seeds remaining	46.6	62.6
Germination period		
Non-germinating due to:		
fungal attack	13.1	17.3
seed dormancy	12.7	10.5
Total not germinating	25.8	27.8
Seedlings	20.8	34.8
Post-germination period (1 year after germination)		
Mortality due to:		
fungi	5.4	12.8
other causes	7.3	4.6
Total mortality	12.7	17.4
Seedlings surviving	8.1	17.4

herbivores do, indeed, appear to have an effect on parental output, since the recruitment of seedlings, one year after germination, was significantly increased in their absence. The second point, however, is that this effect is rather less than might be expected from a consideration of vertebrate granivores alone. Other sources of mortality, particularly fungal attack at various stages, appear to act in a density-dependent, compensatory fashion. Thus the herbivores, like the minks preying on Errington's musk-rats, are, to some extent, removing individuals that are already doomed.

We have, of course, already met a similar example of granivory in Section 4.3: Brown & Davidson's (1977) granivorous ants tended to consume seeds that would otherwise have been taken by rodents, and vice versa. Yet the two guilds, together, caused a measurable reduction in the size of the seed population; and since germination and subsequent establishment depends on the occupation of a favourable micro-site (which is largely a matter of chance), this, in turn, must have caused a reduction in the size of the populations of adult plants in subsequent generations. Granivores seem, therefore, to adversely affect plants, but not necessarily to the extent suggested by superficial examination.

When we turn from seeds and fruits to seedlings and adult plants, the assessment of herbivore-damage becomes, if anything, even more difficult. In some cases there is indisputable fitness-reduction. Haines (1975), for instance, obtained frequency distributions of seedling populations in different areas whilst investigating the activities of leaf-cutting ants (*Atta colombica tonsipes*). (These ants cut leaf, flower and fruit material, which is then transported to their nests where the harvested plant material is decomposed by a fungus. The fungus is eaten by the ants, and the degraded plant remains dumped at 'refuse tips'.) The survivorship and size of seedlings of forest plants are inversely related to the frequency of visitation by the ants (Fig. 5.8). Seedlings in parts of the forest away from nests and dumps are present in all size classes, but on the nests and dumps themselves most of the seedlings present are very small. Thus, the chance of a seedling maturing into the forest canopy appears to be zero if it arises in areas regularly frequented by ants.

Such simple destruction of whole plants, however, is rather rare; and by contrast, the consequences of *partial* defoliation to the subsequent survivorship of a plant are generally difficult to unravel, while the demonstration of proximate death by defoliation is often elusive. This is because a frequent effect of partial defoliation is to disturb the integrated nature of the whole physiology of the plant. Loss of leaf means not only the loss of photosynthetic area, but also the mobilization of stored reserves for leaf replacement through bud growth; new growth demands protein, carbohydrate and minerals—the high quality components required by predators. Thus, persistent defoliation of young, actively growing leaves may constitute a rapid drain of stored reserves. Moreover, such changes in physiology may result in an alteration in the rate of root growth, and in some cases in root dieback. And such reductions in root volume, photosynthetic area and stored reserves necessarily

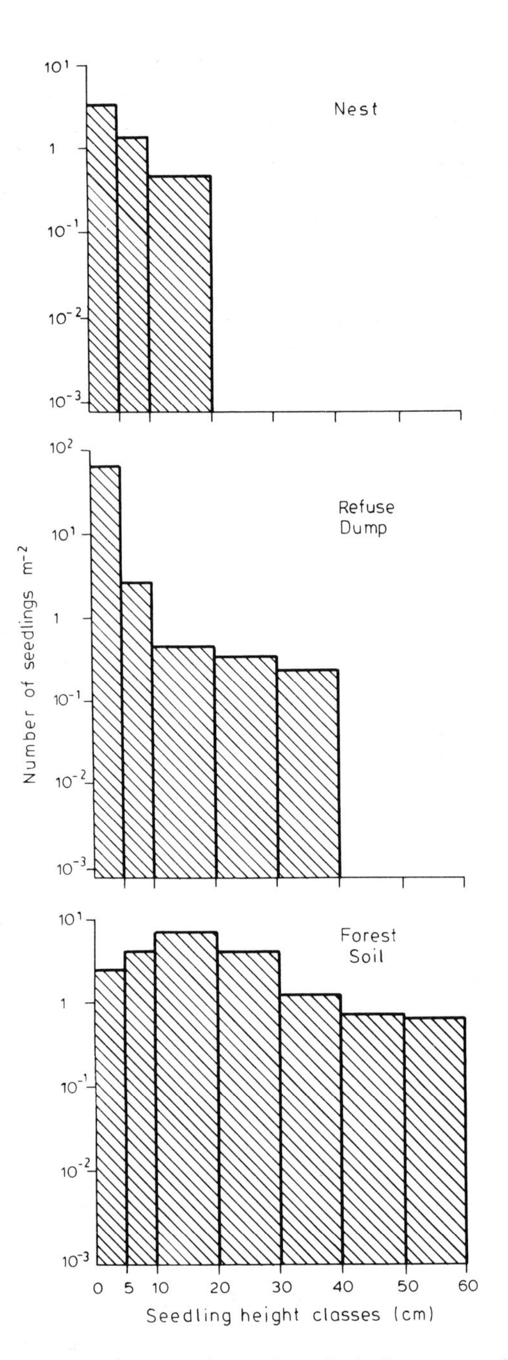

Fig. 5.8. Seedlings do not survive to a large size in the presence of ants. The size-structure of seedling populations in tropical rain forests on nests and on refuse dumps of leaf-cutting ants *Atta colombica tonsipes* and on forest soil. The density of seedling height classes are presented. (Data from Haines 1975.)

place a defoliated individual at a disadvantage in any competitive struggle—a disadvantage that would obviously be intensified still further if buds, the regenerative organs, were subject to predation as well.

Defoliation may also have important repercussions on plant fecundity, i.e. seed production. Regular removal of vegetative growth in perennial plants tends to reduce flowering, but the time at which defoliation occurs is critical in determining the actual response of the plant. If the inflorescence is formed prior to defoliation, the response to leaf removal is seed abortion, or for individual seeds to be smaller; whereas defoliation before inflorescence-production is likely to halt or severely constrain flower formation.

Some plant predators do not defoliate their host, but extract food requirements from within the plant. The detailed work of Dixon (1971a, b), for instance, has clearly demonstrated that aphid infestations on lime saplings cause at least a tenfold reduction in the rate of total dry weight increase (from 7.7 down to 0.6 g week^{-1}), even though average leaf area, and total leaf area per sapling remained unaltered. Yet, examination of the root systems of the infested plants revealed that no growth had occurred below ground subsequent to infestation; and the energetics of this interaction suggest that, on average, 30 aphids per leaf during the growing season is sufficient to completely drain the annual net production of the tree. This average is obviously an oversimplification, but the figure agrees reasonably well with observed natural aphid densities (Dixon 1971b), implying that although they cause no immediate visual damage, aphids may substantially limit the growth of lime trees, and they may, in consequence, affect survivorship.

In discussing herbivory so far, we have been dealing with the effect of herbivores on whole plants. Yet, in Section 2.5.4 we recognized that some plant populations (e.g. grasses) were more conveniently considered as populations not of genets but of independent plant parts (i.e. ramets or tillers). Such plant populations suffer herbivore predation in the process of grazing—a subject we shall be considering in more detail below (Section 5.13). At this point, however, it is sufficient to realize that the effect of predation on tiller populations is partly analogous to that on seeds. Plant physiologists have shown that light penetration of a canopy decreases exponentially below the surface (Donald 1961). In a dense canopy, therefore, leaves at the base receive little sunlight and are not able to photosynthesize sufficiently for respiratory purposes. If they are not to die, they must import energy from other parts of the canopy (nearer the surface), and thus they make a negative contribution to the *net* assimilation rate of the whole canopy. There is, then, an optimum leaf area at which the whole canopy is making a maximum net photosynthetic contribution—a state at which leaf growth-rate will also be at its maximum (Fig. 5.9). Herbivores, by removing excess growth as it accumulates, might actually maintain a pasture population at its optimum leaf area: in such a case, herbivory would not, on balance, be having an adverse effect.

Overall, therefore, we can see, in both seeds and adult plants, that although herbivores commonly harm their food plants, our initial impression of the inflicted damage may be highly misleading. Sometimes it is much more than appearances suggest; often it is much less. In rare cases, grazing may actually benefit a plant population.

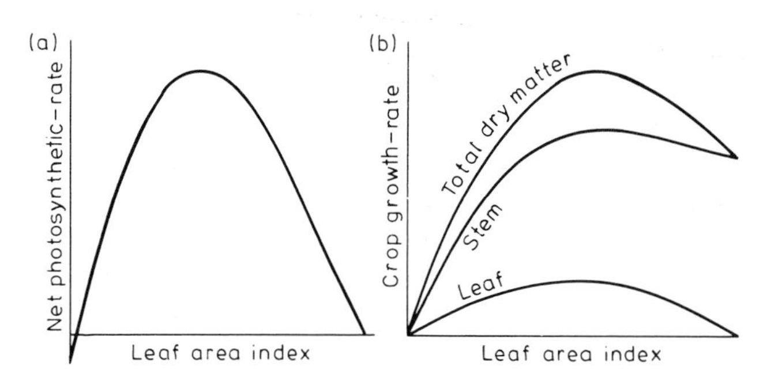

Fig. 5.9. (a) Idealized relationship between crop growth-rate and leaf area index. (b) Actual relationship found in subterranean clover by Davidson & Donald (1958). (After Donald 1961.)

5.6 The effects of predation-rate on predator fitness

5.6.1 Thresholds

The effects of predation-rate on predator fitness—like the effects of predation-rate on prey fitness—appear, superficially, to be straightforward; and in the case of parasitoids they certainly are: every host successfully attacked by a parasitoid represents an incremental increase in parasitoid fitness. In all other cases, however, there is an added complication. Predators, herbivores and parasites all require a certain quantity of 'prey' tissue for basic maintenance; and it is only when their intake exceeds this threshold that increases in predation-rate lead to measurably increased benefits to the predator. This is illustrated with typical examples in Fig. 5.10a (predator growth rate) and Fig. 5.10b (predator fecundity). The consequences of this threshold for the stability of predator–prey interactions are fairly obvious. There is a tendency at low prey densities for predation-rate to fall below the threshold, causing predator fitness to slump to zero. The adverse effects of low prey density on predator fitness are, therefore, exaggerated, and the interaction generally is *destabilized.*

5.6.2 *Food quality*

There is another factor complicating the relationship between predation-rate and the fecundity and survivorship of predators, however, which is of greater importance, namely *food quality*. It is not the case that each item (or even each gram) of food consumed by a predator or herbivore is equivalent. The chemical composition of food, and its accessibility via digestion to the predator, both have a considerable bearing on the way in which food consumption affects predators. This is particularly apparent amongst herbivores (see White 1978; Lawton & McNeill 1979).

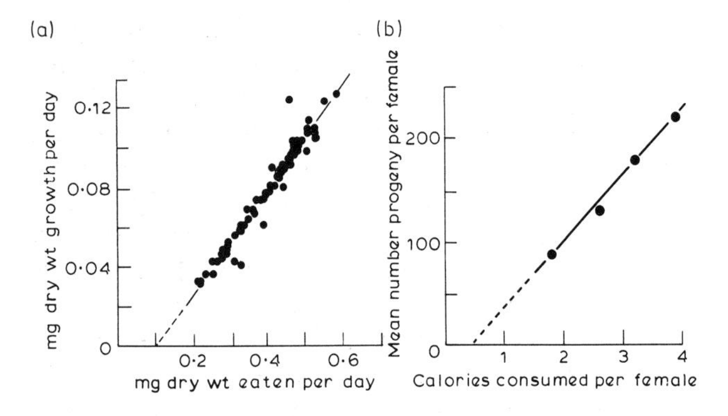

Fig. 5.10. Prey thresholds for predators. (a) Growth in the spider, *Linyphia triangularis* (Turnbull 1962). (b) Reproduction in the water flea, *Daphnia pulex* var. *pulicaria* (Richman 1958). (After Hassell 1978.) 1 calorie (non-SI unit) = 4.186 joules.

In particular, herbivores are greatly affected by the nitrogen content of their food. One has only to consider the honeydew (excess carbohydrate) excreted by aphids, to realize that many herbivores must ingest vast quantities of plant tissue in order to consume sufficient amounts of amino acids. Moreover, there is good evidence that herbivore-abundance can be limited by nitrogen content (i.e. food quality). Many herbivores only

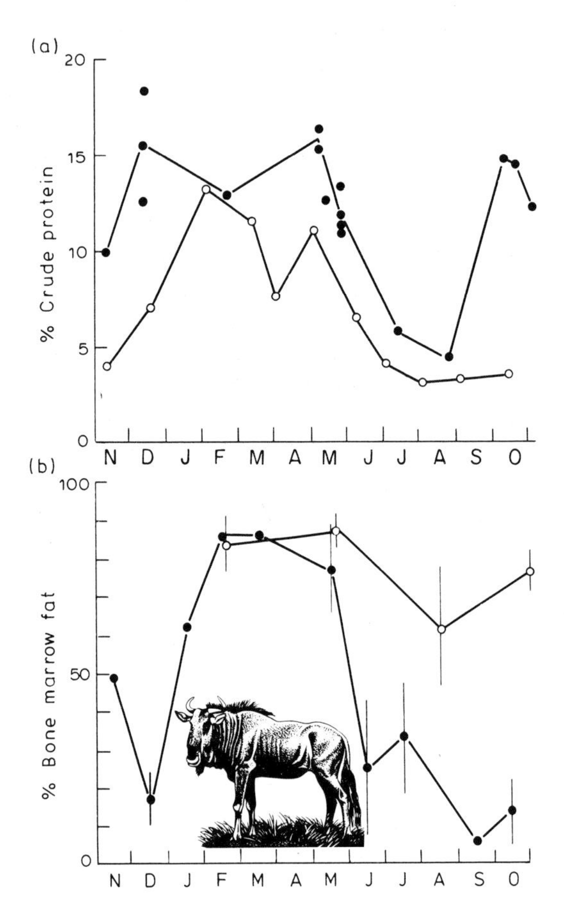

Fig. 5.11. (a) The quality of food measured as % crude protein available to (O) and eaten by (●) wildebeest in the Serengeti during 1971. During the dry season, food quality fell below the level for nitrogen balance (5–6% crude protein) despite selection. (b) The fat content of the bone marrow of the live male population (O) and those found dead from natural causes (●). Vertical lines, where present, are 95% confidence limits. (After Sinclair 1975.)

have access to low-quality food, and they have insufficient time and energy to digest enough of this to provide them with protein for maintenance, let alone growth and reproduction. This was shown, for instance, by Sinclair (1975) who noted the protein content of the food available to the wildebeest in Serengeti (Tanzania) during 1971, and compared this with the protein content of the food they ate (Fig. 5.11a). He also monitored (Fig. 5.11b) the fat reserves in the bone marrow of live males, and of males that died of natural causes (these reserves being the last to be utilized).

It is clear from Sinclair's results that, despite selecting nitrogen-rich plants and plant-parts, the wildebeests consumed food in the dry season which was below the level necessary even for maintenance (5–6% crude protein); and, to judge by the depleted fat reserves of dead males, this was an important cause of mortality. Moreover, when we consider that during late-pregnancy and lactation (December–May in the wildebeest) the food requirements of females are three or four times the normal (Agricultural Research Council 1965), it becomes obvious that *shortage of high-quality food* can have drastic effects on herbivores.

A similar conclusion can be drawn from the work of McNeill (in McNeill & Southwood 1978). As Fig. 5.12 shows, seasonal peaks in the densities of insects feeding on the grass *Holcus mollis* are related to peaks in food quality (measured as soluble nitrogen) in the leaves and the stems—the latter approximating phloem flows to the flowers and developing seeds in the middle of the summer. Once again, food quality appears to be having an extremely important effect on predators.

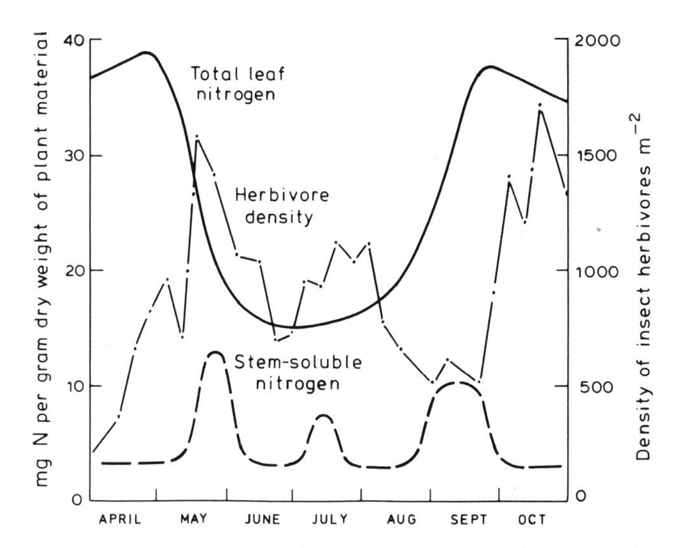

Fig. 5.12. Seasonal changes in the mean numbers of insects on the grass *Holcus mollis* related to changes in food quality in the leaves and stems (McNeill & Southwood 1978). (After Lawton & McNeill 1979.)

Moreover, plant quality does not only affect herbivores because of what plants lack (in terms of nutrients), but also because of the toxic or digestibility-reducing 'secondary compounds' that many plants contain by way of protection. However, while this may have an important evolutionary effect on herbivores, causing them to specialize, it presumably has relatively little effect on those herbivores that are specifically adapted to feed on the plants producing these compounds (Lawton & McNeill 1979).

5.7 The functional response of predators to prey availability

5.7.1 *The 'type 2' response*

We now turn, for our next component, to the way in which the predation-rate of predators, herbivores or parasitoids is influenced by prey availability, and we begin with the simplest aspect of availability: prey density. A fairly typical response is illustrated in Fig. 5.13, which depicts the numbers of *Daphnia* (of a particular size) eaten during a 24-hour period by tenth-instar nymphs of the damselfly, *Ischnura elegans* (Thompson 1975). Fig. 5.13 clearly shows that, as prey density increases, the predation-rate responds less and less and approaches a plateau (approximately 16 *Daphnia* per 24-hour period). A similar result is shown in Fig. 5.14 for a herbivore: slugs feeding on *Lolium perenne* (Hatto & Harper 1969). Such *functional responses* of predators to changes in prey density were described first by Solomon (1949), but discussed more extensively by Holling (1959), who attributed the form taken by curves like the ones in Figs 5.13 and 5.14 (which he called 'type 2' responses) to the existence of the predator's *handling time*. (This, as we have seen, is the time the predator spends pursuing, subduing and consuming each prey

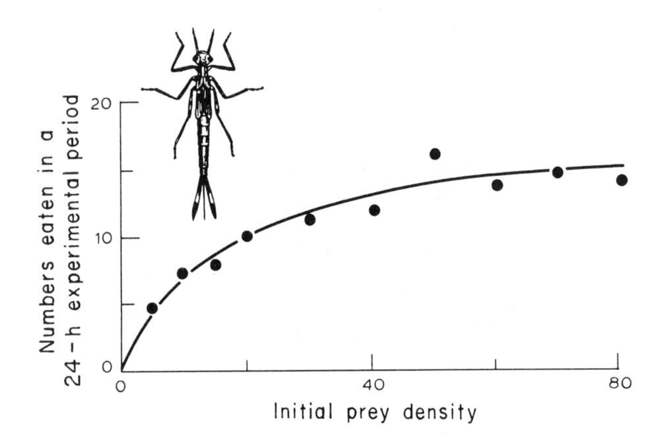

Fig. 5.13. The functional response of tenth instar damselfly nymphs to *Daphnia* of approximately constant size. (After Thompson, 1975.)

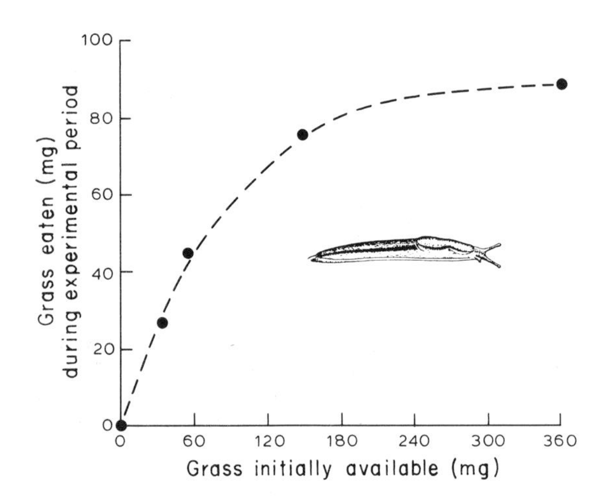

Fig. 5.14. The functional response of single slugs to changes in the amounts of the grass *Lolium perenne* available to be eaten. (Data from Hatto & Harper 1969.)

item it finds, and then preparing itself for further search.) Holling argued that as prey density increases, 'search' becomes trivial, and handling takes up an increasing proportion of the predator's time. Thus, at high densities the predator effectively spends all of its time handling prey, and the predation-rate reaches a maximum, determined by the maximum number of handling-times that can be fitted into the total time available.

This view of the type 2 functional response is confirmed and illustrated in Fig. 5.15, which summarizes some of the results obtained by Griffiths (1969) in his work on the ichneumonid *Pleolophus basizonus* parasitizing the cocoons of the European pine sawfly, *Neodiprion sertifer*. Taking parasitoids of different ages, Griffiths plotted the number of ovipositions per parasitoid over a range of host densities, but he also calculated the *actual* maximum oviposition-rate by presenting other parasitoids of the same ages with a *superabundant* supply of host cocoons. It is clear from Fig. 5.15 that the type 2 functional response curves did indeed approach their appropriate maxima. Yet, while these maxima (of around 3.5 ovipositions per day) suggest a handling-time of around 7 hours, further direct observation indicated

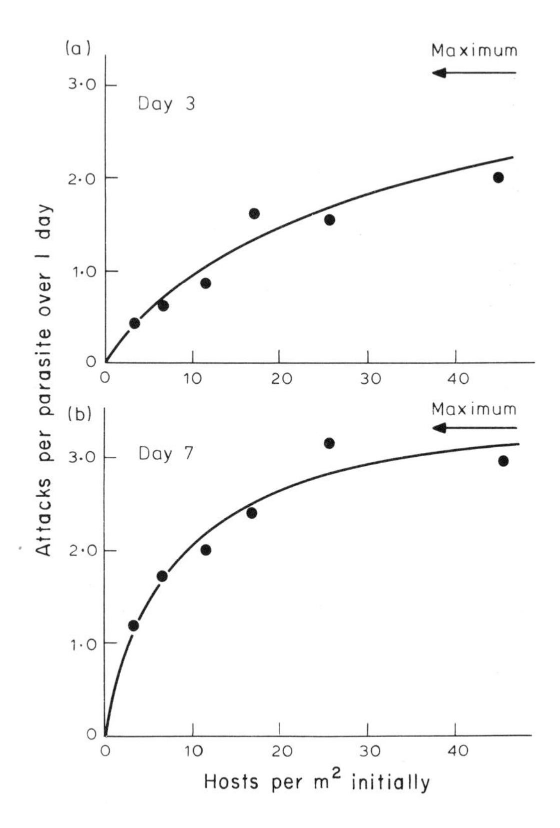

Fig. 5.15. The functional responses of the ichneumonid parasitoid *Pleolophus basizonus* to changes in the density of its host *Neodiprion sertifer*; arrows indicate maxima observed in the presence of excess hosts. (a) parasitoids on their third day; (b) parasitoids on their seventh day. (After Griffiths 1969.)

that oviposition takes, on average, only 0.36 hours. The discrepancy is accounted for, however, by the existence of a 'refractory period' following oviposition during which there are no eggs ready to be laid. 'Handling-time', therefore, includes not only the time actually taken in oviposition, but also the time taken 'preparing' for the next oviposition. Similarly, the handling-times suggested by the plateaux in Figs 5.13 and 5.14 will almost certainly include time devoted to activities, peculiar to the damselflies and slugs, other than the direct manipulation of food items. It is in this (non-literal) sense that handling-time must be understood.

A further point to note from Fig. 5.15 is that, while the plateau level (and thus the handling-time) remains approximately constant with increasing age, the rate of approach to that plateau is much more gradual in the younger parasitoids. In other words, it is apparent that the younger parasitoids search less effficiently, or attack at a slower rate. Thus, at low host densities they oviposit less often than their older conspecifics; but at high densities there is such a ready supply of hosts that even they are limited only by their handling time. Hence, the form taken by a type 2 functional response curve can be characterized simply in terms of a handling time and a *searching efficiency* (or attack rate); and Hassell (1978) discusses the methods by which these parameters can be obtained from the data. The value of estimating these parameters is illustrated by the work of Thompson (1975) who fed *Daphnia* of a variety of sizes to tenth-instar damselfly nymphs (Fig. 5.16). Fig. 5.16b shows that the various functional responses in Fig. 5.16a are the result of the ways in which both handling-time and attack rate change with prey size. It appears, quite reasonably, that damselfly nymphs need more time to handle larger prey, and that they are most efficient and effective at catching *Daphnia* of size D, with attack rate declining rapidly as prey size increases.

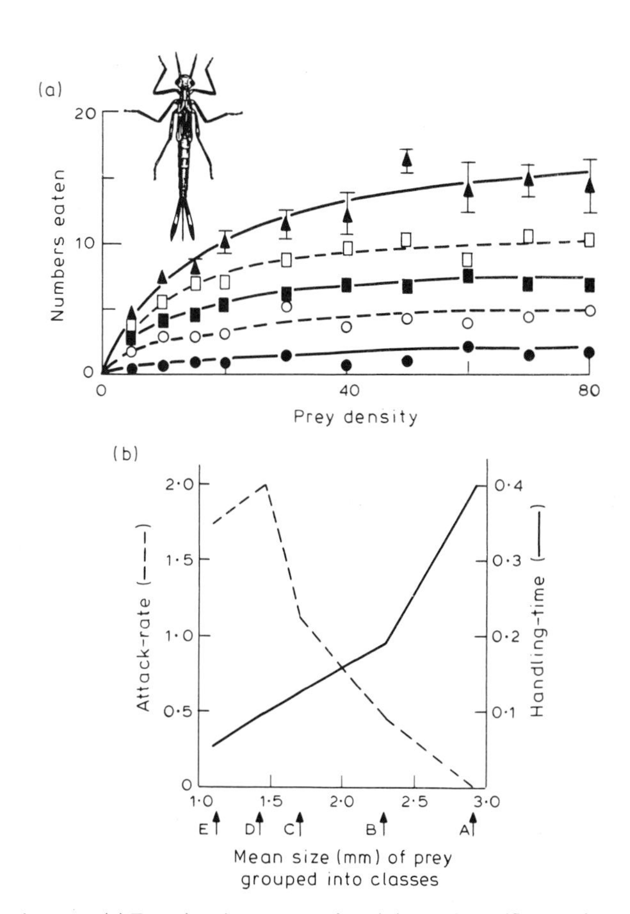

Fig. 5.16. (a) Functional responses of tenth-instar damselfly nymphs to *Daphnia* of various sizes. Size A prey (●); B (O); C(■); D (□); E (▲). Standard errors are fitted to the top line. (b) The attack rates and handling-times of these functional responses. (After Thompson 1975.)

5.7.2 *The 'type 1' response*

A Holling 'type 2' functional response is commonly observed in herbivore–plant, predator–prey and parasitoid–host interactions. Much less common, by comparison, is his 'type 1' response, an example of which is shown in Fig. 5.17. This figure describes work

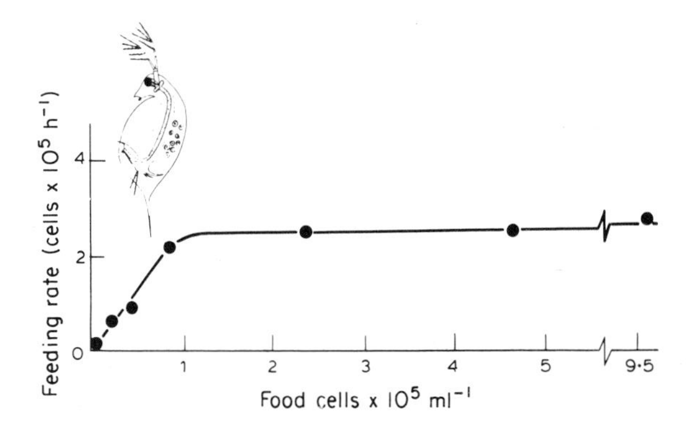

Fig. 5.17. The functional response of *Daphnia magna* to different concentrations of the yeast *Saccharomyces cerevisiae*. (After Rigler 1961.)

carried out by Rigler (1961) on the feeding rate of *Daphnia magna* with the yeast *Saccharomyces cerevisiae* as its 'prey'. *D. magna* is a filter feeder, extracting yeast cells at low density from a constant volume of water washed over its filtering apparatus. Below 10^5 yeast cells ml^{-1} the predation-rate is directly proportional to the food concentration. But the *Daphnia* must also swallow (i.e. handle) their food. At low concentrations this does not interfere with the predation-rate, because it happens sufficiently quickly to remove all the food accumulated by filtration. Above 10^5 cells ml^{-1}, however, the *Daphnia* are unable to swallow all the food they filter. At all such concentrations, therefore, they ingest food at a maximal rate, limited by their 'handling' time. The type 1 response is, therefore, an extreme form of the type 2 response in which the handling-time exerts its effect not gradually but suddenly.

It is important to note, in both the type 1 and particularly the type 2 response, that the *rate* of predation on the prey gets less as prey density increases. This is, in other words, a case of inverse density-dependence (Section 2.6), in which large prey populations suffer proportionately less mortality than small. The consequent effect on the predator–prey interaction is clearly *destabilizing*.

5.7.3 *Variation in handling-time and searching efficiency: 'type 3' responses*

It is instructive, at this point, to note explicitly the various components of the handling-time and searching efficiency (following Holling 1965, 1966). Handling-time is determined by:

(i) the time spent pursuing and subduing an individual prey;

(ii) the time spent eating (or ovipositing in) each prey;

(iii) the time spent resting or cleaning or fulfilling any other essential function (like digestion) prior to the act of predation itself.

Searching efficiency (attack rate) will depend on:

(i) the maximum distance at which a predator can initiate an attack on a prey;

(ii) the proportion of these attacks that are successful;

(iii) the speed of movement of the predator and prey (and thus their rate of encounter);

(iv) the 'interest' taken by a predator in obtaining prey as opposed to fulfilling other essential activities.

We have already seen that at least some of these components are likely to change with predator age (Griffiths 1969; Fig. 5.15) and prey size (Thompson 1975; Fig. 5.16); and the length of time since the predator's last meal (its hunger) is also likely to modify its response (i.e. its 'interest' in food, and thus its attack rate, will be altered). Of particular importance, however, is the way in which these components vary with prey density, or, more generally, relative and absolute prey availability.

An example which effectively shows *attack rate varying with prey density* is illustrated in Fig. 5.18a (Hassell *et al.* 1977): when host densities are low, the parasitoid *Venturia canescens* spends a relatively large proportion of its time in activities other than probing for larvae (see component (iv) above). The consequence of this is shown in Fig. 5.18b (Takahashi 1968). At low host densities, there is an upward sweep in the functional response curve (Fig. 5.18b region A), because an increase in density elicits an increased amount of probing and thus an increased effective rate (Fig. 5.18a). At higher host densities, on the other hand, the attack rate (as well as the handling-time) is relatively constant, leading to the decrease in slope (region B) which is typical of a type 2 response. Overall, the resulting functional response is S-shaped or sigmoidal, and is, in Holling's (1959) terminology, 'type 3'. Clearly, it will occur whenever attack rate increases or handling-time decreases with increasing prey density.

5.7.4 *Switching and 'type 3' responses*

There are other (related) circumstances, however, which will also lead to a type 3 functional response. These are the cases of predator switching considered in Section 5.3.2. The resemblance between Fig. 5.18b and Figs 5.4a and 5.5a is obvious. The major difference is that in the latter case there were two types of prey (rather than one), so that the numbers of prey eaten varied with relative prey availability; we cannot be sure that a type 3 functional response would have resulted had the alternative prey been absent. Nevertheless, the effective result of predator switching is a type 3 functional response: the numbers of prey consumed varies with prey density in a sigmoidal fashion.

Type 3 functional responses, then, resulting from switching or from changes in handling-time or attacking efficiency, can occur in predators, parasitoids or herbivores; although it remains to be seen how common and widespread they are. Whenever and for whatever reason they occur, however, the effect on the stability of the interaction will be essentially the same. Throughout region A (Fig. 5.18b)—the upward-sweeping part of the curve—increases in prey density lead to increases in the predation pressure on the prey: the process is *density-dependent.* It can, therefore, have a potentially important *stabilizing* effect on the interaction. The *actual* importance, however, depends on:

(a) the concavity of the curve in region A; and

(b) the relevance of the prey densities in region A to a particular field (or laboratory) situation.

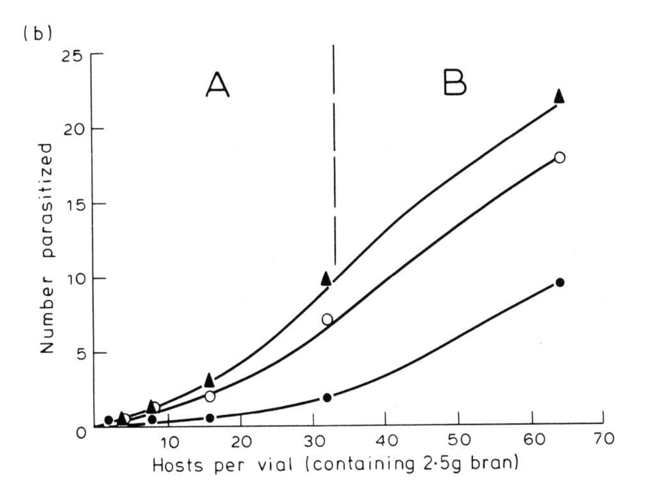

Fig. 5.18. (a) The relationship between the time spent probing by *Venturia canescens* (as a percentage of total observation time) and the density of its host larvae. *Plodia interpunctella.* (b) The sigmoid functional responses of *Venturia canescens* parasitizing *Cadra* larvae of second (●), third (O) and fourth (▲) instars (Takahashi 1968). (After Hassell *et al.* 1977.) For further discussion, see text.

5.8 Aggregated effects

5.8.1 Parasite–host distributions

It was established in Section 5.3.2 that polyphagous predators, by exhibiting preferences, distribute their ill-effects unevenly between prey species. We turn now to the distribution of these ill-effects *within* a *single* species of prey. Consider, to begin with, the distributions of parasites on their hosts shown in Fig. 5.19. In both examples, the observed patterns have been compared with the patterns that would have arisen if the parasites had been distributed at random. Random distributions are the simplest imaginable arrangements in that they occur when all hosts and all parasites are equivalent and independent. (An 'even' distribution would only occur if the parasites positively avoided one another.) Yet the observed patterns are far from random. Instead they are distinctly 'clumped'. There are more hosts than expected supporting large numbers of parasites, but also more than expected with no parasites at all. In effect, there is a *partial refuge* for the hosts: the pattern of distribution ensures that an 'unexpectedly' large number of hosts escape parasitization.

Such patterns are extremely common amongst parasites, and, although proposals for underlying mechanisms are usually speculative (Anderson & May 1978), the effects these patterns have on host–parasite dynamics are fairly obvious. By leaving a large number of hosts unparasitized, even at high levels of infection, the distributions ensure that the host populations are buffered from the most drastic effects of the parasites. This, in turn, tends to ensure that the parasite population has a population of hosts to live on. The basic effect of this partial refuge, then, is to *stabilize* the interaction.

5.8.2 Refuges

Partial refuges, as we shall see, are of considerable importance in a wide variety of predator–prey interactions. Yet there are, in some cases, not partial but total refuges, and these can be of two types.

The first is a refuge for a *fixed proportion* of prey. The parasitoid *Venturia canescens*, for instance, when attacking caterpillars of flour moths (*Ephestia* spp.) can only extend its ovipositor a certain distance into the flour medium. A proportion of the caterpillars—those lying deeply enough to be beyond the ovipositor's reach—are, therefore, protected in an effective refuge (Hassell 1978).

In the second type of total refuge, by contrast, *a fixed number* of prey are protected. This is illustrated by Connell's (1970) work on the barnacle, *Balanus glandula,* in which he found that adults are restricted to the upper zones of the shore, while juveniles are distributed throughout a much broader region (Fig. 5.20). This occurs because in the upper zones there are two exposures to the air at low tide almost every day, which means that the whelks that consume the juvenile barnacles in the lower regions—even *Thais emarginata* (Fig. 5.20)—have only two short, high-tide periods to feed (as opposed to a single long one at lower levels). And this is never long enough to find and eat an adult *B. glandula* in the upper regions. Thus, the *B. glandula* individuals which the upper zones can support are protected from predation, irrespective of *Thais* numbers. There is a fixed-number refuge.

From what has already been noted about partial refuges, it is clear that total refuges will tend to *stabilize* predator–prey interactions; but it should be equally clear that the (density-independent) fixed-proportion refuge will be much less potent than the (density-dependent) fixed-number refuge, since in the latter case the proportion protected increases as prey density decreases.

5.8.3 Partial refuges: aggregative responses

In the living world as a whole, total refuges are probably rather rare. Considerably more common, however, is a tendency for prey to be affected by their herbivores, parasitoids or predators in much the same way as hosts are affected by their parasites (Fig. 5.19): the ill-effects are aggregated so that the prey have a partial refuge. We can illustrate this, initially, in a herbivore — the cabbage

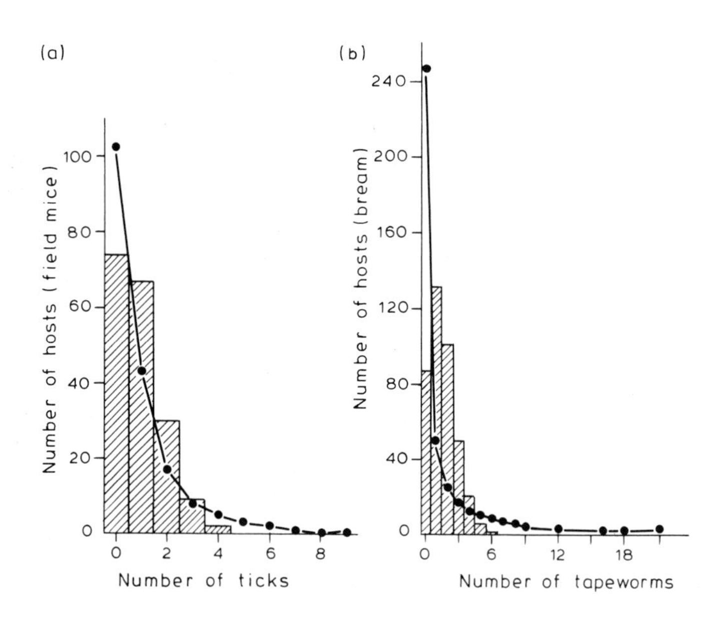

Fig. 5.19. The aggregated distributions of parasites on hosts (points and curve), compared to a random distribution (histograms). (a) Ticks, *Ixodes trianguliceps*, on the wood mouse, *Apodemus sylvaticus* (data from Randolph 1975). (b) Tapeworms, *Caryophyllaeus laticeps* in the bream, *Abramis brama.* (Data from Anderson 1974.)

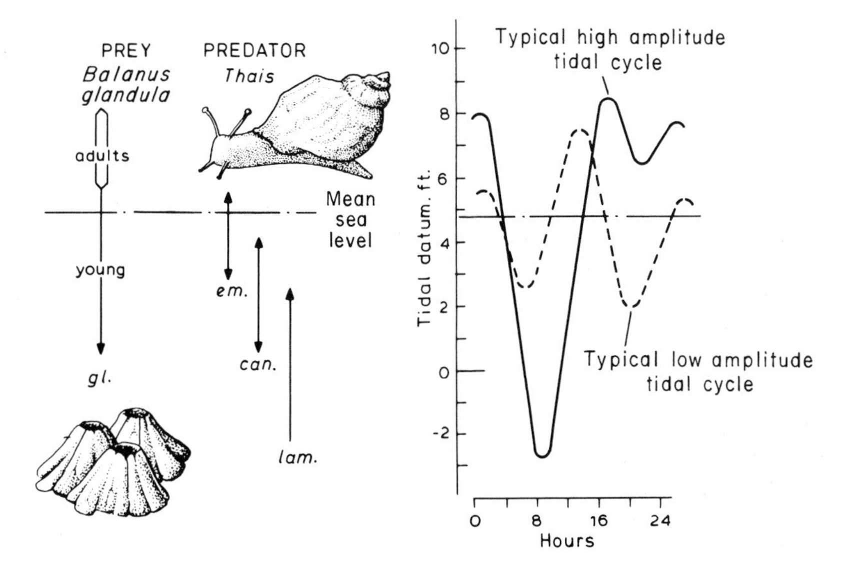

Fig. 5.20. Vertical distributions of barnacle prey (*Balanus glandula*) and whelk predators (*Thais emarginata, Th. canaliculata* and *Th. lamellosa*) in relation to typical tidal fluctuations of large and small amplitude. (After Connell 1970.) For further discussion, see text.

aphid, *Brevicoryne brassicae*. This species, an important pest of cabbage and its relatives, has a marked tendency to form aggregates at two separate levels: nymphs, when isolated experimentally on the surface of a single leaf, quickly form large groups; while populations on a single plant tend to be restricted to particular leaves (Way & Cammell 1970). The effects are illustrated by an experiment (Way & Cammell 1970) in which eight small cabbage plants were each artificially infested by sixteen aphid nymphs. Four of the eight plants had all sixteen nymphs on a single leaf (the normal situation), while the other four had four nymphs on each of the four leaves. Aphid-free leaves were protected from cross-infestation, and colonization by 'outside' aphids was prevented. Productivity was measured by the numbers and weights of adult aphids subsequently produced. Although the differences are small (Table 5.4), it is clear

Table 5.4 Weights of winged adult cabbage aphids produced on small cabbage plants. (From Way & Cammell 1970.)

	Mean weight of adults at time of peak production (mg) (±standard errors)	Mean biomass of adults per plant (mg)
One leaf colonized	0.224±0.012	620
All leaves colonized	0.178±0.009	578

that the normal, aggregated situation, with a single leaf colonized, is the more productive in terms of aphids. Moreover, the uninfested leaves of the 'one-leaf colonized plants' were healthy at the end of the experiment, while the 'four-leaf colonized plants' were virtually dead. The aphids' behaviour, therefore, does more than increase their own productivity: it also provides a partial refuge for the cabbages. Thus, as a by-product of the aphids' behaviour, many more leaves escape destruction than would do so if the aphids were randomly distributed. The cabbage population is partially buffered from the aphids' ill-effects, and the interaction is (relatively) *stabilized*.

Of course, the aggregated cabbage aphids are themselves the prey of other animals and, indeed, their distribution is typical of the 'patchiness' of prey animals generally. The responses of a predator and a parasitoid to their distribution are shown in Fig. 5.21. In both cases (over at least part of the density range) the time spent on a leaf increases with the density of aphids on that leaf. Thus, if we make the reasonable assumption that an increased searching-time leads to an increased proportion of the leaf being searched, which leads to an increase in the proportion of aphids attacked, then clearly the aphids at the lower densities have a smaller probability of being attacked. Once again, therefore, there is a partial refuge, but this time it is the aphids themselves

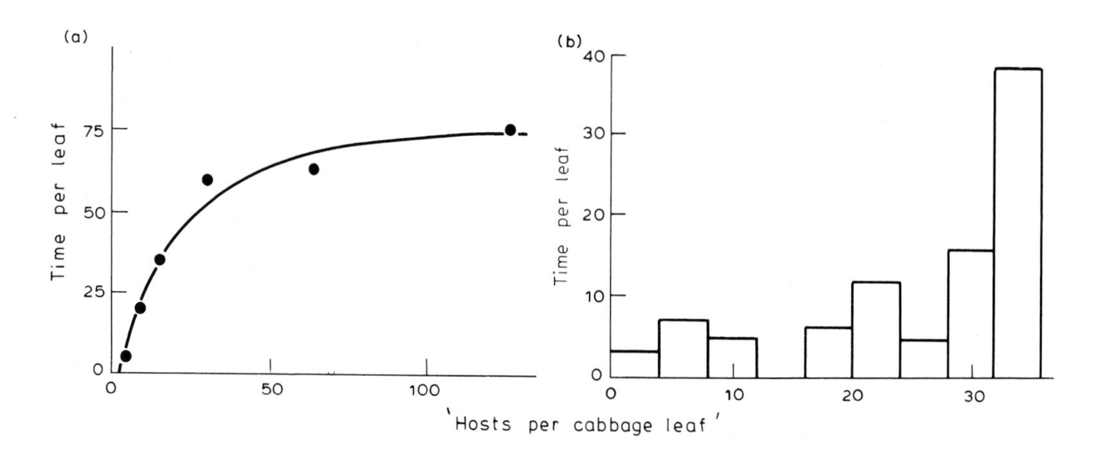

Fig. 5.21. Aggregative responses by a parasitoid and a predator of the cabbage aphid *Brevicoryne brassicae*. (a) Searching time of the braconid *Diaeretiella rapae* at different densities of its host (Akinlosotu 1973). (b) As (a) but using coccinellid larvae, *Coccinella septempunctata*. (After Hassell 1978.)

that are protected: aphids in low-density aggregates are most likely to be ignored. (It is important to note, however, that such effects apply only to those parts of the density axis in which 'time spent' increases. The importance of the effects depends on the relevance of these densities in nature.)

Most herbivores, predators and parasitoids appear capable of exhibiting an 'aggregative response', concentrating their ill-effects on a particular portion or 'patch' of their prey population; and by providing a partial refuge for the prey this tends to stabilize the interaction. But it must be realized that in each case this is essentially a by-product of the response (albeit an extremely important one): there is no question of evolution favouring stable populations. This fundamental point is illustrated in Table 5.4 and Fig. 5.21. The predator and the parasitoid concentrate on particular patches (leaves) and make their individual search more profitable; the aphids aggregate in patches supporting other aphids and increase their individual productivity. In all cases, the aggregative behaviour is *advantageous to the consumer*.

These data also provide further support for the assertion that *predators choose profitable prey* (in this case, profitable patches). Experimental illustration of this is provided for a two-patch situation by the work of Krebs *et al.* (1978). Great tits were offered rewards (food) at two different perches (patches), but the profitabilities of the two perches were unequal. After a period of learning, the birds concentrated (almost exclusively) on the more profitable patch (Fig. 5.22).

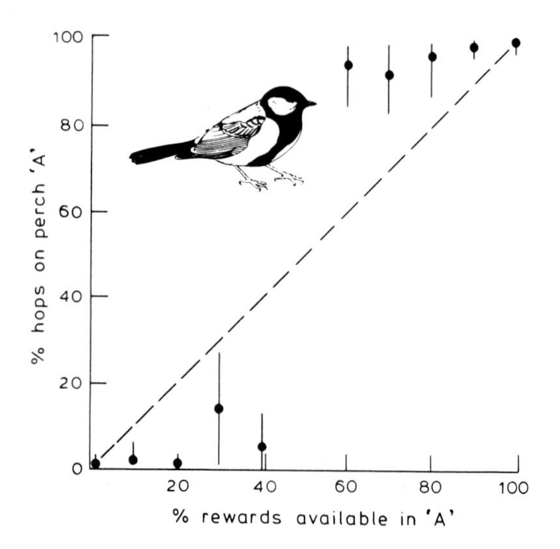

Fig. 5.22. When great tits are faced with a choice of two perches to obtain food, they go for the one with the higher reward rate. The ordinate shows the percentage of responses on one of the perches, and the abscissa is the percentage of rewards available (Krebs *et al.* 1978). (After Krebs 1978.) (Geometric means and 95% confidence limits are shown.)

5.8.4 *Further responses to patchiness*

Of course, by blandly asserting that consumers concentrate on profitable patches we beg the question of what actually constitutes a patch. In the examples shown in Fig. 5.21, for instance, the aphids' predators and parasitoids appear to treat each leaf as a patch; while in the further examples in Fig. 5.23 (which once again show concentration on profitable 'patches') the term 'patch' applies to a whole plant in the case of the braconid *Apanteles glomeratus* (Fig. 5.23a), and simply to a unit area in the case of the ichneumonid *Diadromus pulchellus* (Fig. 5.23b). Similarly, while the cabbage aphid appears to treat a leaf (or part of a leaf) as a patch, the appropriate scale for many herbivores is the whole plant. One example of this is the moth *Cactoblastis cactorum,* which has been used in Australia to control the prickly pear cacti *Opuntia inermis* and *O. stricta. C. cactorum* and its food plants now live at an apparently stable low level of abundance, which can be explained in terms of aggregation by considering some of the results obtained by Monro (1967) for *C. cactorum* and *O. inermis* (Table 5.5). The distribution of *C. cactorum* egg-sticks (each containing 70–90 eggs) on prickly pears is distinctly clumped, and the very limited mobility of the larvae means that the 'unexpectedly' high number of plants without eggs are indeed protected. Moreover, the interaction is also stabilized by the death of larvae on plants with two many egg-sticks (roughly more than two per plant). These plants become 'overloaded': they are completely destroyed by the dense aggregation of sedentary larvae, but the larvae themselves then have insufficient food to complete development. This, therefore, is an extreme example of a widespread phenomenon: the predators' ill-effects are aggregated, tending to stabilize the interactions between them and their prey.

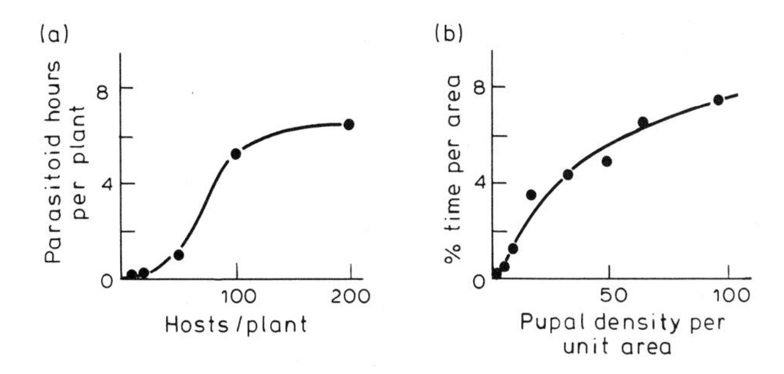

Fig. 5.23. Aggregative responses of (a) the braconid *Apanteles glomeratus* to plants of different *Pieris brassicae* density (Hubbard 1977), and (b) the ichneumonid *Diadromus pulchellus* to different densities of leek moth pupae per unit area (Noyes 1974). (After Hassell 1978.)

Table 5.5 Aggregation in a plant–herbivore interaction. Comparison of observed distributions of *Cactoblastis* egg-sticks per *Opuntia* plant with corresponding Poisson distributions. (After Monro 1967.)

Site	Mean density (egg-sticks/ segment)	Egg-sticks/plant Mean	Egg-sticks/plant Variance	Comparison of distribution with Poisson distributions of the same mean by χ^2 test
1	0.398	2.42	6.16	†P < 0.001
2	0.265	2.09	22.40	†P < 0.001
3	0.084	1.24	5.09	†P < 0.001
4	0.031	0.167	0.247	0 P > 0.005
5	0.112	0.53	1.47	†P < 0.01
6	0.137	1.97	18.76	†P < 0.001
7	0.175	0.62	3.55	†P < 0.001
8	0.112	0.34	0.90	†P < 0.05

†Egg-sticks more clumped than expected for random oviposition.
0 Egg-sticks not distributed differently from random.

5.8.5 *'Even' distributions*

Parenthetically within this component of the interaction, it must be stressed that not all 'predator–prey' distributions are clumped. Indeed, Monro (1967) provides a striking example of a herbivore going to the other extreme. Fig. 5.24 shows the distribution of ovipositions by the trypetid fruit fly *Dacus tryoni* in ripe loquat fruit, and also provides an 'expected' random distribution for comparison. It is obvious that the fly spreads its ill-effects much more evenly than expected, so that there are relatively few fruits that escape and few that are overcrowded. Once again, however, as in most of the clumped examples, we can see that the fundamental basis of the pattern is that it is *advantageous to the consumer*—in this case as a result of the reduction in intraspecific competition experienced by each larva.

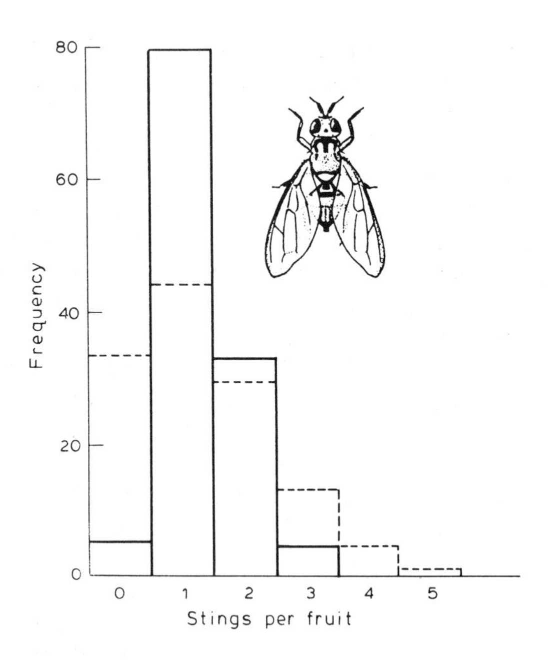

Fig. 5.24. An 'even' distribution: the observed distributions of oviposition-stings made by the fruit-fly *Dacus tryoni* on loquat fruit (——) compared to an expected random distribution (- - - -). (After Monro 1967.)

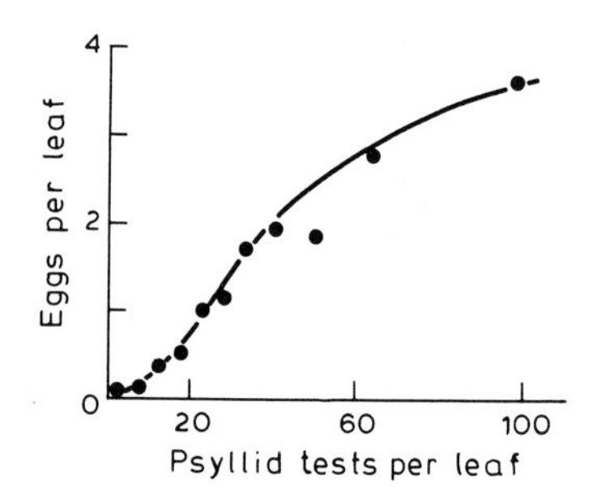

Fig. 5.25. Distribution of *Syrphus* eggs in relation to the number of tests of the psyllid *Cardiaspina albitextura* per leaf surface (Clark 1963). (After Hassell 1978.)

5.8.6 *Underlying behaviour*

There are various types of behaviour underlying the aggregative responses of predators, but they fall into two broad categories: those involved with the *location* of prey patches, and those that represent the response of a predator once *within* a prey patch. Within the first category we can include all examples of predators perceiving, at a distance, the existence of heterogeneity in the distribution of their prey. Rotheray (1979), for instance, found that the parasitoid *Callaspidia defonscolombei* was attracted to concentrations of its syrphid host by the odours produced by the syrphids' own prey: various species of aphid.

Within the second category—responses of predators within prey patches—there are three distinct types of behaviour. The first is illustrated in Fig. 5.25: a female predator of one generation tends to lay her eggs where there are high densities of prey so that her relatively immobile offspring are concentrated in these profitable patches. This response, as far as the predatory individuals themselves are concerned, is essentially passive. By contrast, the second type of behaviour involves a change in a predator's searching pattern in response to encounters with prey items. In particular, there is often an increased rate of predator turning immediately following an intake of food, which leads to the predator remaining in the vicinity of its last food item. Increased turning causes predators to remain in high-density patches of food (where the encounter-rate and turning-rate is high), and to leave low-density patches (where the turning-rate is low). Such behaviour has been demonstrated in a number of predators, and is illustrated in Fig. 5.26 for coccinelid larvae feeding on aphids (Banks 1957).

The third type of behaviour is demonstrated by the data in Table 5.6 (Turnbull 1964), referring to the web-spinning spider *Archaearanea tepidariorum* preying on fruit flies in a large experimental arena. The spiders tend, simply, to abandon sites at which their capture-rate is low, but remain at sites where it is high. In this case, therefore, the spiders modify their leaving-rate (rather than their turning-rate) in response to prey encounters, but the result, once again, is that predators congregate in patches of high prey density.

5.8.7 *'Hide-and-seek'*

We have seen, then, that the distributions of predators and prey can have important effects on predator–prey dynamics, because predators tend to concentrate on

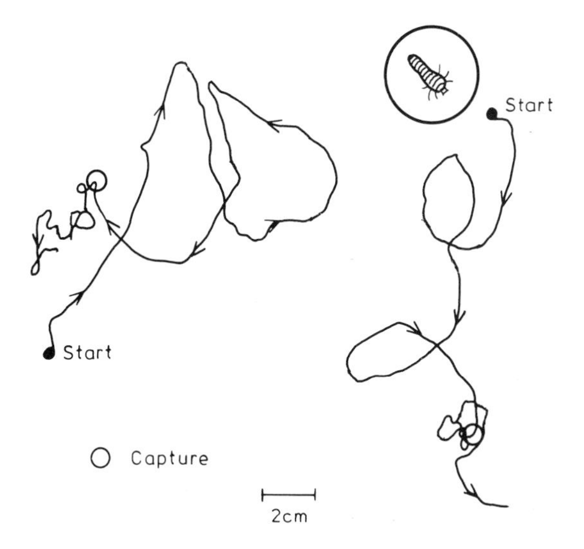

Fig. 5.26. Search paths of hungry, fourth-instar coccinellid larvae before and after capture of a prey (small circle on the path). The rate of turning markedly increases after prey capture (Banks 1957). (After Curio 1976.)

Table 5.6 Site-occupation and feeding-rate of spiders feeding on *Drosophila* in an experimental arena with sites of varying suitability for *Drosophila*. (After Turnbull 1964.)

	Spider					
	1	2	3	4	5	6
Number of sites occupied temporarily	3	2	1	2	4	3
Mean number of days temporary site occupied	2.7	3	5	3.5	3	3
Number of days final site occupied	12	14	15	13	8	11
Mean flies eaten per day at temporary site	0.5	0.2	0	0.1	0.3	0.5
Mean flies eaten per day at final site	3.5	3.4	2.3	2.5	3.4	3.5

profitable patches of prey. There is, however, in some cases, another perspective from which this behaviour can be seen: predators and prey can appear, in effect, to play 'hide-and-seek'. The most famous and illustrative example of this is provided by the experimental work of Huffaker (1958) and Huffaker *et al.* (1963). Their laboratory microcosm varied, but basically consisted of a predatory mite, *Typhlodromus occidentalis*, feeding on a herbivorous mite, *Eotetranychus sexmaculatus*, feeding on oranges interspersed amongst rubber balls in a tray. In the absence of its predator, *Eotetranychus* maintained a fluctuating but persistent population (Fig. 5.27a); but if *Typhlodromus* was added during the early stages of prey population growth, it rapidly increased its own population size, consumed all of its prey and became extinct itself (Fig. 5.27b). The interaction was exceedingly unstable, but it changed when Huffaker made his microcosm more 'patchy'. He greatly increased its size, but kept the total exposed area of orange the same; and he partially isolated each orange by placing a complex arrangement of vaseline barriers in the tray which the mites could not cross. However, he facilitated the dispersal of *Eotetranychus* by inserting a number of upright sticks from which they could launch themselves on silken strands carried by air currents. The overall result was a series of sustained and relatively stable predator–prey oscillations (Fig. 5.27c), probably generated by the following mechanism. In a patch occupied by both *Eotetranychus* and *Typhlodromus*, the predators consume all the prey and then either disperse to a new patch or become extinct. In a patch occupied by the predators alone, there is usually death of the predators before their food arrives. But in patches occupied by the prey alone, there is rapid, unhampered growth accompanied by some dispersal to new patches. Dispersal, however, is much easier for the prey than it is for the predators. The global picture is, therefore, a mosaic of unoccupied patches, prey–predator patches doomed to extinction, and thriving prey patches; with some prey and rather fewer predators dispersing between them. While each patch is ultimately unstable, the spatially heterogenous whole is much less so. Once again, therefore, patchiness has conferred stability; and, in the context of this section, we can see that this example illustrates the effect of temporary 'temporal refuges'.

A rather similar, and ultimately more satisfying example (since it comes from the field), is provided by the work of Landenberger (1973; in Murdoch & Oaten 1975), who studied the predation by starfish of mussel clumps off the coast of Southern California. Clumps which are heavily preyed upon (or are simply too large) are liable to be dislodged by heavy seas so that the mussels die: the predators are continually driving patches of prey to extinction. Yet the mussels have

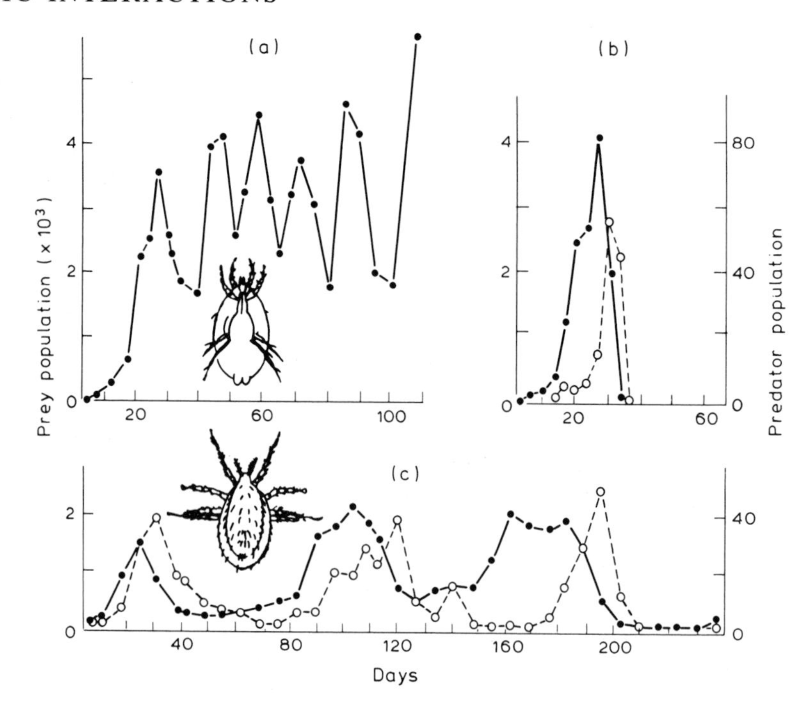

Fig. 5.27. Predator–prey interactions between the mite *Eotetranychus sexmaculatus* (●) and its predator *Typhlodromus occidentalis* (O). (a) Population fluctuations of *Eotetranychus* without its predator; (b) a single oscillation of predator and prey in a simple system; (c) sustained oscillations in a more complex system (Huffaker 1958). (After Hassell 1978.)

planktonic larvae which are continually colonizing new locations and initiating new clumps. The starfish, on the other hand, disperse much less readily. They tend to stay wherever the clumps are, and there is a time-lag before they leave an area when the food is gone. The parallel with Huffaker's mites is quite clear: patches of mussels are continually becoming extinct, but other clumps are growing prior to the arrival of starfish. The starfish show aggregative behaviour, concentrating on large, profitable clumps, and allowing the initially small, protected clumps to become large and profitable themselves. 'Hide-and-seek' and aggregative behaviour are, therefore, essentially indistinguishable, and both illustrate the important stabilizing effects of aggregation on the predator–prey interaction.

We have one further, important aspect of this 'aggregative behaviour' component to consider, but before we do so it is necessary to examine another component: the interactions that occur *between* predators.

5.9 Mutual interference amongst predators

Predators commonly reserve particular aspects of their behavioural repertoire for interactions with other predators of the same species: herbivorous (nectar-feeding) humming-birds, for instance, actively defend rich sources of food (Wolf 1969); badgers patrol and visit the 'latrines' around the boundaries between their territories and those of their neighbours; and females of *Rhyssa persuasoria*, an ichneumonid parasitoid of wood wasp larvae, will threaten and, if need be, fiercely drive an intruding female from the same area of tree trunk (Spradbery 1970). On a more quantitative level, Kuchlein (1966) has shown that an increase in the density of the predatory mite *Typhlodromus longipilus* (and thus an increase in the number of predator–predator encounters) leads to an increase in the rate of emigration from experimental leaf discs containing prey mites; and a similar situation is shown for a parasitoid of leek moth pupae in Fig. 5.28. In both qualitative and quantitative examples, then, the essential effect is the same: the time available to the predator (or herbivore, or parasitoid) for 'prey'-seeking is reduced by encounters with other predators. And the importance of this effect increases with predator density, because this increases the rate of predator–predator encounters.

The precise characteristics of such *mutual interference* will vary from species to species, but all examples can be reduced to a common form in the following manner

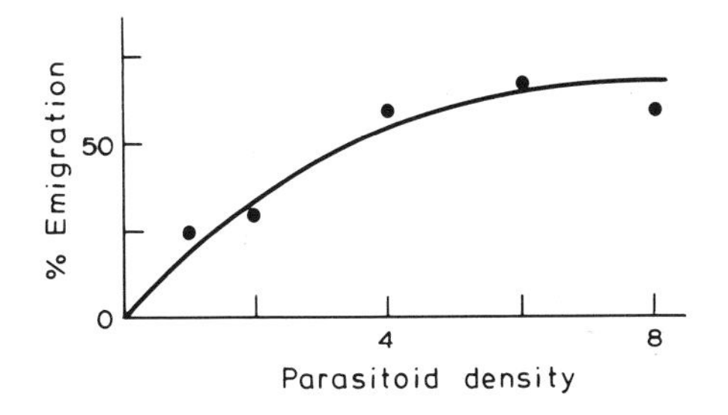

Fig. 5.28. The effect of female parasitoid density on parasitoid emigration from an experimental cage; the ichneumonid *Diadromus pulchellus* (Noyes 1974). (After Hassell 1978.)

(Hassell & Varley 1969; Hassell 1978). The *end-result* of mutual interference is that each predator consumes less than it would otherwise do. Consumption-rate *per predator* will, therefore, decline with increasing predator density. Thus, if we ignore the fact that search-time is reduced by mutual interference as predator density increases (i.e. assume that search-time remains constant), it will *appear* as if searching (or attacking) *efficiency* is declining. Mutal interference can then be demonstrated by plotting apparent attacking efficiency (calculated from data on consumption-rates on the assumption of *random* search; see Section 5.10, below) against predator density. This has been done in Fig. 5.29, on logarithmic scales. As expected, the slope in all cases is negative, and may be denoted by *-m*, where *m* is termed the *coefficient of interference.* The general form of this relationship is probably represented by Fig. 5.29a,b, in which *m* remains constant at high and moderate predator densities, but decreases at low densities. This indicates that apparent attacking efficiency cannot continue to rise as predators become increasingly scarce (moving from *right* to *left*). As Fig. 5.29c–e shows, however, the coefficient of interference often remains constant throughout the range of densities actually examined. This important observation will be utilized in Section 5.13.

We have just seen that as a result of mutual interference amongst predators, attacking efficiency decreases as predator density increases. There is, in other words, a density-dependent reduction in the consumption rate per individual, and thus a density-dependent reduction in predator fitness, which will have a *stabilizing* effect on the predator–prey interaction. The coefficient of interference, *m*, is a measure of this stabilizing effect.

5.9.1 *A similar effect amongst parasites*

Interestingly, although parasites are not subject to the same sort of mutual interference as predators, herbivores and parasitoids, they are affected by another process which has a rather similar result. Immunological responses by hosts can play an important role in parasite mortality, and the strength of the response is often directly related to the size of the parasite burden. Thus, the probability of host-induced parasite mortality will increase with greater parasite density, as Fig. 5.30 illustrates. There is no direct interference, of course, but an

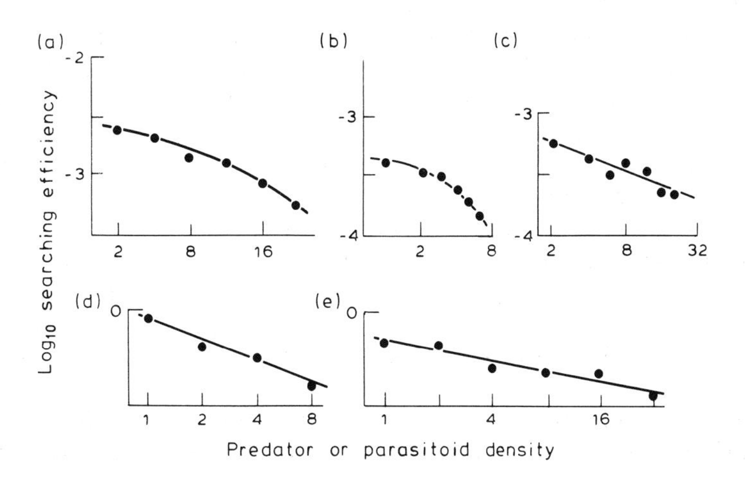

Fig. 5.29. Interference relationships between the searching efficiency (log scale) and density of searching parasitoids or predators. (a) *Encarsia formosa* parasitizing the whitefly, *Trialeurodes vaporariorum* (Burnett 1958). (b) *Chelonus texanus* parasitizing eggs of *Anagasta kühniella* (Ullyett 1949b). (c) *Cryptus inornatus* parasitizing cocoons of *Loxostege stricticalis* (Ullyett 1949a). (d) *Coccinella septempunctata* feeding on *Brevicoryne brassicae* (Michelakis 1973). (e) *Phytoseiulus persimilis* feeding on deutero-nymphs of *Tetranychus urticae* (Fernando 1977). (After Hassell 1978.)

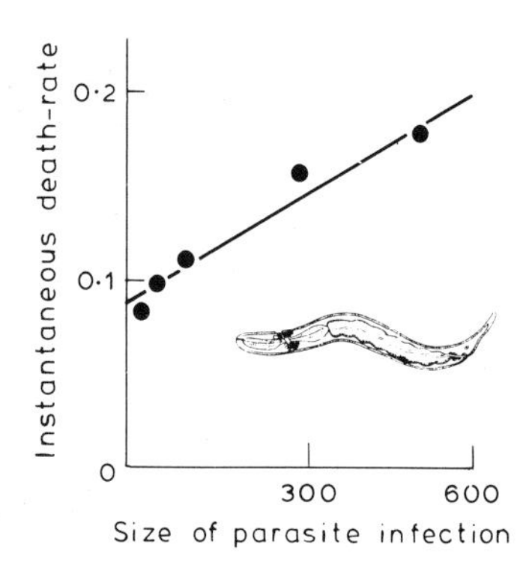

Fig. 5.30. The relationship between parasite death-rate and parasite density within individual hosts for chickens infected with the fowl nematode *Ascaridia lineata* (Ackert *et al.* 1931). (After Anderson & May 1978.)

individual parasite's fitness is reduced by the presence of other parasites. Once again, therefore, the general effect is to *stabilize* the interaction (see Anderson & May 1978).

5.10 Interference and pseudo-interference

We can return now to aggregative behaviour, and consider an important alternative approach to its effects. This was introduced in a mathematical way by Free *et al.* (1977); but for our purposes their ideas can be discussed verbally. By concentrating on profitable patches, a single predator increases the number of prey it eats per unit time; the same predator searching randomly would eat less. This means that the apparent attacking efficiency (calculated from consumption-rates on the *assumption* of random search) is higher in the 'aggregated' predator. Yet the same predator, by removing prey from the profitable patches, will marginally reduce those patches' profitability. This, in time, will make aggregated search more like random search, and the apparent attacking efficiency will decline. And if there are many predators, all concentrating on profitable patches, then this decline may be immediately perceptible. Thus, in general, we can expect the apparent attacking efficiency to decrease as predator density increases. Yet this is the very relationship that we saw resulting from mutual predator interference in the previous section. In the present case, therefore, this *consequence of aggregative behaviour* may be described as 'pseudo-interference' (Free *et al.* 1977).

The importance of pseudo-interference, along with the nature of mutual interference itself, is amply illustrated in the following example (Hassell 1971a,b, 1978) in which larvae of the flour moth *Ephestia cautella* were parasitized by *Venturia canescens*. The larvae were confined in small containers at densities ranging from 4 to 128 per container, and exposed to 1, 2, 4, 8, 16 or 32 parasitoids for 24 hours. Line A in Fig. 5.31a was obtained in the manner of the previous section (by counting the total number of hosts parasitized and assuming there was random search): there was, apparently, a high degree of mutual parasitoid interference (m large). In fact, continuous observation showed *explicitly* that the time spent by parasitoids on host containers did decrease with increasing parasitoid density (Fig. 5.31b). In other words, mutual interference caused parasitoids to leave host patches, and thus spend less time searching. The true searching time, then, is not 24 hours, but the proportion of 24 hours indicated by Fig. 5.31b; and taking this into account increases the apparent searching efficiency and yields line B in Fig. 5.31a. Yet there is clearly some interference remaining after these effects have been removed.

A partial explanation of this is provided by the existence of an additional aspect of behavioural interference. Observation showed that parasitoid–parasitoid encounters often led not to departure from a patch, but simply to an interruption of probing (lasting about one minute). This, too, can be used to reduce the 'true' searching-time appropriately, because the number of such encounters was observed and noted; and recalculating the apparent attacking efficiency this time yields line C in Fig. 5.31a.

Lines A, B and C have negative slopes (i.e. coefficients of interference) of 0.67, 0.45 and 0.27 respectively. Thus, 33% of the total interference [(0.67–0.45)/0.67×100] is accounted for by parasitoids leaving host patches, and a *further* 27% by parasitoids simply interrupting their searching. However, this leaves 40% of the total unaccounted for by actual mutual interference. The preceding argument (Free *et al.* 1977) would suggest

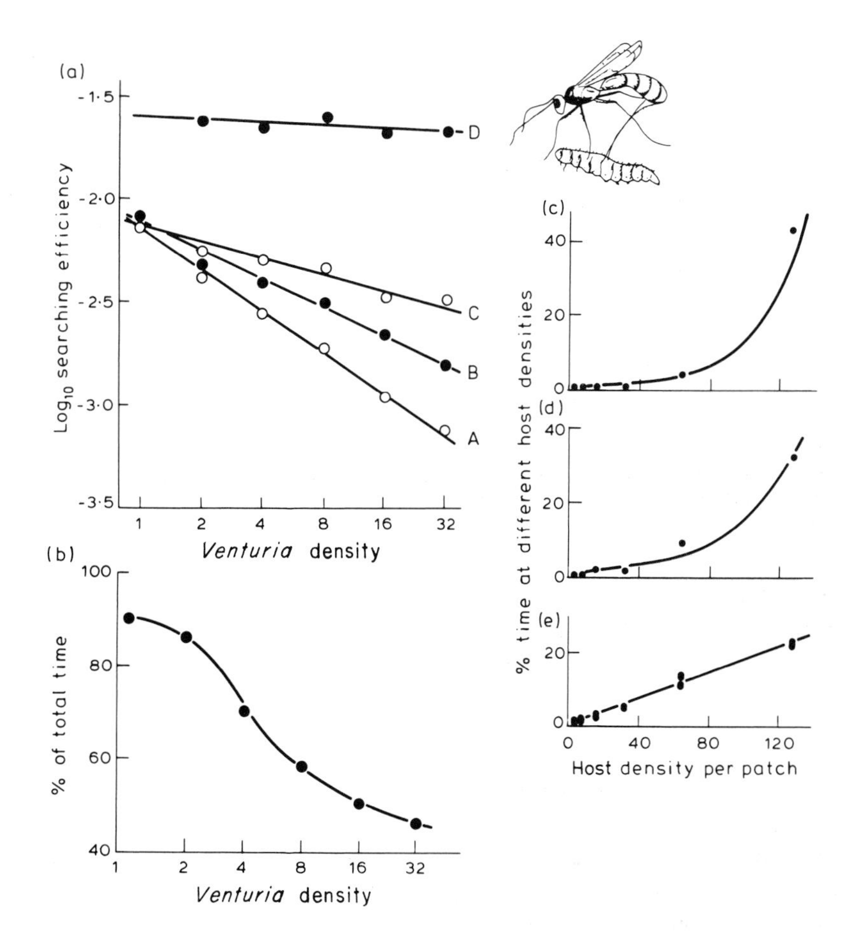

Fig. 5.31. (a) Relationships between searching efficiency and density of searching *Venturia canescens*; (b) relationship between the proportion of time spent searching and *Venturia* density; and the percentage of time spent at different host densities with (c) one parasitoid, (d) two parasitoids and (e) four, eight, sixteen and thirty-two parasitoids. (After Hassell 1978.) For further discussion, see text.

that this is pseudo-interference, and the aggregative behaviour shown in Fig. 5.31c–e certainly supports this. Much more persuasive, however, is line D in Fig. 5.31a. This has been obtained by abandoning the assumption of random search, and using instead the observed data on search-pattern from Fig. 5.31c–e. The slope, −0.03, indicates that there is no longer any significant amount of interference. The interference of line C, in other words, was purely a consequence of the aggregative behaviour. It was, indeed, pseudo-interference.

Perhaps the initial conclusion to be drawn from this work is that, in the absence of the detailed observational data, it would have been impossible to partition the total interference of line A into mutual interference and pseudo-interference. This is not, however, as disappointing a conclusion as it might at first appear. In any natural situation, mutual interference (if it exists at all) is bound to increase with increased aggregation, since aggregation (by definition) leads to a higher probability of encounter. Yet the general effect of mutual interference will be to drive predators from dense aggregates (where encounters are most frequent). This will tend to reduce the level of aggregation, and ultimately reduce the mutual interference itself. There is, in short, a complex and dynamic interaction—which it might be extremely difficult to disentangle—between aggregation (leading to pseudo-interference) and behavioural interference; and it may, therefore, be convenient to encapsulate the effects of both in a single parameter: the coefficient of interference. Plots like the ones in Fig. 5.29 and line A in Fig. 5.31a, in other words, are able to capture the combined, density-dependent, stabilizing effects of aggregative behaviour and mutual predator interference; and there will be many natural situations in which one or both of these effects are extremely significant. Both effects, of course, represent strategies adopted by the predators to increase their own fitness.

5.11 Optimal foraging

One further conclusion that can be drawn from this discussion, however, is that the movement of predators between patches is itself worthy of detailed study—whether this movement is a consequence of interference, or of patch depletion, or is even a means of patch-assessment. Such study would be closely related to investigations of diet width among predators (Section 5.3), and of the way in which predators distribute their effort amongst patches (Section 5.7), because all three are concerned with the ways in which predators tend to maximize their 'profits'. They can be bracketed together as studies of 'optimal foraging'. This is a subject which is reviewed in some detail by Krebs (1978), but we can underline the importance of optimal foraging between patches by means of the following example.

Cook & Cockrell (1978) fed individual, fourth-instar mosquito larvae to adult water-boatmen, *Notonecta glauca,* and as Fig. 5.32a shows, the rate at which nutrients were extracted from a single prey item declined sharply with time. This occurred not because the water-boatmen were satiated, but because the food became increasingly difficult for them to extract. Without bending the rules too much, we can treat each prey item as a 'patch'. From this viewpoint we can see that the profitability of patches to water-boatmen declined rapidly the longer they stayed 'in' them. We can also see that to maximize their profits they would, at some time, have to leave depleted patches and find new, highly profitable ones, i.e they would have to drop the old prey and catch a new one. Yet this in itself is a costly process: the predators take in no food while they are expending energy searching for, capturing and subduing (i.e. handling) the next prey item. Obviously, if they are to forage optimally, the water-boatmen must maximize their profits when they set such costs against the eventual gain from the new patch, and the longer the handling-time the greater these costs will be. We would, therefore, expect optimally foraging predators to spend relatively long periods at a patch (where the rate of profit, even if low, is at least positive) when the handling-time is high; and this is precisely the result obtained by Cook & Cockrell (Fig. 5.32b). The water-boatmen's feeding-times served to increase their profits (Fig. 5.32c i). Prolonged stays (Fig. 5.32c ii) or repeated changes of patch (Fig. 5.32c iii) would clearly have been less profitable.

It must not, of course, be imagined that the water-boatmen consciously weigh up pros and cons and act accordingly. Rather, it is natural selection that has

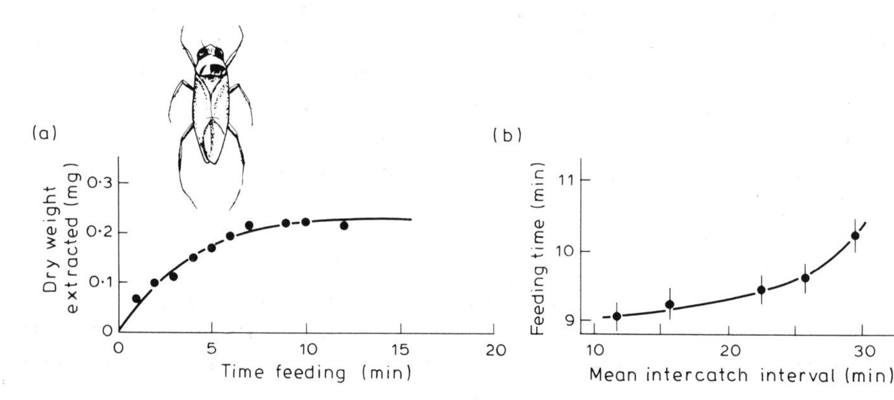

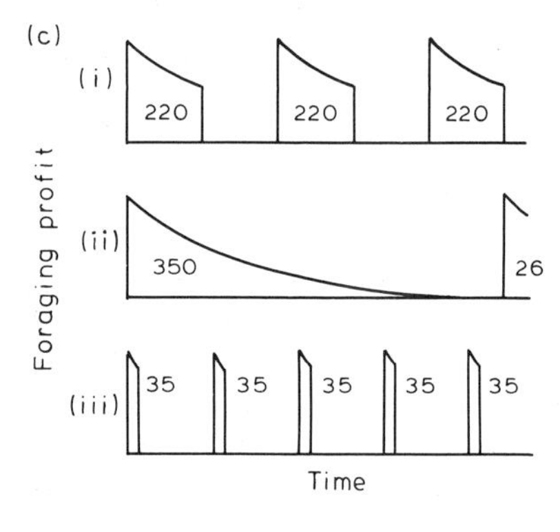

Fig. 5.32. (a) The cumulative dry weight of food extracted from an individual mosquito larva by *Notonecta*, as a function of time spent feeding: as time increases, diminishing returns set in. (b) The relationship between feeding-time (±s.e.) and handling-time (intercatch interval). (After Cook & Cockrell 1978.) (c) The foraging profit of (i) an optimal forager (660 arbitrary units), compared to one with a long stay time (ii) (376 units) and one with a short stay time (iii) (175 units).

favoured individuals that adopt a strategy appropriate for foraging in a patchy environment with marked patch depletion. Nor must it be imagined that animals always adopt the strategy which actually maximizes their foraging profits. As Comins & Hassell (1979) point out, animals must spend (i.e 'waste') time sampling and learning about their environment; and the strategies they adopt will be influenced not only by foraging itself, but also by predator-avoidance, and so on. Nevertheless, we can expect natural selection to exert evolutionary pressures *towards* an optimal foraging strategy; and the recognition of optimal foraging is bound to play an increasingly crucial role in our understanding of predator–prey interactions. Interestingly, Comins & Hassell (1979) suggested, from an analysis of models, that although optimal foraging has important consequences for predator fitness, its effects on predator–prey dynamics are rather similar to those resulting from a much simpler, fixed pattern of aggregation.

5.12 Resumé

Our view of the 'simplest, abstracted, two-species predator–prey system' is, by now, somewhat less naïve than it was. We have seen that intraspecific competition amongst prey, 'type 3' functional responses, aggregated distributions of ill-effects (and, as special cases, spatial and temporal refuges), mutual interference amongst predators, and the increase in host immunological response with increased parasite burden all tend to have a stabilizing influence on the interaction; while time-lags, the increased effects of multiple infections in parasites, the existence of 'maintenance thresholds' and 'type 1' and 'type 2' functional responses all tend to have a destabilizing influence. We have also seen that predators and prey exert a greater influence on the dynamics of one another's populations the more specific (less polyphagous) the predator is; that some prey have relatively little fitness to lose by their death; that the quality as well as the quantity of food can exert important influences on predator populations; and that, in many respects, predators *tend* to forage optimally.

Nevertheless, having examined these components we remain essentially ignorant of their relative importance, and of the patterns of abundance that we might expect them to give rise to, either alone or in combination. In order to go some way towards dispelling this ignorance, we turn, once again, to mathematical models.

5.13 Mathematical models

Of the models that we might consider, Lotka (1925) and Volterra (1926) constructed a classical, simple, continuous-time predator–prey model, which Rosenzweig & MacArthur (1963) developed further in graphical form; Caughley & Lawton (1981) reviewed models of plant–herbivore systems; and Crofton (1971), Anderson & May (1978) and May & Anderson (1978) constructed models of the host–parasite interaction (reviewed by Anderson 1981). However, because of the range of factors which they explicitly and successfully incorporate, we shall concentrate here on difference-equation models of host–parasitoid systems (cf. Section 3.2). These are described in much greater detail by Hassell (1978). Then, in Section 5.13.2, we shall turn briefly to a model of grazing systems (Noy-Meir 1975). The conclusions that we are able to draw throughout will throw important light on predator–prey interactions generally.

5.13.1 Host–parasitoid models

We begin by describing a very simple model, which can be used as a basis for further developments (Nicholson 1933; Nicholson & Bailey 1935). Let H_t be the number of hosts, and P_t the number of parasitoids (in generation t); r is the intrinsic rate of natural increase of the *host*, and c is the conversion rate of hosts into parasitoids, i.e. the mean number of parasitoids emerging from each host. If H_a is the number of hosts actually attacked by parasitoids (in generation t), then, clearly:

$$\begin{aligned} H_{t+1} &= e^{r}(H_t - H_a) \\ P_{t+1} &= cH_a \end{aligned} \qquad (5.1)$$

In other words, ignoring intraspecific competition, the hosts that are not attacked reproduce, and those that are attacked yield not hosts but parasitoids. For simplicity we shall assume that each host can support only one parasitoid ($c = 1$). Thus, the number of hosts attacked in one generation defines the number of parasitoids produced in the next ($P_{t+1} = H_a$).

To derive a simple formulation for H_a we proceed as follows. Let E_t be the number of host–parasitoid encounters (or interactions) in generation t. Then if A is the proportion of the hosts encountered by any one parasitoid:

$$E_t = AH_tP_t$$

and

$$\frac{E_t}{H_t} = AP_t \qquad (5.2)$$

(A can, alternatively, be thought of as the parasitoid's searching efficiency; or the probability that a given parasitoid will encounter a given host; or, indeed, the 'area of discovery' of the parasitoid, within which it encounters all hosts.) Remember that we are dealing with parasitoids. This means that a single host can be encountered several times, but only the first encounter leads to successful parasitization; predators, by contrast, would physically remove their prey, and thus prevent re-encounters.

If encounters occur in an essentially random fashion, then the proportions of hosts encountered nought, one, two, three or more times are given by the successive terms in the appropriate 'Poisson distribution' (described in any basic textbook on statistics). The proportion *not* encountered at all, p_0, is given by:

$$p_0 = \exp\left(-\frac{E_t}{H_t}\right), \qquad (5.3)$$

where $\exp(-E/H_t)$ is another way of writng e^{-E/H_t}. Thus the proportion that *is* encountered (one or more times) is $1-p_0$, and the *number* encountered (or attacked) is:

$$H_a = H_t(1-p_0) = H_t\left\{1-\exp\left(-\frac{E_t}{H_t}\right)\right\}$$

or

$$H_a = H_t\{1-\exp(-AP_t)\} \qquad (5.4)$$

And substituting this expression for H_a into equation 5.1 gives us:

$$H_{t+1} = H_t \exp(r-AP_t)$$
$$P_{t+1} = H_t\{1-\exp(-AP_t)\} \qquad (5.5)$$

This is the Nicholson–Bailey model of the host–parasitoid interaction. Its simplicity rests on two assumptions:

(a) that parasitoid numbers are determined solely by the rate of random encounters with hosts; and

(b) that host numbers would grow exponentially but for the removal of individuals by random encounter with parasitoids.

As Hassell (1978) makes clear, an equilibrium combination of these two populations is a *possibility*, but even the slightest disturbance from this equilibrium leads to divergent oscillations (Fig. 5.33). Thus, our simple model, although it produces coupled oscillations, is highly unstable. Nevertheless, it is clearly a formal restatement of the naïve expectation expressed in Section 5.1: when a single predator and a single prey interact in the simplest imaginable way, coupled oscillations are the result.

Since most observed patterns of abundance are considerably more stable than those produced by the Nicholson–Bailey model, we must be particularly interested in modifications to it which enhance stability. The most obvious modification we can make is to replace the exponential growth of hosts with density-dependent growth resulting from intraspecific competition (Section

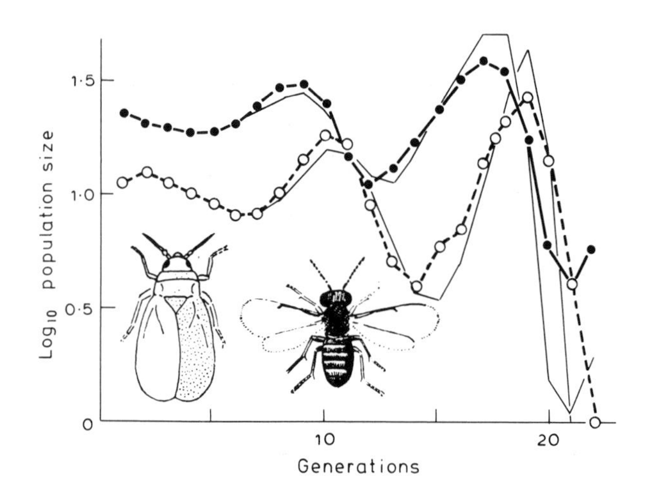

Fig. 5.33. Population fluctuations from an interaction between the greenhouse whitefly *Trialeurodes vaporariorum* (●) and its chalcid parasitoid *Encarsia formosa* (O). The thin lines show the estimated outcome from a Nicholson-Bailey model (Burnett 1958). (After Hassell 1978.)

5.5). Following Beddington *et al.* (1975), this is done by incorporating a term like the one used in the logistic equation (Section 3.3), giving

$$H_{t+1} = H_t \exp\left\{r\left(1-\frac{H_t}{K}\right)-AP_t\right\} \quad (5.6)$$

$$P_{t+1} = H_t\{1-\exp(-AP_t)\}$$

where K is the carrying capacity of the host population in the absence of parasitoids. Fig. 5.34 illustrates the patterns of abundance resulting from this revised model in terms of r and a new parameter, $q(=H^*/K)$ where H^* is the equilibrium size of the host population in the presence of parasitoids. For a given value of r and K, q depends solely on the parasitoids' efficiency, A. When A is low, q is almost 1 ($H^* \cong K$), but at higher efficiencies q approaches zero ($H^* \ll K$).

It is clear from Fig. 5.34 that intraspecific competition amongst hosts can lead to a range of abundance patterns. Moreover, intraspecific competition is obviously a potentially important stabilizing factor in host–parasitoid systems. This is particularly so for low and moderate values of r, and high and moderate values of q; but even with low values of q (high parasitoid efficiency), the fluctuations are not altogether unlike those observed in natural populations. Note, however,

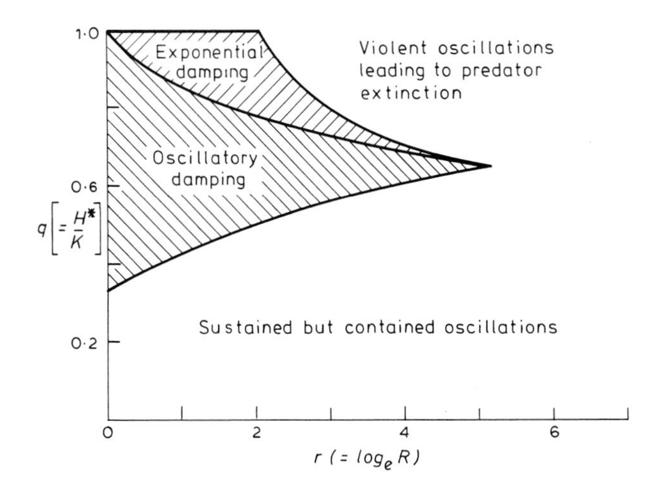

Fig. 5.34. Stability boundaries for the parasitoid-host model with intraspecific host competition, as in equation 5.6. (After Beddington *et al.* 1975.) For further discussion, see text.

that with high reproductive rates and low parasitoid efficiencies we return to the chaotic behaviour characteristic of the single-species populations in Section 3.4.1.

The next modification we can make to the Nicholson–Bailey model is to consider explicitly the parasitoids' functional response to host density (Section 5.7). In the present context, this is described by the relationship between H_a/P_t (the mean number of hosts attacked per parasitoid) and H_t; i.e. the relationship between the number of hosts attacked by a constant number of parasitoids (H_a with P_t constant) and H_t. Until now we have been assuming that this relationship is linear (equation 5.4):

$$H_a = H_t\{1-\exp(-AP_t)\}.$$

In other words, it is implicit in the Nicholson–Bailey model that the 'predation'-rate of the parasitoids continues to rise indefinitely with increasing host density, i.e. handling-time is zero. This is clearly at variance with the data examined in Section 5.7, and is, in any case, impossible. Obviously it is important to replace this linear relationship with the type 2 and type 3 functional responses which are actually observed.

Dealing first with the type 2 response, we saw in Section 5.7.1 that the essential feature underlying it is the existence of a finite 'handling-time'. Thus, we shall let T be the total amount of time available to each parasitoid, T_h the time it takes to deal with *each* host (handling-time), T_s the total amount of time available to each parasitoid for host-seeking, and a the parasitoids' instantaneous rate of search (or attack rate, see Section 5.7). Then, by definition:

$$A = aT_s,$$

so that the total number of hosts handled in generation t by each parasitoid is now aT_sH_t. If we consider that 'total available search-time' is equal to 'total time' minus '*total* handling-time', then:

$$T_s = T - T_h a T_s H_t,$$

which, by rearrangement, gives:

$$T_s = \frac{T}{1+aT_hH_t},$$

so that:

$$A = \frac{aT}{1+aTH_t}.$$

Substituting this into equations 5.4 and 5.5 gives:

$$H_a = H_t\left\{1-\exp\left(\frac{-aTP_t}{1+aT_hH_t}\right)\right\} \qquad (5.7)$$

$$H_{t+1} = H_t\ \exp\left(r-\frac{aTP_t}{1+aT_hH_t}\right)$$

$$P_{t+1} = H_t\left\{1-\exp\left(\frac{-aTP_t}{1+aT_hH_t}\right)\right\} \qquad (5.8)$$

As Fig. 5.35 shows, equation 5.7 generates a type 2 functional response, in which the maximum possible number of hosts attacked is determined (as expected) by T_h/T, and the rate of approach to this asymptote is determined by a (the instantaneous search-rate). Equations 5.8, therefore, represent the Nicholson–Bailey model with a type 2 functional response incorporated into it. Their dynamic properties were examined by Hassell & May (1973) and, as expected, the inverse density-dependence makes this model less stable than the Nicholson–Bailey (recovered from this model when $T_h = 0$). However, increased instability is relatively slight as long as $T_h/T \ll 1$, and we can see in Table 5.7 (Hassell 1978) that this is, indeed, generally the case. This model indicates, therefore, that the destabilizing tendencies of type 2 functional responses are unlikely to be of major importance in nature.

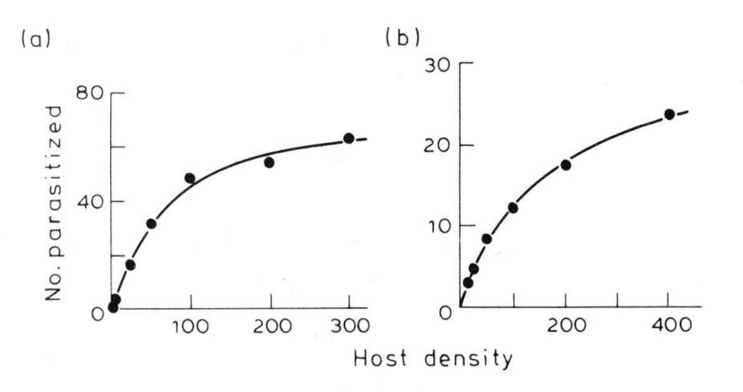

Fig. 5.35. Type 2 functional responses generated by equations 5.7. (a) *Nasonia vitripennis* parasitizing *Musca domestica* pupae (DeBach & Smith 1941) $a = 0.027$, $T_h = 0.52$. (b) *Dahlbominus fuscipennis* parasitizing *Neodiprion sertifer* cocoons (Burnett 1956; $a = 0.252$, $T_h = 0.037$. (After Hassell 1978.)

In order to model the type 3 functional response, we shall follow the pragmatic approach of Hassell (1978), and assume firstly that only the instantaneous search-rate, a, varies with host density, and secondly that it does

Table 5.7 Estimated values of handling-time T_h from equation 5.7 for a selection of parasitoids. The values of T_h/T are based on conservative estimates of longevity. (After Hassell 1978.)

Parasitoid species	Host	Handling-time T_h (hours)	T_h/T	Author(s)
Venturia canescens	*Ephestia cautella*	0.007	< 0.0001	Hassell & Rogers (1972)
Chelonus texanus	*Ephestia kuhniella*	0.12	< 0.001	Ullyett (1949a)
Dahlbominus fuscipennis	*Neodiprion lecontei*	0.24	< 0.003	Burnett (1958)
Pleolophus basizonus	*Neodiprion sertifer*	0.72	< 0.02	Griffiths (1969)
Dahlbominus fuscipennis	*Neodiprion sertifer*	0.96	< 0.01	Burnett (1954)
Cryptus inornatus	*Loxostege sticticalis*	1.44	< 0.02	Ullyett (1949b)

so in the simplest way compatible with the data examined in Section 5.7.3; viz.:

$$a = \frac{xH_t}{1+yH_t},$$

where x and y are constants. Substituting this into equations 5.7 and 5.8 (and rearranging) gives us:

$$H_a = H_t\left\{1-\exp\left(\frac{-xTH_tP_t}{1+yH_t+xT_hH_t^2}\right)\right\} \quad (5.9)$$

$$H_{t+1} = H_t \exp\left(r-\frac{xTH_tP_t}{1+yH_t+xT_hH_t^2}\right)$$

$$P_{t+1} = H_t\left\{1-\exp\left(\frac{-xTH_tP_t}{1+yH_t-xT_hH_t^2}\right)\right\} \quad (5.10)$$

Equation 5.9 generates a type 3 functional response, and equations 5.10 are, therefore, the Nicholson–Bailey model with a type 3 response incorporated into it.

We have already noted (in Section 5.7.4) that the density-dependent aspect of this response is likely to have an essentially stabilizing effect, but, as with the type 2 response, examination of the appropriate model allows us to make an interesting qualification to this informal conclusion. In particular, Hassell & Comins (1978) found that in the situation we have been dealing with—one host and one parasitoid coupled together in a *discrete-generation* model—a type 3 response *alone* is incapable of stabilizing the interaction. On the other hand, there are at least two alternative situations in which it becomes a much more potent stabilizing force. The first is when the time-delay of discrete generations is removed and replaced by the instantaneous reaction of continuous breeding (Murdoch & Oaten 1975). The second is when the parasitoid is polyphagous, and the type 3 response is a result of parasitoid 'switching'. Hassell & Comins (1978) believe that this behaviour would essentially allow the parasitoid to maintain itself at a constant density, so that:

$$P_t = P_{t+1} = P^*.$$

And if this equation is merged with the host part of equation 5.10, then the patterns of abundance are as summarized in Fig. 5.36.

Overall, therefore, these models make it clear that the

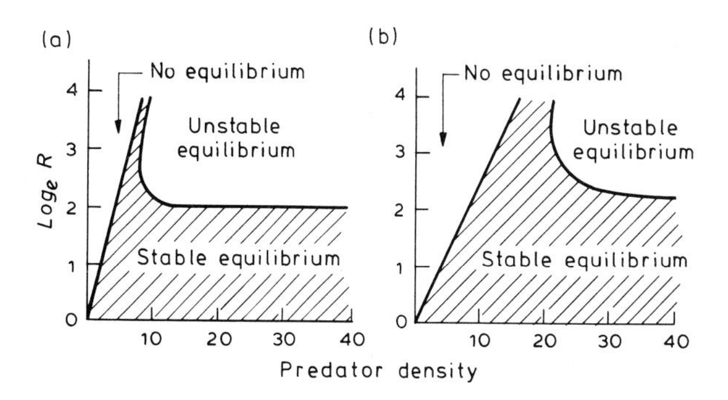

Fig. 5.36. Stability boundaries for the parasitoid–host model with constant parasitoid density (parasitoid switching) and the host as described in equation 5.10a. (a) $y = 0$; (b) $y = 2\sqrt{(xT_h)}$. (After Hassell & Comins 1978.)

apparently simple consequences of the different 'predator' functional responses are subject to significant qualifications.

We turn now to the important question of heterogeneity (Section 5.8), and consider first the special, extreme case of a host 'refuge'. If there is a *constant proportion* refuge such that only a proportion of the hosts, γ, are available to the parasitoids, then equations 5.5 can be replaced simply by:

$$H_{t+1} = (1-\gamma)H_te^r+\gamma H_t \exp(r-AP_t)$$

$$P_{t+1} = \gamma H_t\{1-\exp(-AP_t)\} \quad (5.11)$$

If, on the other hand, there is a *constant number* refuge in which H_0 hosts are always protected, the appropriate modification is:

$$H_{t+1} = H_0e^r+(H_t-H_0) \exp\{r-AP_t\}$$

$$P_{t+1} = (H_t-H_0)(1-\exp\{-AP_t\}) \quad (5.12)$$

The results of these modifications are summarized in Fig. 5.37 (following Hassell & May 1973). It is clear that, while both modifications have a stabilizing effect on the interaction, the constant *number* refuge is by far the more potent of the two. This is no doubt due to the density-dependent effect of having a greater proportion of the host population protected as host density decreases.

By contrast, a very much more general approach to heterogeneity has been taken by May (1978a). He set

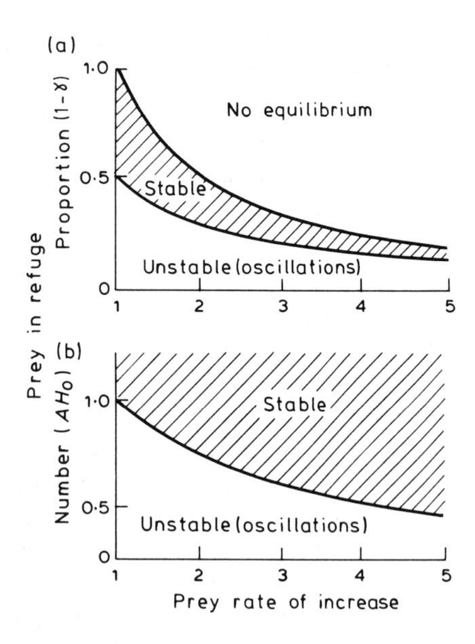

Fig. 5.37. The effects of refuges on stability: (a) a constant proportion refuge (equations 5.11); (b) a constant number refuge (equation 5.12). (After Hassell & May 1973.)

aside the precise nature of host and parasitoid distributions and efforts, and argued simply that the distribution of host–parasitoid *encounters* was not random but aggregated. In particular, he assumed (with some justification—see May 1978a) that this distribution could be described by the simplest and most general of the appropriate statistical models, the negative binomial. In this case, the proportion of hosts not encountered at all is given by:

$$p_0 = \left[1+\frac{AP_t}{k}\right]^{-k}$$

where k is a measure of the degree of aggregation: maximal aggregation at $k = 0$, minimal aggregation at $k = \infty$ (recovery of the Nicholson–Bailey model). The appropriate modification of p_0 in equations 5.6 gives us a model that incorporates both aggregated encounters and intraspecific competition amongst hosts:

$$H_{t+1} = H_t \exp\left\{r\left(1-\frac{H_t}{K}\right)\right\} \left[1+\frac{AP_t}{k}\right]^{-k}$$
$$P_{t+1} = H_t\left(1-\left[1+\frac{AP_t}{k}\right]^{-k}\right) \quad (5.13)$$

The patterns of abundance generated by this model are summarized in Fig. 5.38, from which it is clear that the already moderately stable system described by equations 5.6 is given a marked boost in stability by the incorporation of significant levels of aggregation ($k \leqq 1$). Of particular importance is the existence of stable systems with very low values of q. Of the stabilizing factors so far considered, encounter-aggregation (i.e. heterogeneity) is obviously the most potent.

We consider next a topic that was established in Section 5.9 as being closely connected with the aggregation of encounters: mutual parasitoid interference. Following Hassell & Varley (1969), we shall adopt a simple, empirical approach and derive a form for the searching efficiency, A, which conforms to the log–log plots of Fig. 5.29, i.e.:

$$A = QP_t^{-m}$$

where log Q and $-m$ are the intercept and slope of the plots, and, in particular, m is the 'coefficient of interference'. Note that, because this relationship is empirical, it serves to describe not only interference proper but pseudo-interference as well. The relationship, therefore, incorporates mutual parasitoid interference and aggregation.

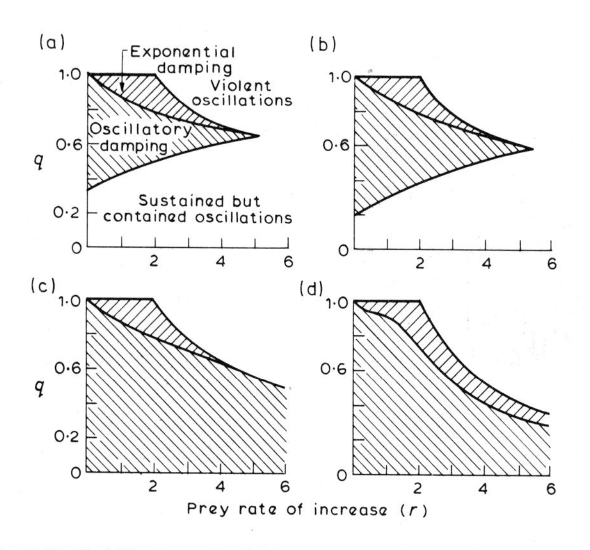

Fig. 5.38. Stability boundaries for the parasitoid–host model with parasitoid aggregations as in equations 5.13: (a) $k = \infty$; (b) $k = 2$; (c) $k = 1$; (d) $k = 0$. (After Hassell 1978.) (See Fig. 5.34.)

The appropriate model is clearly:

$$\begin{aligned} H_{t+1} &= H_t \exp(r - QP_t^{1-m}) \\ P_{t+1} &= H_t\{1-\exp(-QP_t^{1-m})\} \end{aligned} \qquad (5.14)$$

and the patterns of abundance resulting from these equations are illustrated in Fig. 5.39. Not surprisingly, since aggregation of encounters is being incorporated, 'total interference' is shown to be an extremely potent stabilizing force.

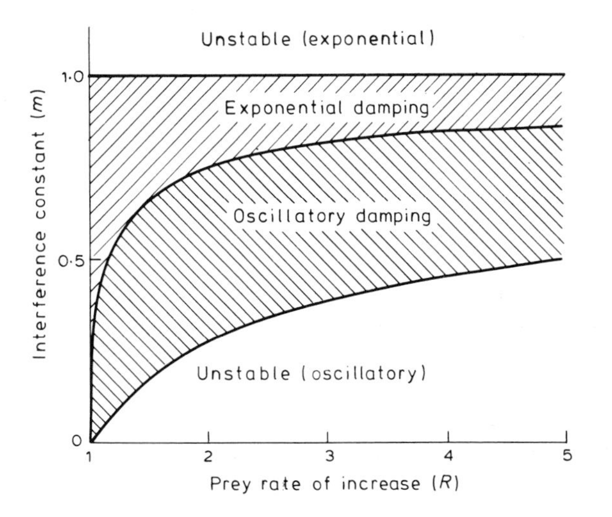

Fig. 5.39. Stability boundaries for the parasitoid–host model with mutual interference as in equations 5.14. (After Hassell & May 1973.)

5.13.2 *A model of grazing systems*

We turn now to a model of grazing systems developed by Noy-Meir (1975). It is typical of a range of models, all of which essentially incorporate the 'Allee effect' (Section 2.6); and all such models indicate that predators and their prey (in this case grazers and their food plants) can coexist at *more than one stable equilibrium* (see May 1977 for a review).

We have already seen (in Section 5.5.1) that plant populations, with the individuals growing as multiply branched units, may not necessarily exhibit a simple pattern of population growth. In particular, we argued that the rate of vegetative growth may be conveniently expressed in terms of photosynthetic assimilation, the *net* assimilation rate being the rate of biomass growth having accounted for respiratory losses. Fig. 5.9 demonstrated that if the rate of growth in biomass is plotted against the leaf area index (the ratio of total leaf area to horizontal ground beneath the canopy), then there is an intermediate, optimum index which maximizes growth-rate—a result of the shifting balance between photosynthesis and respiration as biomass and *shading* increase. Not surprisingly, since total biomass and leaf area index are so closely associated, there is also a humped, curved relationship between growth-rate and biomass. The actual shape of such curves (e.g. Fig. 5.40) will depend on the interaction of many factors (see Blackman 1968 for a review), but in all cases an optimum biomass will exist, yielding a maximum growth-rate.

The removal of vegetation biomass, on the other hand, occurs as a result of grazing, and we have seen that the rate of herbivore consumption depends on a variety of factors. In broad terms, however, consumption-rate is likely to follow the saturation curve of a 'type 2' or, more rarely, a 'type 3' functional response (Section 5.7). Of course, the *total* rate of herbivore consumption will increase with herbivore density. This will result in a family of curves, all of similar shape, but with the height of the maximal, saturation consumption rate itself increasing with herbivore density (see, for instance, Fig. 5.41a).

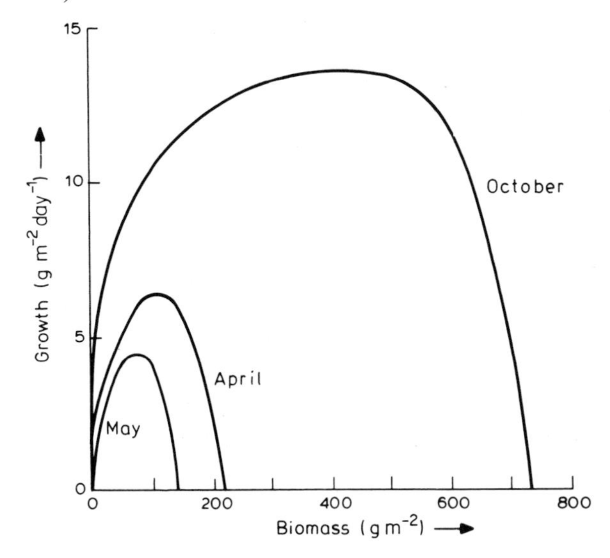

Fig. 5.40. Plant growth in New Zealand ryegrass-clover pastures as a function of biomass (Brougham 1955, 1956). (After Noy-Meir 1975.)

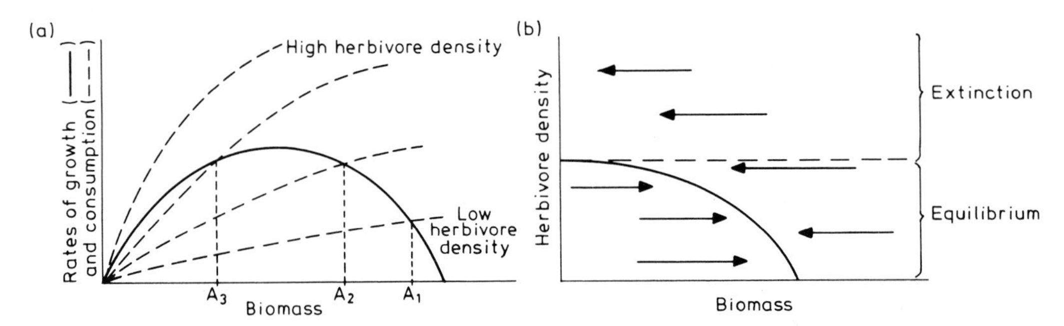

Fig. 5.41. (a) Rates of plant growth and herbivore consumption (at a range of herbivore densities) plotted against plant biomass. A_1, A_2 and A_3 are stable equilibria at which growth and consumption rates are equal. (b) The isocline dividing combinations of plant biomass and herbivore density which lead to biomass increase (arrows left to right) from combinations which lead to biomass decrease (arrows right to left). (After Noy-Meir 1975.) For further discussion, see text.

We now have two families of curves; and they can be used to model the outcome of grazing by superimposing them on one another, because the difference between the rates of growth and consumption gives the *net* change in the growth-rate of vegetation biomass. In Figs. 5.41a, 5.42a, 5.43a and 5.44a, the growth- and consumption-rate are plotted against biomass for four different sets of conditions. These allow the net growth-rates to be inferred for various combinations of herbivore density and biomass, and this information is summarized in the corresponding Figs 5.41b, 5.42b, 5.43b and 5.44b. These illustrate the positions of the biomass zero isoclines, separating circumstances of positive and negative biomass growth (see Fig. 4.22).

The simplest grazing model is shown in Fig. 5.41. For each type 2 herbivore consumption curve (Fig. 5.41a), biomass increases below point 'A' (because growth exceeds consumption), but decreases above point 'A' (consumption exceeds growth). Each point 'A' is, therefore, a stable equilibrium. As herbivore densities get higher, however, (and *total* consumption increases) the level of stable biomass gets lower (shown by the shape of the isocline in Fig. 5.41b), and at herbivore densities exceeding some critical value, consumption is greater than growth for all levels of biomass and the plant population is driven to extinction; this is obviously not unreasonable.

A slightly more complex situation, in which the

Fig. 5.42. Similar to Fig. 5.41 except that in (a), as a result of altered consumption-rate curves, there is an unstable equilibrium at intermediate herbivore densities, B_3; and in (b) there is a shaded area close to the isocline, indicating combinations of plant biomass and herbivore density at which minor changes in either could alter the outcome from stable equilibrium to plant decline and ultimate extinction, or vice versa. For further discussion, see text.

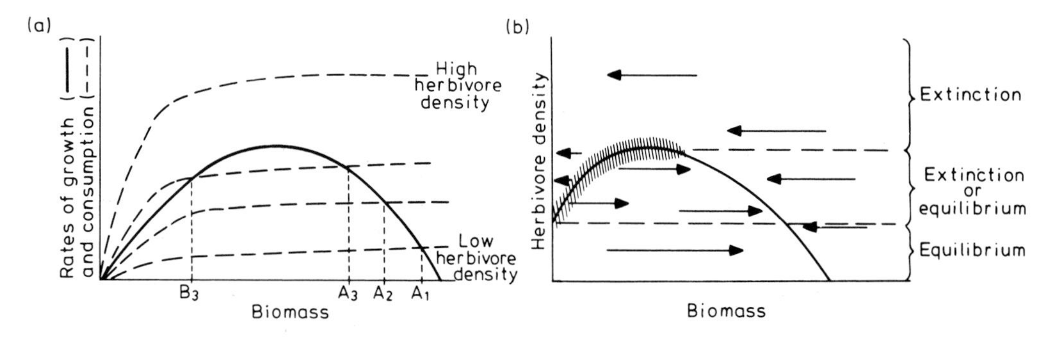

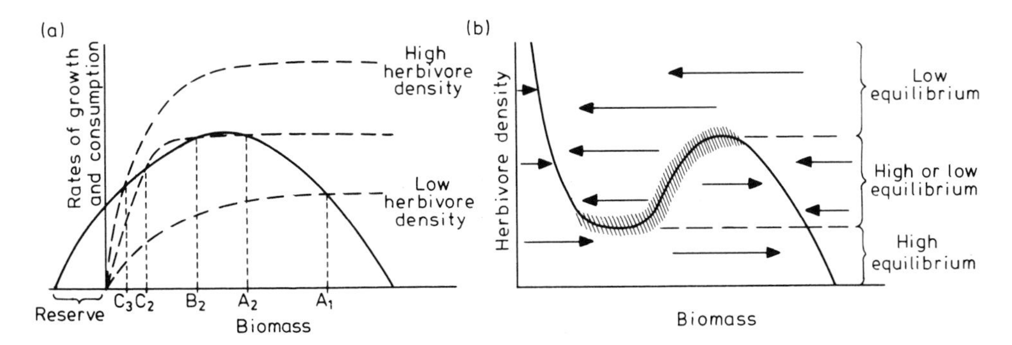

Fig. 5.43. Similar to Fig. 5.42 except that there is a 'reserve' of ungrazable plant biomass. As a consequence there are in (a) low-biomass stable equilibria at high and intermediate herbivore densities (C_2 and C_3). The shaded area in (b), therefore, indicates combinations of plant biomass and herbivore density at which minor changes in either could alter the outcome from low-biomass stable equilibrium to high-biomass stable equilibrium, and vice-versa. For further discussion, see text.

consumption curve reaches saturation more suddenly, is shown in Fig. 5.42. The outcome is unchanged at low herbivore densities (equilibrium) and at high herbivore densities (extinction); but at intermediate densities, the consumption-rate curve crosses the growth-rate curve twice (points 'A' and 'B'). As before, biomass decreases above point 'A' (because consumption exceeds growth), but this is now also true *below* point 'B'; while between points 'A' and 'B', growth exceeds consumption, and biomass increases. Point 'A' is, therefore, still an essentially stable equilibrium, but point 'B' is an unstable *turning point.* A biomass slightly less than 'B' will decrease to extinction, driven by over-consumption; a biomass slightly greater than 'B' will increase to the stable equilibrium at point 'A'. As Fig. 5.42b makes clear, then, there is an intermediate range of herbivore densities at which *either* equilibrium *or* extinction is possible, and small changes in herbivore density or vegetation biomass occurring close to the isocline in this region (hatched in Fig. 5.42b) can obviously have crucial effects on the outcome of the interaction.

In Figs 5.43 and 5.44 two even more complex situations are illustrated, but in general terms the outcomes are the same in both. Fig. 5.43 represents a plant population that maintains a *reserve* of material which is not accessible to grazers (underground storage organs, for example, or plant parts which are inedible). The origin of the growth-rate curve is, therefore, displaced to the left of the origin of the consumption-rate curve. Fig. 5.44 represents a herbivore population exhibiting a 'type 3' functional response. In both cases, the outcome is unchanged at low herbivore densities; there is a single, stable equilibrium (point 'A').

At high herbivore densities, however, the plant

Fig. 5.44. Similar to Figs 5.41 and 5.42 except that the herbivores exhibit a sigmoidal, type 3 functional response. In other respects the figure is like Fig. 5.43. For further discussion, see text.

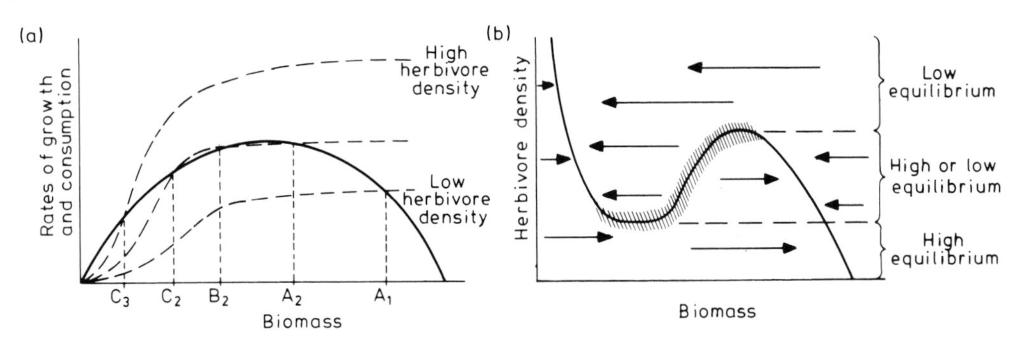

population is no longer driven to extinction (ungrazable reserves and type 3 responses *stabilize* the interaction). Instead there is a point, 'C_3', at which the two curves cross, such that growth exceeds consumption below C_3, while consumption exceeds growth above it. These high-density cases are, therefore, *also* characterized by a single, stable equilibrium. However, while the low herbivore density equilibria were maintained at a high biomass by the self-regulatory properties of the plants, these high herbivore density equilibria are maintained at a *low* biomass by either the ungrazable reserves of the plants or the functional response of the herbivore.

Moreover, at intermediate herbivore densities in Figs 5.43a and 5.44a, the consumption curve crosses the growth curve at *three* points ('A_2', 'B_2' and 'C_2'). As before, point 'A' is a high biomass stable equilibrium, point 'C' is a low biomass stable equilibrium, and point 'B' is an unstable turning point. Populations in the region of point 'B', therefore, might increase to point 'A' or decrease to point 'C', depending on very slight, perhaps random changes in circumstances. These plant–herbivore systems have *alternative stable states*; and a small change in the size of either the plant or the grazer population in the hatched regions of Figs 5.43b or 5.44b can shift the system rapidly from one stable state to the other. The crucial point, therefore, is that systems with ungrazable reserves or type 3 functional responses may undergo sudden drastic changes in population levels, which are nonetheless grounded in the regulatory dynamics of the interaction.

It is difficult to evaluate this model's predictions critically, because much of the data on grazing is not amenable to this form of analysis. Yet the available evidence for Australian pastures certainly suggests that combinations of observed growth and consumption curves may well lead to situations with alternative stable states (Fig. 5.45); and as Noy-Meir (1975) has argued, the predictions of the model are borne out by the experiences of agronomists and range managers. In particular, one practice often recommended in the management of grazing stock is to delay the start of grazing until a minimum of vegetation growth has been exceeded. It is clear that for a given herbivore density this will tend to prevent extinction in Fig. 5.42, while in Figs 5.43 and 5.44 it will favour a high-biomass equilibrium, where the plant growth-rate and thus the food available to the grazers will also be high.

Finally, it must be admitted that this 'one plant/one grazer' model is an unrealistic abstraction of the real, multi-species world. Nevertheless, it provides a simple and effective framework for the comparison and analysis of plant–herbivore dynamics, and also indicates the complex, indeterminate, multi-equilibrial behaviour of predator–prey systems generally.

5.14 'Patterns of abundance' reconsidered

This chapter began, after some preliminary definitions, with an outline survey of the patterns of abundance associated with predation; and, since 'predation' was defined so inclusively, this was effectively a survey of

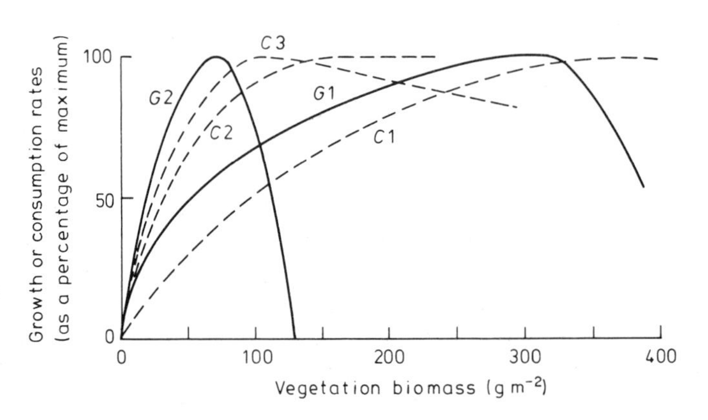

Fig. 5.45. Experimentally obtained curves of growth, G, (ryegrass and subterranean clover) and consumption, C, (by sheep), standardized to maximum rates of 100%. (After Noy-Meir 1975.)

abundance patterns generally. Now, having examined many of the components of 'predatory' interactions, we ought to be in a position to reconsider these patterns. But we must remember that no 'predator–prey system' exists in a vacuum: both species will also be interacting with further species in the same or different trophic levels. Thus, Anderson (1979) considered that parasites probably often play a *complementary* role in the regulation of host populations, increasing their susceptibility to predation (though evidence for this is mainly anecdotal). Similarly, Whittaker (1979) concluded that quite moderate levels of grazing will often produce a severe effect on a target plant when (and because) they are combined with (otherwise ineffective) interspecific plant competition. Lawton & McNeill (1979) concluded that plant-feeding insects stood between the 'devil' of their natural enemies, and the 'deep blue sea' of '. . . food that, at best, is often nutritionally inadequate and, at worst, is simply poisonous'. Finally, Keith (1963) suggested— because of hare cycles on a lynx-free island— that the (supposedly) classical snowshoe hare/Canada lynx predator–prey oscillations (Fig. 5.1a) are actually a result, at least in part, of the hare's interaction with its food. It is clear, in short, that to truly understand the patterns of abundance exhibited by different types of species, we must consider, synthetically, all of their interactions.

Nevertheless, it is apparent from this chapter that we have at our disposal a wealth of *plausible* explanations for the patterns of abundance that occur. Individually, each of the components can give rise to a whole range of abundance patterns, and in combination their potentialities are virtually limitless. What we lack, regrettably, is the sort of detailed field information that would allow us to decide which explanations apply in which particular situations. We are forced, simply, to conclude—and this is especially apparent in the mathematical models—that observed patterns of abundance reflect a state of dynamic tension between the various stabilizing and destabilizing aspects of the interactions.

Beyond this, we can point to food aggregation, and the aggregative responses of consumers to this aggregation, as probably the single most important factor stabilizing predator–prey interactions. Indeed, Beddington *et al.* (1978) noted that the biological control of insect pests is characterized by a persistent, strong reduction in the pest population following the introduction of a natural enemy (i.e. stability at low q); and they suggested, from the analysis of mathematical models, that the mechanism which is most likely to account for this is the differential exploitation of pest patches in a spatially heterogeneous environment. (Even more important than this is the fact that they were able to marshal several field examples in support of this suggestion.) In brief, it appears that aggregation is of very general significance in maintaining prey populations at stable, low densities.

A further insight, stemming from mathematical models, is that predator–prey systems can exist in more than one stable state. Indeed the model in the previous section is only one of a range that possess such properties (see May 1977, 1979 for reviews), and this range covers predator–prey, host–parasitoid, plant–herbivore and host–parasite interactions. We therefore have a ready explanation for parasite epidemics, pest outbreaks, and sudden, drastic alterations in density generally.

Finally, *in general*, we have seen that very similar principles apply to all those interactions included within the blanket term 'Predation'; and in all cases, there has been considerable progress in understanding the details underlying the relationships. Overall, however, it must be admitted that while there is nothing in the observed patterns of abundance that should surprise us by being essentially inexplicable, we are rarely in a position to apply specific explanations to particular sets of field data.

5.15 Harvesting

In this section, we are concerned with examining the dynamics of populations that regularly suffer loss of individuals as a consequence of the deliberate attentions of mankind, through cropping or harvesting. In all predator–prey interactions the predator will profit by maximizing the crop taken while ensuring that the prey does not become extinct; and we have already seen, in this chapter, that in natural populations the survival of a certain number of prey individuals will tend to be ensured as a by-product of certain features of the inter-

Table 5.8 Effects produced in populations of the blowfly, *Lucilia cuprina*, by the destruction of different constant percentages of emerging adults. (After Nicholson 1954.)

Exploitation-rate of emerging adults	Pupae produced per day (a)	Adults emerged per day (b)	Mean adult population (c)	Mean birth-rate (per individual per day) (a/c)	Natural adult deaths per day	Adults destroyed per day (d)	Accessions of adults per day (e = b−d)	Mean adult life-span (days) (c/e)
0%	624	573	2520	0.25	573	0	573	4.4
50%	782	712	2335	0.33	356	356	356	6.6
75%	948	878	1588	0.60	220	658	220	7.2
90%	1361	1260	878	1.55	125	1134	126	7.0

action. In the instances involving man as the harvester, however, the problem of *conscious* harvest optimization remains, i.e. the problem of ensuring neither *over*-exploitation (hastening extinction) nor *under*-exploitation (cropping less than the prey population can sustain). It is with this in mind that many ecologists have devoted considerable attention to the manner in which plant and animal populations can be exploited for the benefit of mankind.

5.15.1 *Characteristics of harvested populations*

From the outset we can see, from common sense alone, that the first, immediate consequence of harvesting is to reduce the size of the population, and this, in turn, will generally affect the life expectancy and fecundity of the survivors in the harvested population. Nicholson (1954), for instance, cultured Australian sheep blowflies (*Lucilia cuprina*) under conditions that restricted adult food supplies but provided larvae with a food excess (ensuring very little larval mortality). His results (Table 5.8) show that, whilst the adult population declined with increasing exploitation, both pupal production and adult emergence *increased*, resulting in an overall increase in the 'birth' rate. Coincident with this rise, moreover, was a decrease in the rate of natural adult death, and this led to extension of the mean adult lifespan. The reduction in population size resulting from the act of harvesting, therefore, brought about two changes: increased fecundity of surviving adults, and reduced adult mortality; and, indeed, we might have expected this from our knowledge of the effects of intraspecific competition

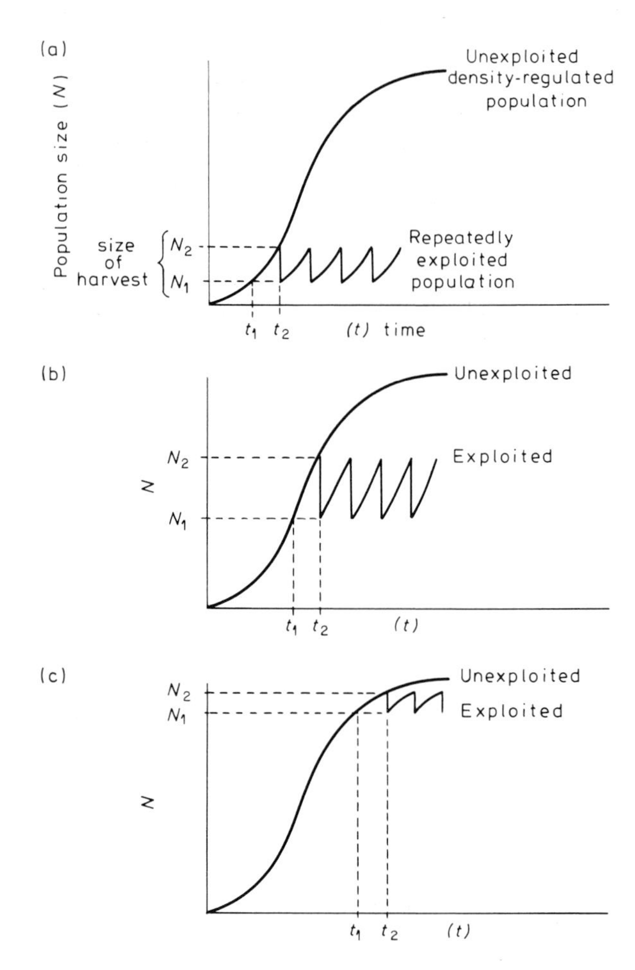

Fig. 5.46. The effects of harvesting at early (a), middle (b) and late (c) stages of population growth with the population growing logistically. For further discussion, see text.

(Chapter 2). Reduction in population density in a resource-limited environment tends to benefit the individual survivors.

The act of harvesting also has an important effect on the *rate of regrowth* of the population. This can easily be seen by considering a population undergoing density-dependent regulation and following a logistic growth curve (Fig. 5.46). At a point in time, t_2 (when the population has reached a size, N_2) we remove some individuals from the population, i.e. we *crop* it. The population will continue to grow, but from the reduced size (N_1) that it had already reached at time t_1. Clearly, the rate of growth of the population after harvesting (the slope of the curve) will depend on the time at which the harvesting occurred. If removal takes place early (Fig. 5.46a), then the rate of subsequent growth will be low, and indeed reduced by harvesting. If removal takes place late (Fig. 5.46c), then the rate of growth, though increased, will also be low. If, however, it takes place when the population is growing most rapidly (Fig. 5.46b), then the rate will be high and largely unchanged by the removal. Moreover, the population size prior to harvest and the rate of subsequent regrowth are also related to the *size* of repeated harvests taken after a unit period of recovery (Figs 5.46 and 5.47). At low population sizes, successive harvests are small (Fig. 5.47a), but with increasing population size they increase steadily to a maximum and then decline; and as Fig. 5.47b shows, there is a parabolic relationship between harvest size and population size. There is, therefore, an *optimal* size at which a population can be maintained, which, on repeated harvesting, ensures a *maximum sustainable harvest*; and repeated harvesting at this size is also followed by the rapid recovery of the population because its rate of regrowth is maximal. Note that in the case of Nicholson's blowflies (Table 5.8), harvest-size ('adults destroyed per day') continued to rise as the exploitation level increased and the mean population

Fig. 5.47. (a) Growth increments in unit periods of time at different stages of growth of a population growing logistically. (b) The parabolic relationship between these increments (or, alternatively, the size of repeated harvests) and population size.

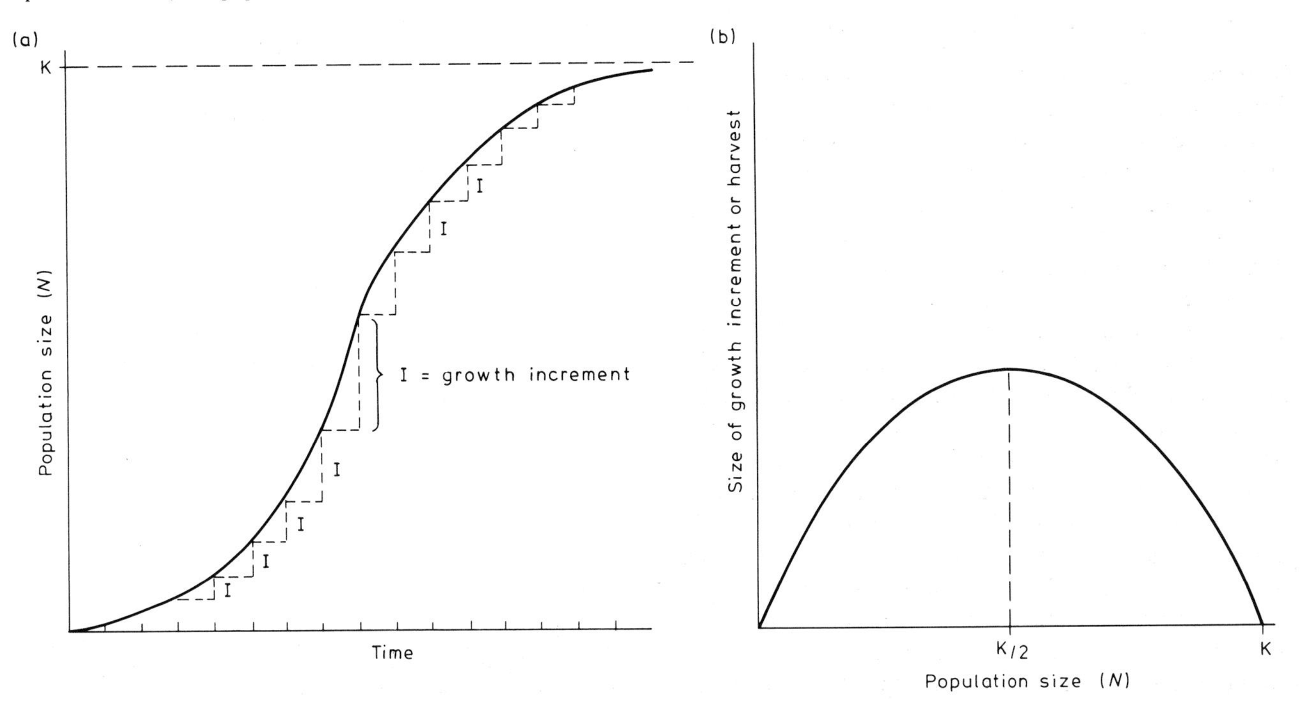

size decreased. There was also a consistent rise in the rate of regrowth (compounded from an increased number of 'adults emerged' and a decreased rate of 'natural adult death'). It seems, therefore, that these populations were towards the right-hand side of the appropriate parabola, where intraspecific competition was fairly intense. For the logistic equation itself, the optimum population size is in fact $K/2$ (where K is the carrying capacity) and we can show this mathematically by differential calculus. (Readers familiar with calculus will know that 'finding the maximum of a function' involves differentiating the appropriate equation—the logistic in this case—setting the derivative to zero, and solving for N.) The important biological point to realize, however, is that the maximum sustainable harvest is not obtained from populations at the carrying capacity, K, but from populations at lower, intermediate densities, where they are growing fastest. The exact shape of the *harvest parabola* depends on the growth function, and the harvest parabolas of different sigmoidal growth curves will vary accordingly. The 'logistic' parabola is symmetrical because the logistic curve itself is symmetrical.

Some justification for this type of model is provided by studies of harvested fish populations. Since it is impossible to estimate the sizes of oceanic fish population numerically, stock sizes (or total biomass) can be used instead. The equation describing the changes must be:

$$S_{\text{year 2}} = S_{\text{year 1}} + Rec + G - D - C$$

where S is stock size, and Rec is recruitment of stock, G growth, D loss by natural mortality and C stock caught by fishing. If stock size remains constant from one year to the next then:

$$Rec + G = D + C.$$

In other words, at *any* stock size, if the stock is in a *steady state*, then the gains from recruitment and growth will be exactly offset by the losses from natural mortality and fishing. So, if the stock is growing logistically, and we wish to *maintain steady-state conditions*, then we must exploit the stock according to the harvest parabola. Fig. 5.48a shows the catch sizes of yellow fin tuna from 1934 to 1955 plotted against fishing intensity (a measure recorded by the fishing industry that may easily be related to stock size). Interestingly, the actual recorded catches do seem to lie on the estimated harvest parabola; and indeed the size of the catch from 1934 to 1950 was increased by 100 million pounds *without* destroying the fish population, and from 1948 onwards large catches *could* be sustained. A fairly similar situation is shown for a lobster fishery in Fig. 5.48b. It does appear, therefore, that at least some fisheries are regulated in a density-dependent and possibly even logistic fashion.

It would be gratifying to be able to claim that the management and conservation of all exploited fish populations are based on detailed knowledge and careful modelling. Yet, while substantial research in the fishing industry has deepened our understanding of the factors influencing the stability and yield of exploited populations (Beverton & Holt 1957; Gulland 1962), for three main reasons we are still unable to fully comprehend their dynamics. The first reason is a genuine ignorance of population dynamics: obtaining the data for long-lived species demands extensive and often expensive study over many seasons. The second reason is that many populations are structured, and we shall consider this problem presently. The third reason is environmental variability. The environment, particularly climate, frequently affects natural populations independently of density, and environmental fluctuations may, therefore, create considerable difficulty in the evaluation of the effects of harvesting on population size. This is illustrated by data on the Pacific sardine, *Sardinops caerulea*, off the coast of California, where the annual catch exhibits extreme fluctuations (Fig. 5.49a). There has been considerable controversy over the relationship between the population sizes of (i) the 'spawning stock', (ii) the mature breeding fish (2 years old and over) and (iii) the 2-year-old recruits; and although Murphy (1967) has established that the number of recruits is partly determined by the water temperature when the fish were spawned (Fig. 5.49b), there are other environmental variables that play a highly significant role. The movements of ocean currents within the coastal nursery grounds, for instance, are often seasonally erratic, and young planktonic fish may be swept into the deep ocean where they are unlikely to survive. Temperature, currents and other factors, then, are all likely to blur any relationship between the size of the breeding population and the number of offspring born to it. However, Radovich (1962) has argued that the recruitment relationship is in fact curvilinear with a recognizable optimum. Thus, in Fig. 5.49c three curves have been drawn: II gives the best statistical fit to the data, while I and III have been drawn by inspection to encompass the variability observed. I represents the relationship we might expect when environmental factors combine to give the 'best' possible conditions, and III the relationship for the 'worst' conditions. The clear consequence of this underlying curvilinearity is that recruitment tends to be little affected by changes in spawning stock size near the optima, but is more drastically affected at the extremes. The utility of these curves stems from the ability they give us to assess the levels of fishing that safeguard against extinction even under the poorest circumstances. If for instance we wish to ensure a consistent recruitment X (Fig. 5.49c), then the bare minimum of stock that must be left after harvesting under the 'average' conditions is S_2. The difference, $S_1 - S_2$, represents the additional stock that should not be harvested if we wish to guarantee recruitment when recruitment is lowest, in the 'poorest' environment. Overall, the problems of incorporating

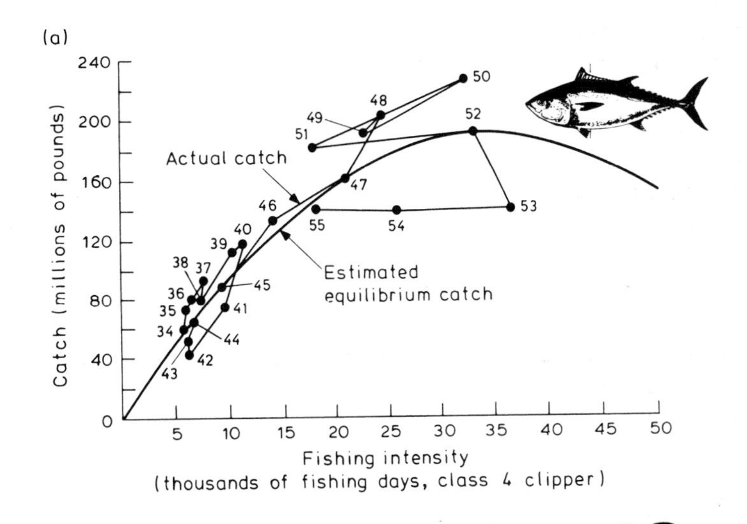

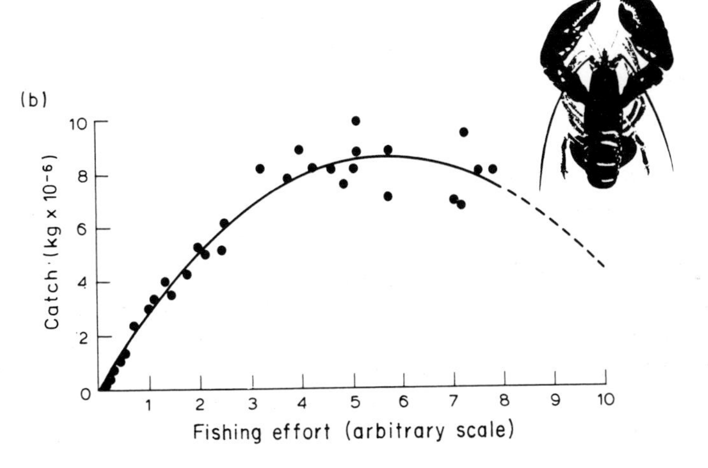

Fig. 5.48. (a) The relationship between fishing intensity and yield of yellow fin tuna in the eastern pacific between 1934 and 1955. The curved line is the estimated harvest parabola (Schaefer 1957). (After Watt 1968). (b) Similar relationship for the western rock lobster *Palinurus cygnus* (Hancock 1979). (After Beddington 1979.)

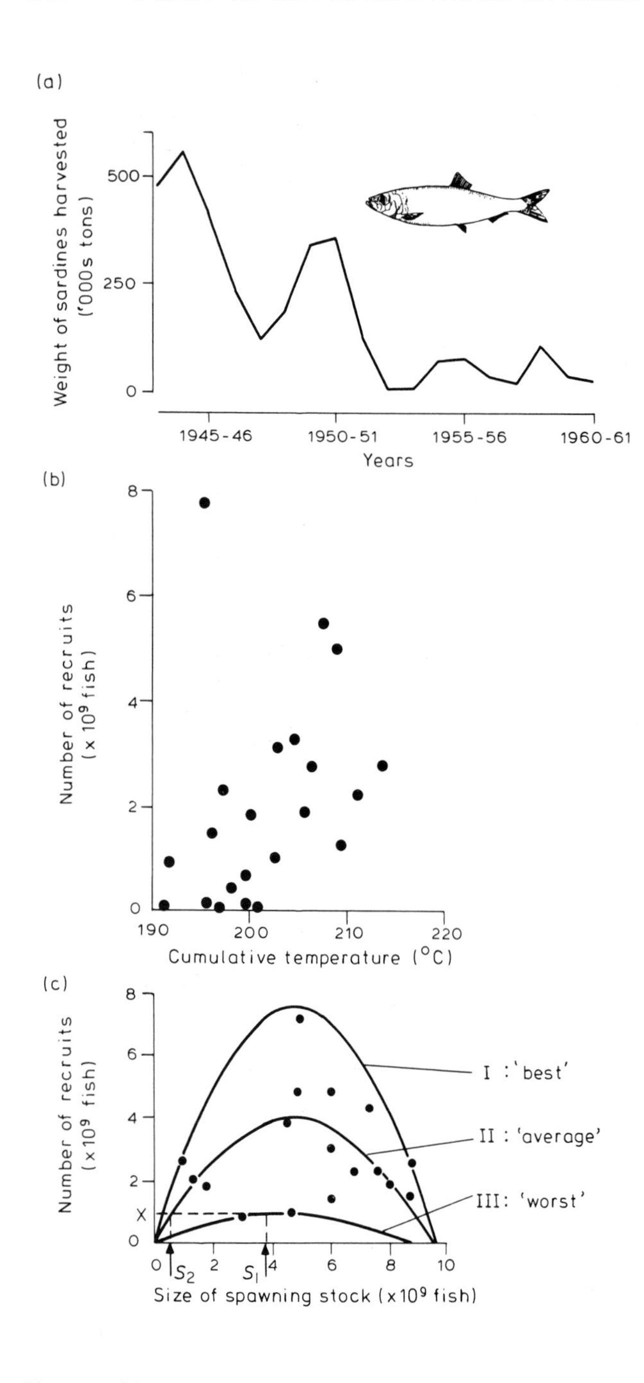

Fig. 5.49. (a) The annual catch of the Californian sardine, and (b) the effect of sea temperature on the recruitment of the sardine (Watt 1968). (c) The average parabola relating the number of recruits to the size of the spawning stock, together with the maximum and minimum parabolas (Radovich 1962). For further discussion, see text. (After Usher 1973.)

environmental variability into a predictive harvesting model are readily apparent (but see Iles (1973) for a more sophisticated attempt to overcome these problems).

We are now in a position to review some of the consequences of harvesting. Firstly, it is quite clear that populations can be systematically exploited with a consequent reduction in population size. It is also clear that harvests can be optimized (i.e. the size of *repeated* harvests *maximized*) by harvesting at some *intermediate* density where the population growth-rate is greatest. However, such harvesting is only sustainable at an exploitation-rate that allows sufficient time for the replacement of cropped individuals. This period of time will depend on the fecundity and generation time of the species in question. It is easy to imagine an intensity of harvesting (an exploitation-rate) in excess of the replacement-rate, such that the population declines to extinction ('stepping down' the curve in Fig. 5.47a). We can also conclude (from the *Lucilia* population, and from a consideration of intraspecific competition) that the fitness of *survivors* in a harvested population is often increased.

5.15.2 *Harvesting in structured populations*

In practice, many harvesting procedures deliberately *select* individuals to be cropped. In fishing, for instance, many nets permit small fry to escape so that large (and possibly older) individuals are captured. In other situations (seal culling for instance) small or young members are the principal object of the harvest. We must therefore consider the age and size composition of populations being exploited.

To investigate the properties of such structured populations in the field under particular harvesting regimes is an immensely arduous task, and, in consequence, many workers have utilized relatively simple laboratory systems. A good example is the work of Slobodkin & Richman (1956) using water fleas (*Daphnia pulicaria*), which breed continuously throughout the year and possess a prodigious capability for population increase by virtue of the very large numbers of eggs they lay. They are also particularly convenient for harvesting studies because they have a short lifespan; and in their population structure we can recognize young, adolescent and adult forms. Fig. 5.50a,b shows the consequences of taking harvests every 4 days from *Daphnia* populations which were maintained with a constant supply of food at a constant temperature for about 9 months. Harvesting was aimed specifically at the smallest-size class which (apart from a few young adolescents) contained *Daphnia* which were less than 4 days old. The population structure in the absence of harvesting is shown in Fig. 5.50a, '0%': the numbers per class decline to a minimum at the large adolescent stage

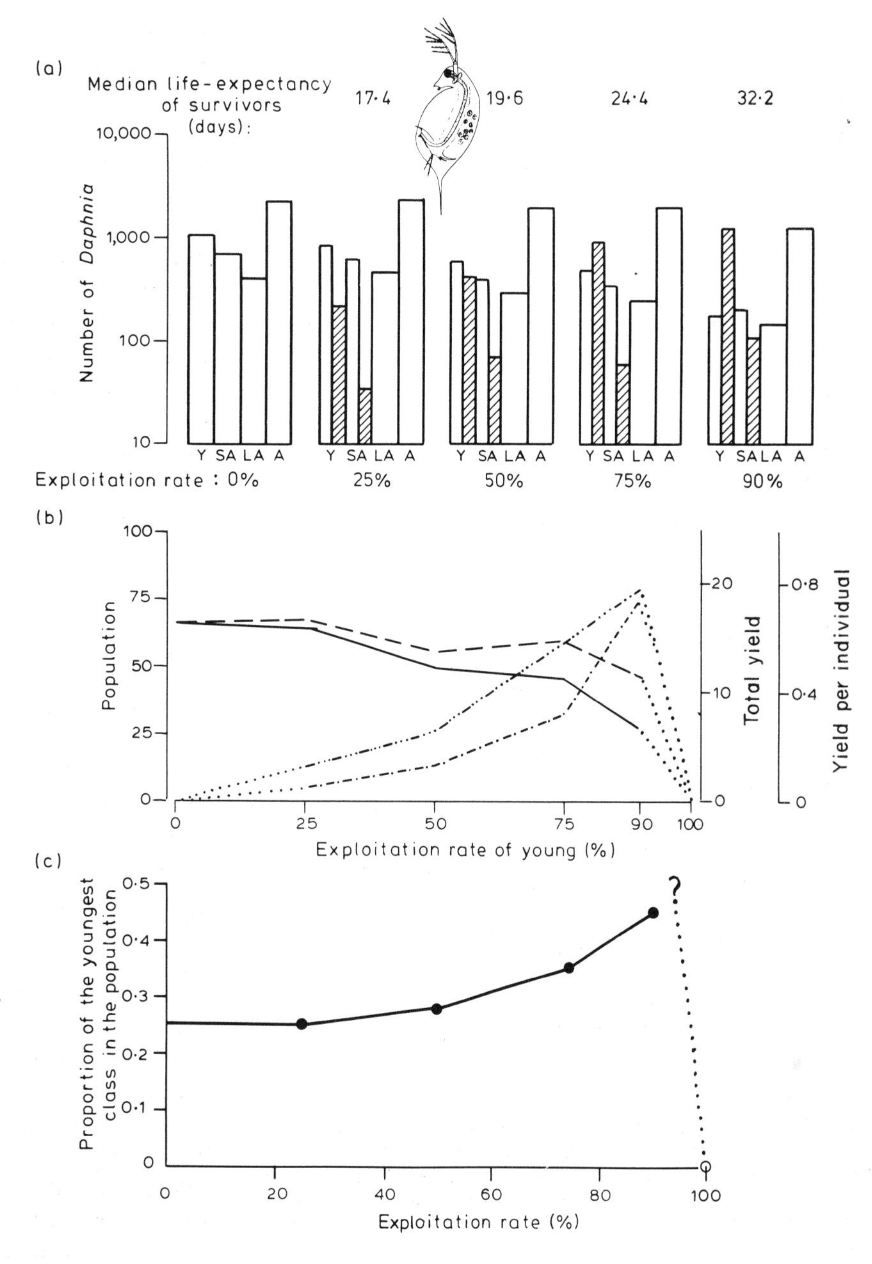

Fig. 5.50. (a) The effect of harvesting on the size structure of a *Daphnia* population, where the size-classes are: Y = young, SA = small adolescent, LA = large adolescent and A = adult. The shaded halves of the columns represent those removed and the open halves those left after harvesting. (b) The effect of harvesting on mean population size before harvest (– – – –), mean population size after harvest (——), total yield (– ··· — ··· –) and yield per individual (– · – · – · –). (c) The effect of harvesting on the proportion of the youngest class in the total population (all Slobodkin & Richman 1956). For further discussion, see text. (After Usher 1974.)

(LA), and the adult class is the most abundant (the data are presented on a *logarithmic* scale). Increasing the proportion of young fleas harvested had two main effects. The first was to consistently reduce the total population size (Fig. 5.50a,b). The second was to alter the population structure (Fig. 5.50a): as the harvesting of small *Daphnia* increased from 25% to 90% of their number, the discrepancy between young, small and large adolescent frequencies diminished. We may note in addition, however, that 90% exploitation did not push the population towards extinction, and indeed, there was a consistent rise in total yield as exploitation-rate increased (Fig. 5.50b). This occurred because the yield *per individual* rose quickly enough to more than compensate for the decline in population size (Fig. 5.50b); and this consistent rise in the fitness of survivors (cf. Nicholson's blowflies) was also reflected by a steady increase in life-expectancy averaged over all size-classes (Fig. 5.50a). Such increases were associated, moreover, with a consistent rise in the proportion in the population of the exploited, youngest class *prior* to harvesting (Fig. 5.50c), i.e. a further indication of the changing population structure. This presumably occurred because exploitation reduced the numbers available to enter the older classes, and also reduced intraspecific competition, leading to an increased *per capita* production of the youngest class such that its *proportion* in the population increased. For the exploitation-rates examined here, these effects apparently more than compensated for the losses due to harvesting, in that there was a consistent increase in yield. Beyond some higher point, however, increases in productivity would be unable to compensate for reductions in numbers, and the population would decline to extinction. In the present case, this point is apparently between 90% and 100% exploitation.

Parenthetically, we should note that once we have accepted that populations are structured, it is important to determine whether 'yield' should be measured in terms of numbers or biomass, since the two are not necessarily equivalent. Fig. 5.51 (Usher *et al.* 1971) illustrates this problem in populations of an invertebrate herbivore, the collembolan *Folsomia candida*. The populations were subjected to regular harvesting over all size-classes at rates of 0%, 30% and 60% of the total population every 14 days, and the animals were always given excess food. Harvesting reduced the numerical size of the *Folsomia* populations far less than it reduced their biomass. Moreover, the crop taken with 30% exploitation yielded a higher biomass than with 60% exploitation, while the actual numbers of arthropods taken were approximately equal in the two regimes.

It is legitimate at this point to ask whether field populations of vertebrates behave in a similar way to these laboratory populations, and, regrettably, studying fish populations in the same detail presents almost intractable experimental problems. Silliman & Guttsell (1958), however, have succeeded in modelling the process of commercial fishing practices, with density-regulated populations of guppies (*Lebistes reticulatus*). Selective fishing, permitting small fish to escape, was applied every 3 weeks (corresponding to the reproductive periodicity of the fish), and the catch-size was carefully controlled but changed after 40-week periods, so that the effects of a range of exploitation from 10% to

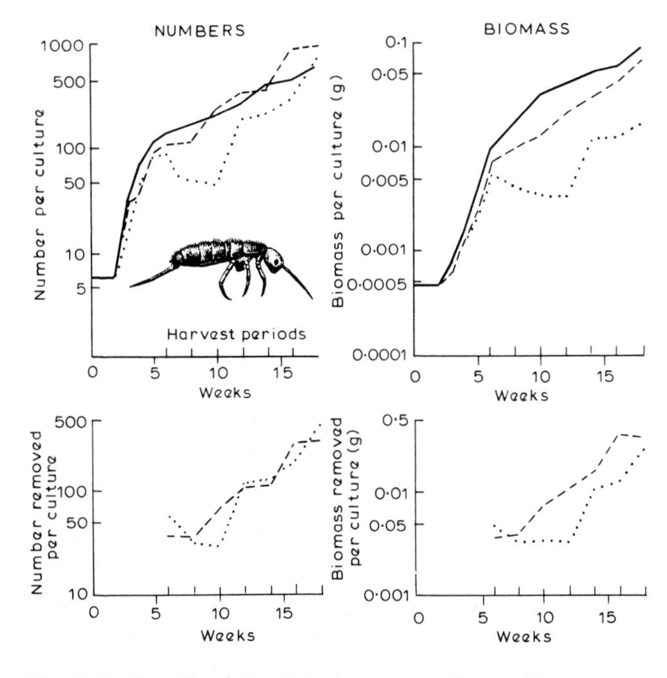

Fig. 5.51. The effects of exploitation on a population of the collembolan *Folsomia candida*; control, 0% exploitation (———), 30% explotation (– – – –), 60% exploitation (····). (After Usher *et al.* 1971.)

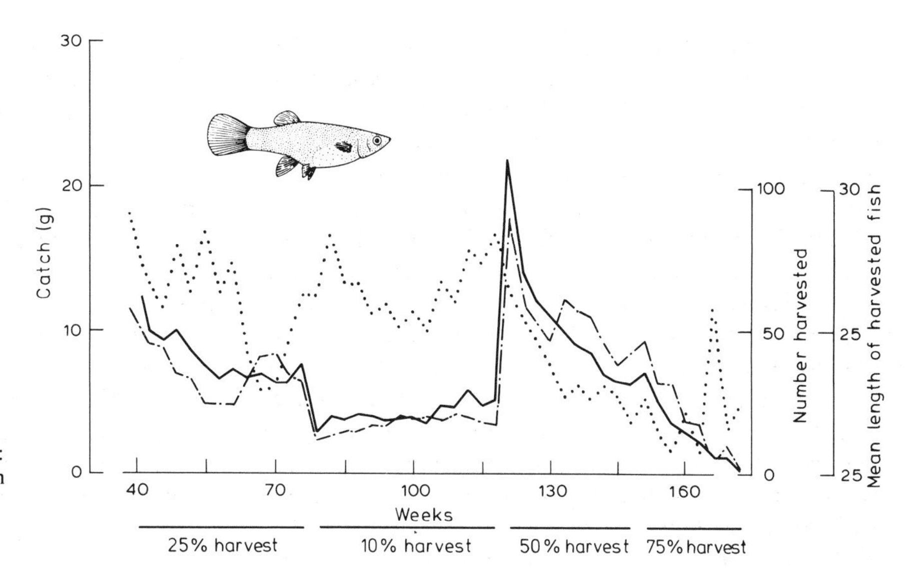

Fig. 5.52. Exploitation of guppy populations: biomass of the catch (———), number of fish caught (–·–·–·–), average length of harvested fish (······). (After Silliman & Gutsell 1958.)

75% could be assessed. Although the results are by no means as clear as those we have examined so far, two points of interests are reinforced (Fig. 5.52). Firstly, as we might expect, the size of the catch, measured as numbers *or* total biomass, was related to the level of exploitation; it was higher (though decreasing slightly) at 25% than it was at 10%, but at 50% it decreased sharply, and at 75% it led to the extinction of the population. Thus, we have confirmation that the maximum sustainable biomass yield must be attained at some *intermediate* level of exploitation—probably close to or just below 25%. The second feature is that harvesting had a discernible effect on both the size-structure of the guppy population (since the mean length of the harvested fish declined noticeably with increasing exploitation), and the age-structure (Fig. 5.53). This, like the data on *Daphnia,* illustrates the general conclusion that the *structure of populations changes* when specific classes are exploited.

The guppy data also illustrate the process of over-exploitation, i.e. of harvesting at a level in excess of the maximum which is sustainable. In contrast to the *Daphnia* data (Fig. 5.50c), there appears to be a consistent *decline* in the proportion of the exploited, adult class between 25% and 75% exploitation (Fig. 5.53). These higher exploitation-rates apparently result in a lowering of the adult proportion to a level that is unable to replenish the fish removed by harvesting. Increases in the fitness of survivors are insufficient to compensate for the decreases in numbers, and extinction would inevitably follow if harvesting at these levels (i.e. over-exploitation) were to continue.

Comparisons between animals as different as *Daphnia* and guppies must obviously be carried out with caution. Nevertheless, it is interesting to note that when a very *juvenile* portion of the *Daphnia* populations is exploited, the maximum sustainable yield is obtained at a very high rate of exploitation (greater than 90%); but

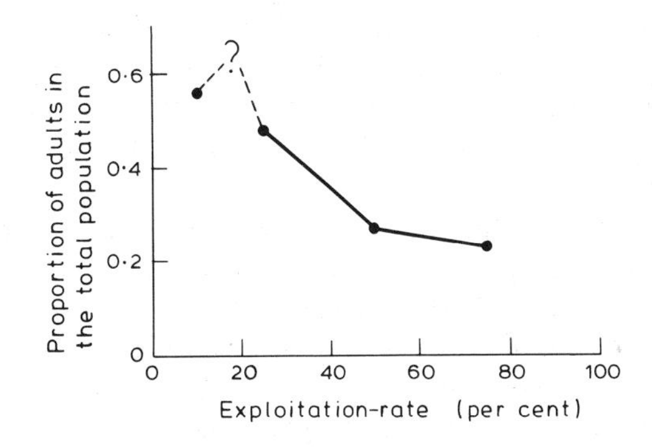

Fig. 5.53. Adult proportions in Silliman & Gutsell's (1958) guppy populations at different rates of exploitation, with the presumed relationship interpolated between the points.

when *adult* guppies are exploited, the peak in sustainable yield occurs at a much lower exploitation-rate. It appears that in a structured population, higher exploitation-rates can be sustained if juveniles are being exploited, because some of these juveniles would fail to reach maturity anyway. To generalize, we might suggest that maximal yields can be obtained from structured populations by removing those individuals that are likely to contribute least to the production of progeny in future.

5.15.3 *Incorporating population structure: matrix models of harvesting*

In the logistic model of harvesting, we did not distinguish between age-classes of individuals; and this may justifiably be seen as a neglect of the significance of population structure and of age-specific fecundity and survival. Age-determined parameters can be incorporated into a model of a harvested population, however, by use of the matrix model from Chapter 3, and Usher (1972) has illustrated the practical application of this approach in an examination of the potential of the blue whale (*Balaenoptera musculus*) for harvesting. He collated data into a transition matrix describing the fecundity and survival of *female* blue whales, but as Table 5.9 shows, the paucity of these data necessitated the structuring of the population into only six 2-year age-classes up to the age of 12 and a single age-class thereafter (12+).

As explained in Section 3.5, the subdiagonal elements of the matrix are measures of the probability of survival from one class to the next (0.77 for each of the first six 2-year periods). Similarly, the lower right-hand corner element of the matrix is the 2-yearly survival-rate for females over 12. Its value of 0.78 gives individuals entering the 12+ class a mean life-expectancy of 7.9 years; female blue whales can live to ages of between 30 and 40 years. The top row of values in the matrix give the fecundity terms for females: they reflect the fact that breeding does not start until the fifth year, and that full sexual maturity does not arrive until females are seven.

Table 5.9 Transition matrix for a population of female blue whales. (After Usher 1972.)

Age class	0–1	2–3	4–5	6–7	8–9	10–11	12+ years
	0	0	0.19	0.44	0.5	0.5	0.45
	0.77	0	0	0	0	0	0
	0	0.77	0	0	0	0	0
	0	0	0.77	0	0	0	0
	0	0	0	0.77	0	0	0
	0	0	0	0	0.77	0	0
	0	0	0	0	0	0.77	0.78

The fecundity terms of the 8–9 and 10–11 year classes are equal to 0.5, because females produce only one calf every 2 years and the sex ratio of the population is 1:1. The value for the 12+ class has been reduced to 0.45 to take account of irregular breeding in older whales.

Using the iterative procedure outlined previously (Section 3.5), we can calculate the net reproductive rate of the population (R) when it has achieved a stable age structure. This equals 1.0072, which, being very close to 1, indicates that the whale population can grow at only a very slow rate. We can now calculate the level of exploitation that the population can withstand, without entering a decline. In percentage terms this is $\{(R-1)/R\}\times 100$, or 0.71%. In other words, the sustainable yield of the population every year is approximately 0.35% of every age-class. Harvesting-rates in excess of this would lead to losses that the species could not counteract unless homeostatic mechanisms acted to alter the fecundity and survival values in the matrix. In view of this, and the intensity with which whaling has been carried out, it is not surprising that blue whale numbers declined significantly in the early 1930s; and the species has been threatened ever since.

Further models, taking into account the effects of differentially exploiting different classes of a structured population, are beyond the scope of this book. The interested reader can consult Law (1979a) and Beddington (1979).

PART 3 SYNTHESES

Chapter 6 Life-History Strategies

6.1 Introduction

The earlier sections of this book have been concerned with the spatial and especially the temporal patterns of abundance in single, and mixed-species populations; and we have seen, particularly in Chapter 1, that these patterns are, ultimately, the consequences of the schedules of fecundity and survivorship exhibited by the individuals in the population. Everything we know about natural selection would indicate that those individuals with the fecundity and survivorship schedules most suited to maximize fitness in their environment will have been favoured in the course of evolution. In this sense, many crucial aspects of these schedules represent life-history 'strategies', exhibited by individuals, that have been favoured by natural selection (though other aspects will simply reflect the immediate environment of the organism). In this chapter we shall examine (or re-examine) the components of these strategies: the age and size at which reproduction begins; the relative 'efforts' devoted to reproduction, growth, survivorship, predator avoidance and so on; the apportionment of reproductive effort between many small or a few large offspring; the distribution of this effort over an individual's life-time; and the diversion of energy to migration or dispersal. ('Reproductive effort' can be defined as the proportion of the available resource input that is allocated to reproduction, though it is often assessed by some index such as 'gonad weight : body weight'.) These components of the strategies are, hopefully, of interest in themselves as parts of each organism's general biology. But they are also the building blocks with which 'Population Ecology' is constructed. Each individual, each population and each species must, clearly, have its own unique life-history strategy. We shall be trying to discern pattern within this multiplicity.

6.2 Allocation of energy

6.2.1 *The necessity for compromise*

It is not difficult to describe a hypothetical organism with exceptionally high (not to say maximal) fitness. It reproduces almost immediately after its own birth; it produces large clutches of large, protected offspring on which it lavishes parental care; it does this repeatedly and frequently throughout a long life; and it outcompetes its competitors, avoids its predators and catches its prey. But while easy to describe, such an organism is rather more difficult to imagine. This is essentially because an organism putting the maximum amount of the energy at its disposal into reproduction cannot also put maximal energy into survivorship; and an organism which takes great care of its offspring cannot also continue to produce large numbers of them. In other words, common sense alone tells us that each real organism's life-history strategy must, to some extent, be

Table 6.1 Average palatability to slugs, *Arion* and *Agriolimax*, of herbs from various successional stages. (From Cates & Orians 1975, after Southwood 1976.)

Plant community	Number of plant species tested	Palatability index* for *Arion*	Palatability index* for *Agriolimax*
Early successional annuals and biennials	18	0.99	0.96
Early successional perennials	45	0.69	0.77
Later successional and climax plants	17	0.40	0.46

*Palatability index is defined as log (amount of test material eaten)/log (amount of control eaten).

a compromise. Support for this notion is provided, for instance, by the data in Table 6.1 (Cates & Orians 1975). It is the plants that come later in succession, and put relatively little effort into early, explosive reproduction, that make a significant investment in energetically expensive, quantitative defensive chemicals, which make them relatively indigestible. Similarly, Cody (1966) noted that birds made greater reproductive efforts (produced larger clutches) in habitats (like islands, for instance) that were relatively free from competitors and predators (and therefore demanded relatively little effort for 'overcoming' competitors and predators).

The necessity for life-history compromise is more strongly supported, however, by data that show that an *individual* devoting increased energy to one aspect of its strategy must pay for this by devoting less to some other aspect. Fig. 6.1, for instance, shows that if female fruit flies, *Drosophila subobscura,* (many of which fly considerable distances in the field) are subjected to a simulated migration, then their subsequent fecundity is immediately depressed, and it never quite returns to the levels exhibited by non-migrant 'controls' (Inglesfield 1979).

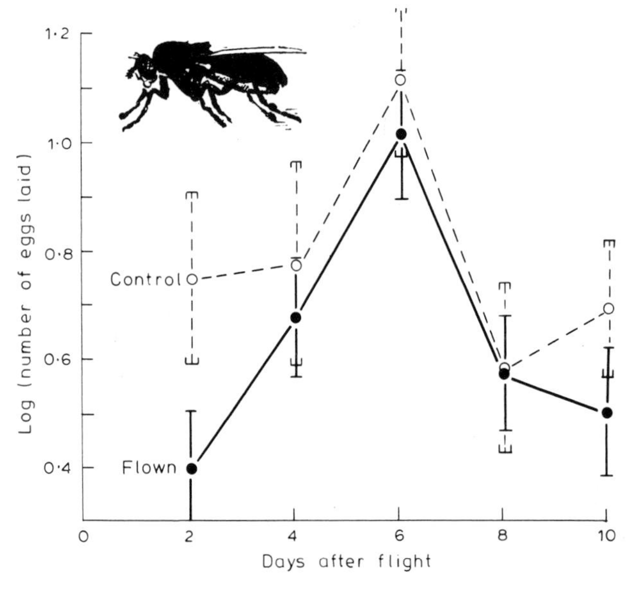

Fig. 6.1. The effect of an induced flight on the fecundity of female fruit flies, *Drosophila subobscura* (plus standard errors). (After Inglesfield 1979.)

6.2.2 *The cost of reproduction*

The most important aspect of life-history compromise, however, is illustrated in Fig. 6.2, which summarizes data from a variety of sources indicating that there is a particularly apparent cost involved in reproduction itself. Fig. 6.2a,b are both examples from mammals in which there is an increase in mortality amongst adults during and after reproduction. But the effects are most marked amongst those organisms that continue to grow throughout pre-mature and mature adult life. Fig. 6.2c shows that in annual meadow grass (*Poa annua*) the price of high reproduction rates early in life is smaller plant size subsequently (and also a reduction in reproductive-rate and survivorship; Law 1979b). A negative correlation between future survivorship and current fecundity is also shown for the predatory rotifer *Asplancha brightwelli* in Fig. 6.2d (Snell & King 1977). While Fig. 6.2e,f shows that in the barnacle *Balanus balanoides*, and Douglas fir, increased current fecundity leads to decreased rates of growth and therefore decreased potential future fecundity (Barnes 1962; Eis *et al.* 1965). It is clear, in short, that *by increasing its current reproductive effort (or by reproducing at all) an individual is likely to decrease its survivorship and/or its rate of growth, and therefore decrease its potential for reproduction in future.* That this results from individuals having only limited total amounts of resources at their disposal is emphasized by the work of Lawlor (1976) on the common pill-bug, *Armadillidium vulgare.* As Table 6.2 shows, despite an overall minor increase in growth-plus-reproduction energy-allocation on the part of reproductive females (resulting, perhaps, in decreased survivorship), the major effect of reproduction is to *divert* a considerable quantity of energy away from

Fig. 6.2. The cost of reproduction. (a) Increased survival-time (with standard deviations) in non-breeding female field voles, *Microtus agrestis* (Clough 1965). (b) Increased rates of mortality following increased rates of birth in female red deer, *Cervus elaphus* (Lowe 1969). (c) The effect of increased breeding in year 1 in *Poa annua* on plant size reduction in year 2 (rank correlation probability < 0.001; Law 1979b). (d) Increased fecundity leading to a decreased probability of survival in the rotifer *Asplancha brightwelli* (Snell & King 1977). (e) Reproduction leading to decreased growth in the barnacle, *Balanus balanoides* (Barnes 1962). (f) Increased reproduction associated with reduced annual growth increment in Douglas fir trees (*Pseudotsuga menziesii*) (Eis *et al.* 1965).

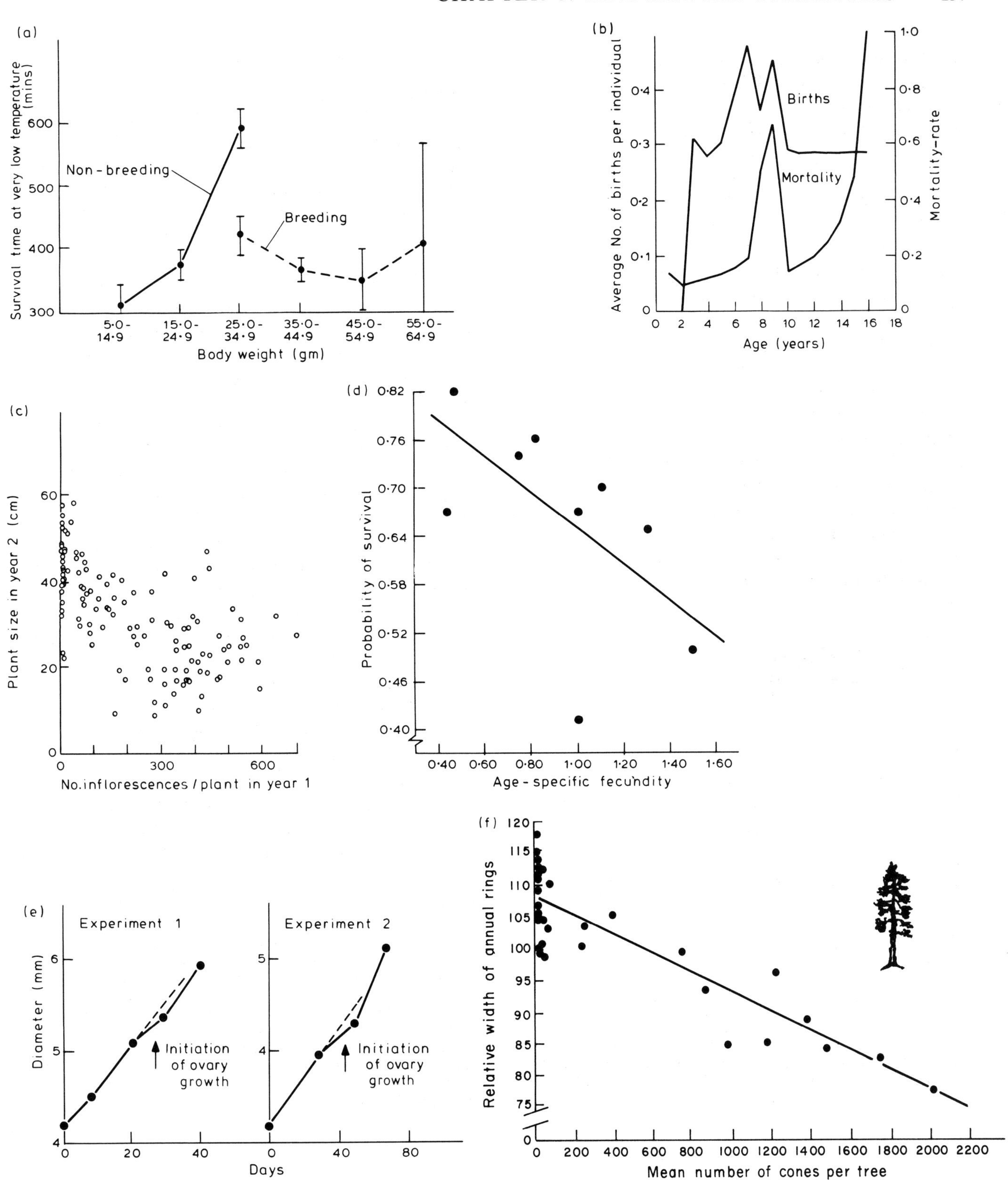
(a)
Survival time at very low temperature (mins)
Non-breeding
Breeding
Body weight (gm)
5·0-14·9
15·0-24·9
25·0-34·9
35·0-44·9
45·0-54·9
55·0-64·9
(b)
Average No. of births per individual
Births
Mortality
Mortality-rate
Age (years)
(c)
Plant size in year 2 (cm)
No. inflorescences / plant in year 1
(d)
Probability of survival
Age-specific fecundity
(e)
Experiment 1
Experiment 2
Diameter (mm)
Initiation of ovary growth
Days
(f)
Relative width of annual rings
Mean number of cones per tree

Table 6.2 Estimated energy allocated by *Armadillidium vulgare* to growth and reproduction. Values are in calories expended during one complete moult cycle. (From Lawlor 1976.)

	Size class (♀ in mg)			
	Reproductive		*Non-reproductive*	
	25–59	60–100	20–59	60–100
Growth	10.0	11.9	24.1	30.5
Reproduction	16.0	26.4	—	—
Total production	26.0	38.3	24.1	30.5

growth. The importance of reduced growth stems from the existence of a strong positive correlation between size (past growth) and clutch size (Fig. 6.3, Paris & Pitelka 1962).

Overall we can see that the strategy adopted by an animal or plant is a compromise allocation of energy to the various aspects of its life-history, each of which contributes to total fitness. The result is a co-adapted

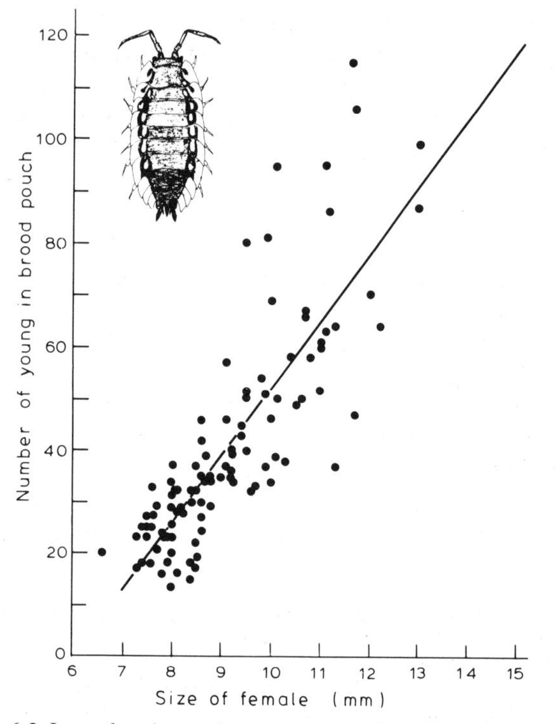

Fig. 6.3. Large females produce a greater number of young in the terrestrial isopod *Armadillidium vulgare*. (After Paris & Pitelka 1962.)

'suite' of characteristics which natural selection has favoured. Two crucial questions arise from this:

(a) Are there discernible patterns in these suites, in which certain characteristics are generally associated with certain others? and

(b) Are there discernible patterns in which particular types of life-history strategy are associated with particular types of habitat?

6.3 The effects of size

In searching for patterns in the construction of life-history strategies, however, it must be recognized, firstly, that there are certain traits which do not result simply from the effects of selection acting on life-histories; and secondly that these traits themselves exert causal influences on other life-history characteristics. The most important such trait is size.

Size is undoubtedly to some extent an inherent property of particular grades of organization (insects are small, etc.). Yet there is a strong positive correlation between the size of an organism and its generation time (Fig. 6.4a; Bonner 1965), and an equally strong negative correlation between its size and its intrinsic rate of increase (r) (Fig. 6.4b; Fenchel 1974). Size itself, therefore, has an important influence on life-history traits, and a plausible mechanism through which this influence might act has been suggested by Southwood (1976). As size decreases, the rate of metabolism per unit weight increases (Fig. 6.4c; Fenchel 1974) leading to a decrease in longevity. This inevitably leads to a decrease in generation-time, and thus to an increase in r. And while this explanation cannot possibly be all-encompassing, there is enough truth in it to warn us against a comparison of the life-histories of an elephant and a fly couched simply in terms of strategic adaptations to different habitats. Size, and its *physiological* correlates can exert major influences on life-history strategies.

6.4 Habitat classification: the organism's view

Southwood (1977) saw the search for patterns in the way life-histories are suited to habitats as an attempt by

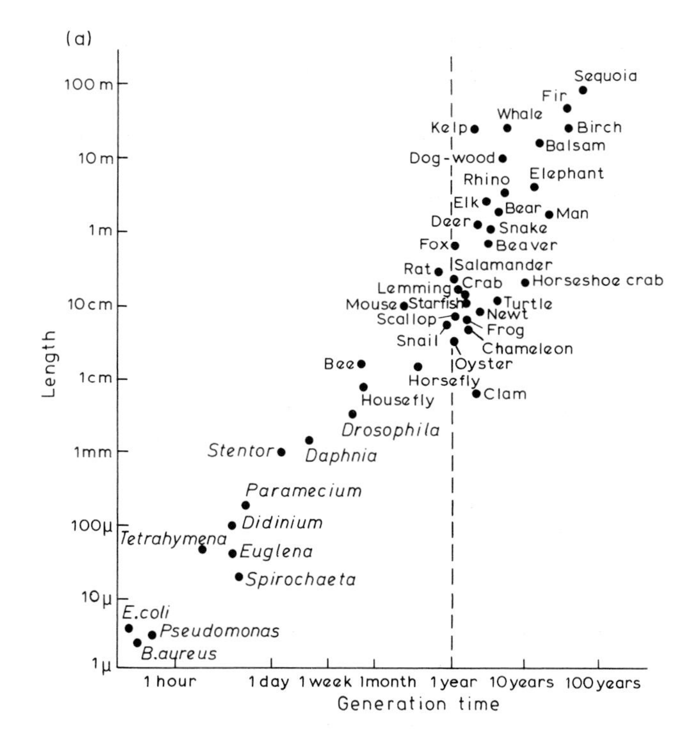

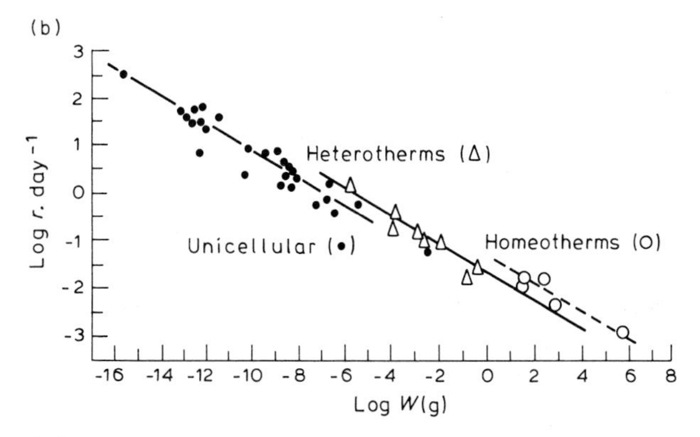

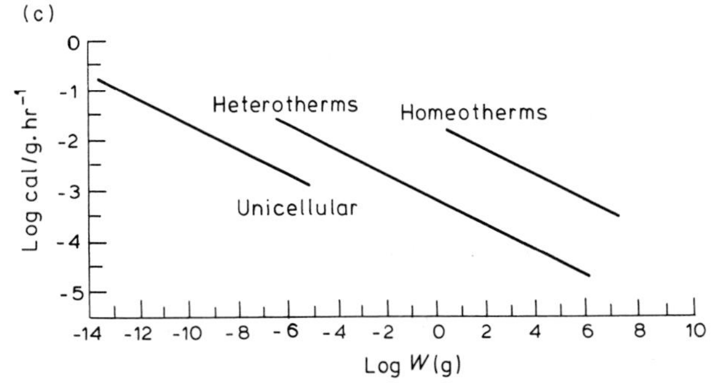

Fig. 6.4. (a) The positive correlation between size and generation time (Bonner 1965). The negative correlations between weight (size) and (b) the intrinsic rate of increase, (r) and (c) the metabolic rate in various animals (Fenchel 1974). (After Southwood 1976.) (Note when considering (a) and (b) that generation time is itself inversely related to r: for a given reproductive rate per generation, r—measured in time rather than generations—will increase as generation time decreases.)

ecologists to construct their equivalent of the chemists' Periodic Table. For the eighteenth-century chemist, '. . . each fact had to be discovered for itself and each had to be remembered in isolation', and a similar fate may await the ecologist unless he can discern, in outline at least, the way in which an organism's habitat acts as a template for its life-history strategy. Of course, if this is to be done, then despite the fact that each organism's habitat is essentially unique, habitats must be classified in terms that apply to them all. Moreover, these terms must not describe the way in which we ourselves see a habitat. Forest fires, for instance, although they appear to *us* to occur unpredictably and irregularly, occur far less unpredictably within the time-frame of the forest trees. Similarly, consider the generalist, fruit-eating orang-utan with a spatial range encompassing many types of tree from which it feeds relatively indiscriminately: its total habitat is relatively predictable and stable. For a fruit-fly larva, however, the 'same' habitat has very different properties. Oviposition will limit its experience to just one (unpredictable) type of fruit, and the destruction of a single fruit body could be a catastrophe for the larva while being of absolutely no consequence to the orang-utan (Southwood 1976). In short, an organism's habitat must be described and classified from the point of view of the organism itself.

Bearing this in mind, we can proceed to a simple classification of habitats based on that suggested by Southwood (1977). Time (Fig. 6.5a) and space (Fig. 6.5b) have, for simplicity, been separated, but in both cases the inherent 'favourableness' (F) of habitats has been quantified. Alternatively, F may be thought of, from the organism's point of view, as the *intrinsic* rate of increase that an individual might expect to attain in that habitat, i.e. the difference between birth- and death-rates in the absence of conspecifics.

In Fig. 6.5a, H measures the length of time that a location permits existence, i.e. the length of time during

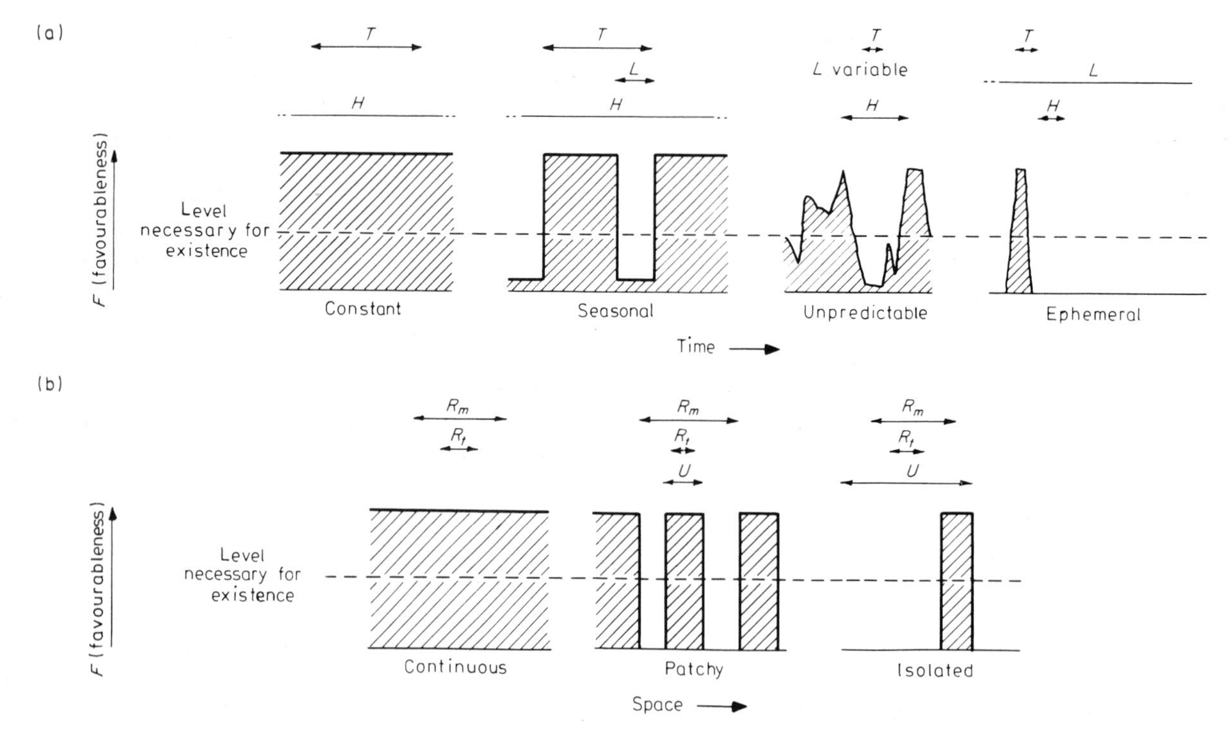

Fig. 6.5. A classification of habitats in time (a) and space (b). (Modified from Southwood 1977.) For further details, see text.

which the balance between birth and death *each generation* is favourable enough to avoid extinction (this clearly depends on generation-time, T); and L measures the length of time that a location remains unsuitable for breeding. In predictable environments, L-periods, if they occur, are of fairly constant duration, and short enough compared to T for H to be very long. F-levels may be *constant*, or they may fluctuate in a predictable pattern, in which case the environment is *seasonal*. In *unpredictable* environments, F-levels are far more variable, and L-periods are too, though they are short enough compared to T for H to be at least moderate. In *ephemeral* environments, L-periods are predictably very long; H is therefore short.

In Fig. 6.5b, R_t measures the 'trivial' range of an organism over which it gathers food on an essentially 'day-to-day' basis, while R_m measures its migratory range over which it may locate favourable habitats by the use of some 'special' dispersive mechanism (e.g. the production of propagules by plants, or of pelagic larvae by sessile marine invertebrates). The size of unfavourable areas between favourable patches is measured by U. In *continuous* environments, F-values are either continuously high or else they vary on a scale that is small compared to R_t. In *patchy* environments U-values exceed R_t, but are small compared to R_m. In *isolated* environments, U-values generally exceed even R_m.

6.5 Diapause, dormancy, migration and dispersal

If these time and space classifications of habitat are combined, we get twelve possible habitat types, of which two could not support life: the continuous-ephemeral and the isolated-ephemeral habitats. Amongst the remaining ten, however, we can begin to discern particular strategies associated with particular habitats. Dormancy will be favoured by natural selection in habitats in which there is either *seasonal* adversity, or *unpredictable* but moderately common adversity. Migratory or dispersive strategies will be favoured by natural selection in *patchy* environments, especially those that are *ephemeral* or *unpredictable*, but as we shall see, they can also evolve in a variety of other environments.

6.5.1 Diapause and dormancy

The costs associated with dormancy are of three types: the risks of mortality during the extended dormant phase, the costs of diverting energy away from reproduction towards the maintenance of a well-adapted, dormant stage in the life-history, and the delay in reproduction leading to a reduced reproductive rate. Benefits come simply from the increased chances of establishment and reproduction that pertain after the environment has improved. We can take it as axiomatic that dormancy will evolve when benefits outweigh costs; but it is nevertheless useful to remember, when we examine dormancy more closely, that there are costs as well as benefits. Analogous comments apply to migration and dispersal.

Dormancy has two important aspects. The first is that one or more of the organism's life-history stages are resistant to the extreme, adverse conditions. The second is the synchronization of the resistant and non-resistant stages with the appropriate adverse or benign periods of time. Such strategies can either be *predictive* or *consequential* (Müller 1970). The predictive strategy occurs (obviously) in advance of the adverse conditions, and is adapted to predictable, seasonal environments. Predictive dormancy is generally referred to as diapause in animals and is often called 'innate dormancy' (Harper 1977) or 'primary dormancy' in plants. The consequential strategy, by contrast, occurs as a direct result of the adverse conditions. As we shall see, such consequential (or 'secondary') dormancy may be either enforced or induced (Harper 1977). Note, however, that all these distinctions are somewhat blurred at the edges, and intermediate examples defying exact classification do occur.

Diapause has been most intensively studied in insects, and examples occur in all developmental stages, more or less throughout the Class. A fairly typical example is the field grasshopper, *Chorthippus brunneus*, discussed in Section 1.3.2. This annual species passes through an *obligatory* diapause in its egg stage, when it is resistant to cold, north-temperate winter conditions that would quickly kill the nymphs and adults. Obligatory diapause is, by definition, predictive; but synchronization is enhanced by the egg's requirement for a long cold period, around 5 weeks at 0°C (Richards & Waloff 1954), before development is able to recommence. This ensures that the egg is not affected by a short, freak period of warm winter weather which may then be followed by normal, dangerous, cold conditions. Thus, this species, like many others, has evolved a strategy that is adapted to maximize average survivorship through a predictably adverse period, even though this precludes the ability to make use of freak favourable conditions.

Diapause is also commonly shown by species with more than one generation per year. The fruit fly, *Drosophila obscura*, for instance, passes through four generations per year in England and diapauses during only one of them (Begon 1976). This *facultative* diapause shares many essential features with obligatory diapause: it, too, is an adaptation for maximizing average survivorship during a predictably seasonal, adverse (winter) period; and it is experienced by *resistant*, diapausing adults with arrested gonadal development and large reserves of stored abdominal fat. Synchronization in this case, however, is achieved not only during diapause, but also prior to it. Emerging adults react to the short day-lengths of autumn (and, to some extent, to low temperatures) by laying down fat and entering the diapause state; they recommence development in response to the longer days (and higher temperatures) of spring. Thus, by relying, like so very many organisms, on the utterly predictable *photoperiod* as a cue for seasonal development, *D. obscura* is largely unaffected by freak weather conditions, and diapause, although predictive, is confined to those generations that would otherwise pass through the adverse conditions.

Diapause or innate dormancy in plants is, in essence, the same as diapause in animals, although predictive, facultative diapause is apparently unknown (presumably because so few plants pass through several generations per year in a seasonal environment). Seed diapause is very common amongst annuals (though not 'winter annuals' that germinate in the late-summer or autumn and pass the winter as established plants);

perennials regularly pass through a dormant resting phase (strictly seasonal in temperate regions, much less so in the tropics); cues are often provided by reliable seasonal stimuli like photoperiod; and the dormant seeds of many perennials are induced to germinate by the light and long days associated with spring. Probably the most significant aspect of diapause in plants, however, is the existence in many species (especially weeds) of an adaptive variability in germination requirements amongst the seeds produced by a single individual.

This is seen, for example, in *Xanthium* species, where the seeds are borne in pairs with a large and a small seed joined and dispersed together. The dormancy breaking requirements of the two are normally different, so that at least 12 months separates their germination. Thus, the parent exhibits a mixed strategy, releasing offspring into at least two (temporally) different environments (Harper 1977). The situation is similar in *Rumex crispus* (Cavers & Harper 1966). In constant light with the temperature alternating between 10°C and 20°C, almost all seeds from all plants germinate. But as Fig. 6.6 shows, in other regimes the responses vary greatly: between nearby sites, within sites and, once again, within the progeny of individual plants. The advantages of these mixed strategies in seasonal environments with an unpredictable level of future adversity are obvious: the plants are almost literally refraining from putting all their eggs in one basket.

Consequential, enforced dormancy is clearly a strategy that requires less by way of evolved foresight than does predictive dormancy. Nevertheless, it is easy to see why it should evolve in an unpredictable environment, or, very often, in a seasonal environment with an *unpredictable* level of adversity. In such circumstances, there *will* be a disadvantage in responding to adverse conditions only after they have appeared. But this may very well be *outweighed* by the advantages:

(a) of responding to *favourable* conditions immediately they reappear, and

(b) of only entering a dormant state if adverse conditions *do* appear.

Thus, when many mammals enter hibernation they do so (after an obligatory preparatory phase) in direct response to the adverse conditions. And, having

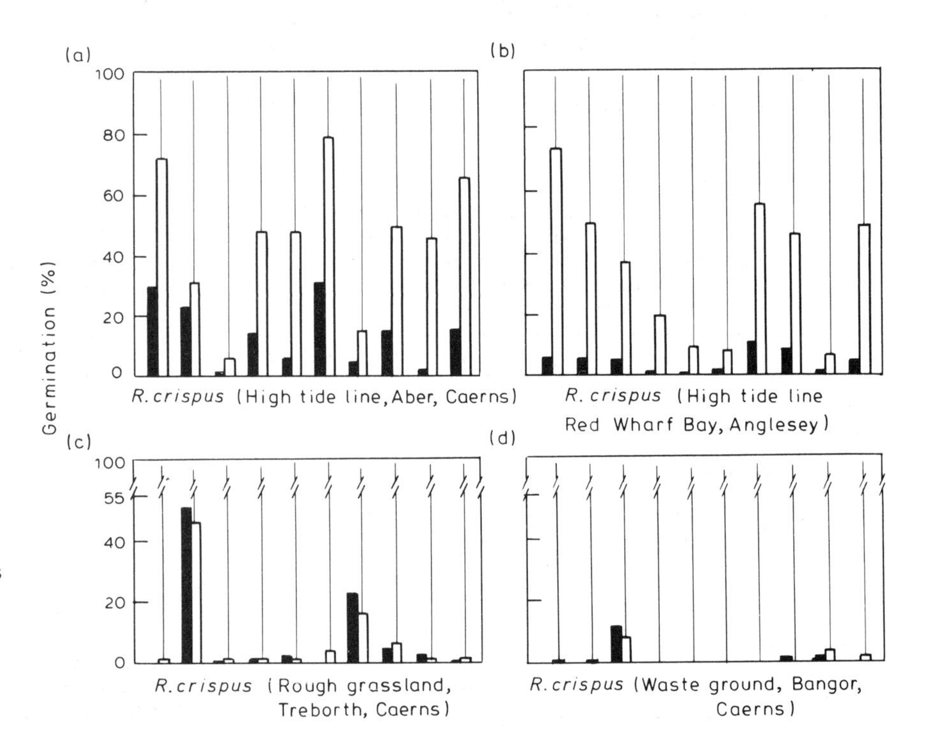

Fig. 6.6. The percentage germination of seeds from ten different plants in each of four different habitats tested under contrasting germination conditions. Black bars: darkness, constant 20°C; white bars: darkness, alternating 10°C and 20°C. (After Cavers & Harper 1966.)

achieved 'resistance' by virtue of the energy they conserve at a lowered body temperature, and having periodically emerged and sensed their environment, they eventually cease hibernation whenever the adversity disappears (Precht *et al.* 1973). Similarly, many insects—e.g. *Drosophila subobscura*, a southerly sibling species of the diapausing *D. obscura*—simply enter a state of 'arrested' development in cold weather; development then *recommences whenever* the temperature rises (Begon 1976). Plant seeds, too, often appear to be held in a state of enforced dormancy by the absence of the conditions necessary for early seedling growth (Harper 1977).

There is, however, a rather more complex, consequential strategy adopted by some plants, namely induced dormancy (Harper 1977), the nature of which is well illustrated by the work of Wesson & Wareing (1969). They found, by a combination of field and laboratory experimentation, that seeds of several species that did not require light for germination when fresh, had this requirement (and thus dormancy) induced in them if they were buried. Moreover, the extent to which burial inhibited germination depended on the depth and water-content of the medium. This, in strategic terms, is clearly a consequential response to an unpredictable environment, but it contains, in addition,

Table 6.3 Examples of species of insects able, or unable, to migrate or to fly, in relation to the permanence of their habitat. (Permanent: lakes, rivers, streams, canals, trees, bushes, salt marshes, woods. Temporary: ponds, ditches, pools, annual plants, perennial plants but not climax vegetation, arable lands.) (After Johnson 1969.)

	Permanent habitats	Temporary habitats
British Anisoptera		
Total number of species	20	23
Migrant or occasionally migrant	6	13
British Macrolepidoptera		
Total number of species	594	181
Migrant	14	42
British Water Beetles		
Total number of species	52	127
Able to fly	13	81
Unable to fly	20	9
Flight variable	19	37

a distinction between the cue for dormancy (in this case, burial) and the induced cue for germination (light).

6.5.2 *Migration and dispersal*

Superficially, we might imagine that there is a very close association between migratory or dispersive strategies and patchy environments, especially those that are ephemeral or unpredictable. This is certainly shown by the insects in Table 6.3 (Johnson 1969); by 'super-tramp' bird species (Diamond 1973), which are able to reach newly colonizable islands early, but have their stay curtailed by the greater competitive abilities of later arrivals (Fig. 6.7); and by the large numbers of tiny, wind-dispersed 'dust seeds' produced by many parasitic plants in their search to colonize new hosts, the positions of which they are unable to predict (Harper 1977). There are, however, many other circumstances in which migration or dispersal are of great importance.

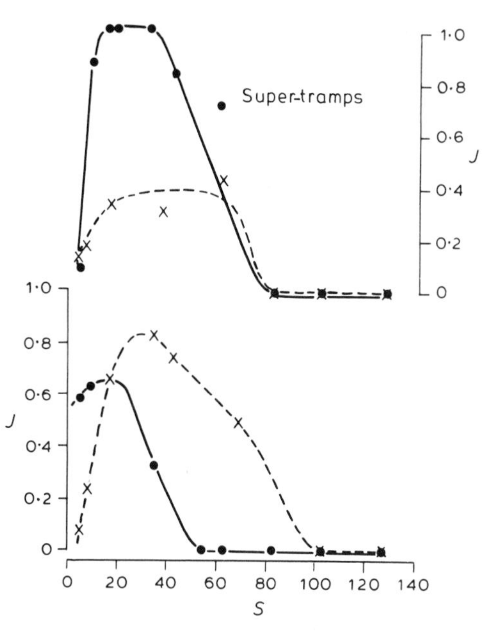

Fig. 6.7. 'Super-tramp' bird species occurring in the Bismarck islands with a high frequency (J) only where the number of other species (S) is low. Above: ●——— the flycatcher, *Monarcha cinerascens;* ×– – – the honeyeater, *Myzomela sclateri.* Below: ●——— the honeyeater, *Myzomela pammelaena*; ×– – – the pigeon, *Ptilinopus solomonensis.* (After Diamond 1975.)

Long-distance migration is exhibited in seasonal environments, especially by many birds, as an alternative to predictive dormancy. Most species move towards the longer days nearer the poles in the summer, returning to warmer, more equatorial regions in the winter. There are certain other species, moreover, that utilize migration as an alternative to consequential dormancy. The snowy owl, for instance, moves from its normal arctic residence into the northern United States and adjacent Canada every 4 years on average, in response to a shortage of prey on its normal wintering grounds (Gross 1947).

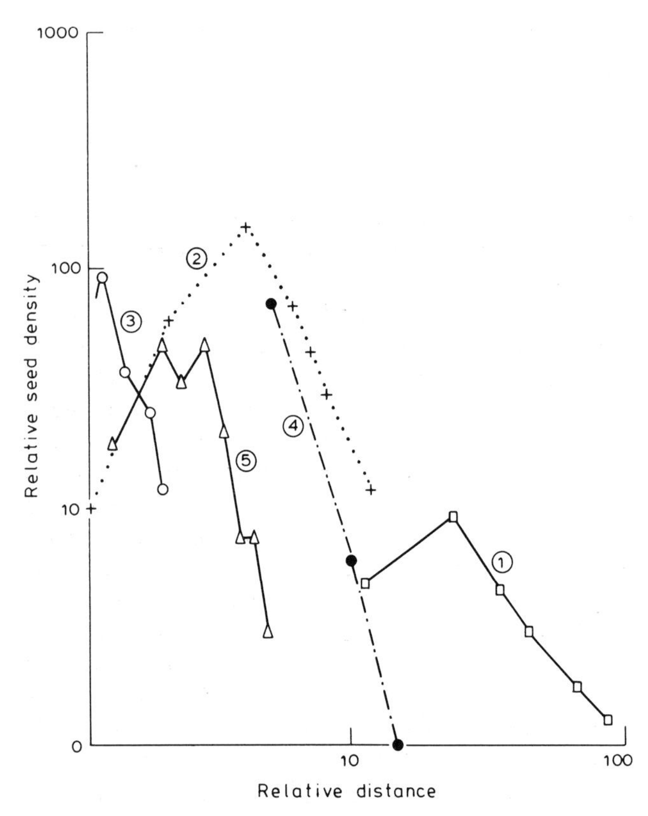

Fig. 6.8. The relationship between seed number and distance of seed dispersal from isolated plants plotted on log/log coordinates. (1) *Eucalyptus regnans* (Cremer, 1965); (2) *Verbascum thapsus* (Salisbury 1961); (3) *Papaver dubium* (Salisbury 1942); (4) *Dipsacus sylvestris* (Werner 1975); (5) *Dactylis glomerata* (Mortimer 1974). (After Harper 1977.)

Much more important than this in numerical terms, however, are the migratory and dispersive strategies exhibited by many species in environments that are apparently predictable and continuous. These can probably be understood most readily in plants that do not use animals as dispersive agents, and some typical patterns of seed dispersal from individual plants are illustrated in Fig. 6.8. The two important facets of these patterns are:

(a) the general tendency for the density of shed seed to decline sharply with increasing distance from the parent; and

(b) the reversal of this trend in the immediate vicinity of the parent.

This can be interpreted as an adaptation by the plants to small-scale heterogeneities in an environment that is basically continuous and predictable. They appear to shed their seed where the environment is most likely to be favourable: close to the parent, which by its existence has proved favourability. However, they appear to avoid the area most likely to be crowded: the immediate vicinity of the parent. Thus, their dispersive strategy seems to be a compromise between locating areas of *inherent* favourability and avoiding intraspecific competition. In other words, even continuous habitats seem patchy when we recognize that:

(a) for each individual, conspecifics are part of the environment, and

(b) aggregated distributions are the general rule (see Section 5.8).

6.6 '*r*'- and '*K*'-selection

Diapause, dormancy, dispersal and migration are all important strategies in their own right. The most dominant feature in the study of life-history strategies, however, has been a tendency to divide suites of characteristics, and the environments with which they are associated, into two contrasting types: '*r*' and '*K*' (MacArthur & Wilson 1967; Pianka 1970). These letters refer to parameters of the logistic equation (Section 3.3), and indicate that *r*-selected individuals have a strategy which tends to maximize fitness by enabling them to reproduce rapidly in an uncrowded environment (i.e. they have a high value of *r*), whilst *K*-selected

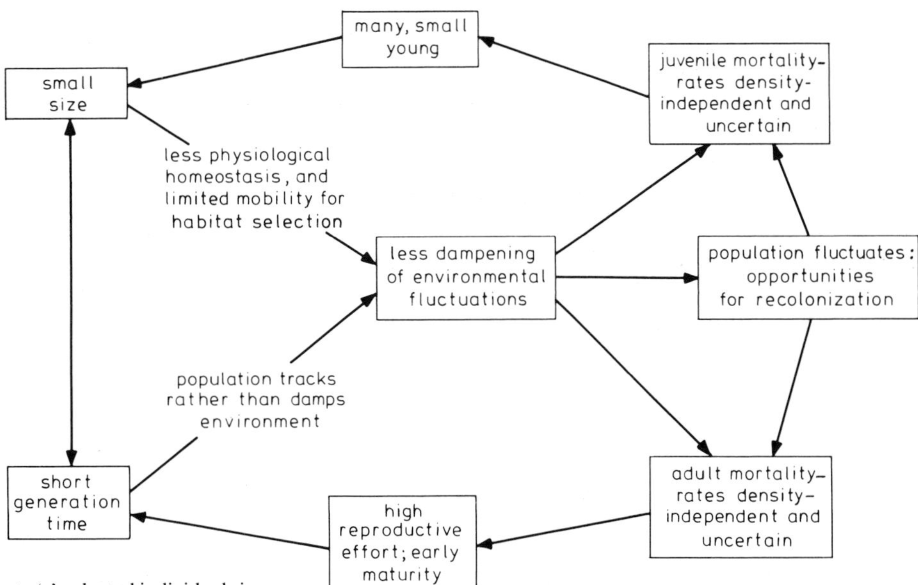

Fig. 6.9. The chain of causality giving rise to '*r*'-selected individuals in '*r*'-selecting environments. (Modified from Horn 1978.)

individuals maximize fitness by making a large proportional contribution to a population that remains close to its carrying capacity. We shall see below that this simple dichotomy is an inadequate classification of the natural world; but the attention the classification has received, its pedigree, and the substantial element of truth in it, all suggest that we should proceed by examining the r/K concept. Bearing in mind that the environment must be seen from the organism's point of view, we can begin by looking in more detail at the pattern of causality which is supposed to give rise to the r- and K-selected individuals. (Note that discussions of 'juvenile mortality' (below) refer to all deaths occurring before organisms are established in the population as free-living individuals potentially competing with mature adults. Juvenile mortality is, therefore, demographically indistinguishable from other elements in the birth process.)

The r-selected population (Fig. 6.9) lives in an environment which is either *unpredictable* in time, or *ephemeral* ('through the eyes of the organism') and it therefore experiences significant environmental fluctuations. As a consequence, the population itself fluctuates widely in size (either within a site, or as a result of repeated colonization), and juvenile and adult mortality-rates are unpredictable and often independent of density. Density-independent juvenile mortality tends to favour the production of many small (as opposed to a few large) young. This is because, with the *proportion* surviving remaining essentially unchanged, numbers-surviving increases with numbers-produced. And small young tend to lead, in their turn, to small adults. Density-independent mortality amongst *adults* tends to favour early and explosive reproduction (high reproductive effort), since efforts towards maintenance and current growth are worthless in the face of catastrophe; and this explosive reproductive effort leads to short lifespans and generation times. As Fig. 6.9 shows, however, small size and short generation time both influence the organism's frame of reference such that environmental fluctuations are all the more likely. The pattern of influence has, therefore, moved full circle: the selective pressures favouring r-type individuals—*with earlier maturity; more, smaller young; smaller size; larger reproductive effort and shorter life*—have been reinforced.

The K-selected population, by contrast (Fig. 6.10), lives in an environment which is either *constant* or *pre-*

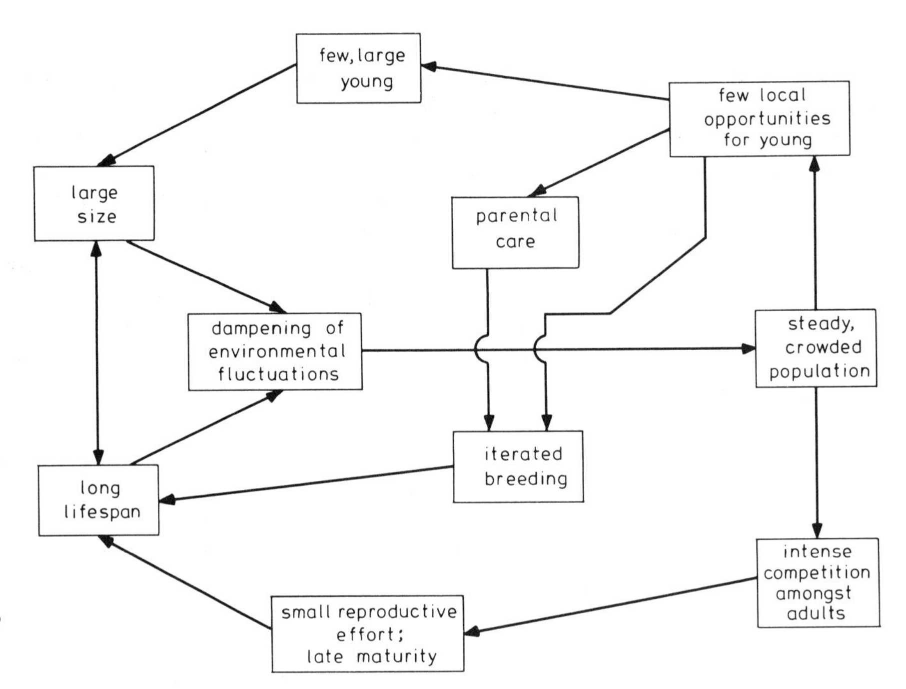

Fig. 6.10. The chain of causality giving rise to '*K*'-selected individuals in '*K*'-selecting environments. (Modified from Horn 1978.)

dictably seasonal in time, and it therefore experiences very little in the way of random environmental fluctuations. As a result, a crowded population of fairly constant size is established in which there is intense, density-dependent competition amongst adults and very few opportunities within the habitat for young to become established. Intense competition amongst adults favours successful competitors, and these will tend to be individuals devoting considerable effort to maintenance and growth, and therefore making a small reproductive effort and *delaying* maturity. The density-dependent difficulties experienced by small, young individuals will, likewise, ensure that only the most successful competitors survive. Thus, the production of a few large (as opposed to many small) young will be favoured (leading to large adults) as will devotion to the young of significant amounts of parental care. Moreover, this necessity for parental care, along with the necessity to reduce the density-dependent effects on the young, will lead to reproduction being extended over time. And this iterated breeding, along with the small reproductive effort, will tend to give rise to a long lifespan and generation time, which, combined and compounded with large size, will influence the organism's frame of reference such that environmental fluctuations are all the more damped. Once again, therefore, the circle has been completed. This time, *K*-selected individuals—*with later (more delayed) maturity; fewer, larger young; larger size; smaller reproductive effort; longer life; iterated breeding and parental care*—are the predicted result.

An alternative, though related, dichotomy was considered by Schaffer (1974). He suggested that all environments fluctuated and led to random variations in mortality-rates, but distinguished between those in which the major effect was on juvenile mortality and those in which it was on adult mortality. He found (mathematically) that in the former case the favoured life-history characteristics are essentially '*K*'-type, while in the latter they are essentially '*r*'-type. Yet as Horn (1978) has pointed out, there need be no contradiction between these two classifications. Unpredictable mortality of established adults (one end of Schaffer's dichotomy) is, perhaps, the major distinctive feature of *r*-selecting environments; while variable juvenile mortality (the other end) merely compounds the uncertainties associated with the high mortality-rates of non-established juveniles in a crowded, *K*-selecting environment. All in all, therefore, Schaffer's classification collapses into the *r*/*K* dichotomy. We are left with a circular causal system of positive feedback through which *r*- (or *K*-) type environments lead to *r*- (or *K*-) type strategies, which in turn lead to the environment being, *from the organism's point of view,* even more *r*- (or *K*-) selecting. This association together of *r*-selected traits in an *r*-selecting environment and *K*-selected traits in a *K*-selecting environment will be referred to as the 'accepted scheme' (Stearns 1977): we shall see that such uncritical 'acceptance' is probably unwarranted.

It must be noted, however, that several ecologists (notably Pianka 1970 and Southwood 1976) have argued that '*r*' and '*K*' are not branches of a dichotomy but ends of a continuum. Yet, while this view is essentially sensible, there is a major problem in accepting it fully. The elements of life-history strategies are, for two reasons, very difficult to quantify. In the first place, reproductive (and other) 'efforts' are made over long periods of time *using processes of differing energetic efficiency*, and in animals they include important but intractable behavioural components. 'Effort' is, therefore, almost impossible to measure in absolute terms. In the second place, while a blackbird's young, for instance, are larger than a snail's or a buttercup's, this comparison tells us nothing about '*r*-ness' or '*K*-ness',

(a) because each organism is subject to the design constraints of being a bird, a mollusc or a plant, and

(b) because the various organisms' frames of reference are so different that their environments are, essentially, incomparable.

In other words, studies of life-history strategies must involve organisms of the same type; and, because absolute measurements are near-impossible, they must be *comparative.* A continuum between and beyond the organisms is imaginable, but it will be qualitative rather than quantitative and will consist mainly of hypothetical strategies which are never found; and the gaps cannot be filled by organisms of another type because they are on a *different* continuum. In practice, therefore, if we compare two or more species (or populations), we can *only* know which are more '*r*' or more '*K*' than the others. The hypothetical continuum need rarely concern us.

6.7 Some evidence for *r*- and *K*-selection

It is possible, to a limited extent, to bring together these ideas on life-history strategies with those on fecundity and survivorship schedules discussed in Chapter 1. For instance, we might consider *K*-selected species to be characterized by Pearl's type I survivorship curve (generally high survivorship until late in the lifespan), and a fecundity schedule with a long pre-reproductive period and either a long, low plateau or a moderately long series of small peaks. Similarly, *r*-selected species might be expected to have a type III survivorship curve, a short pre-reproductive period and a single, high peak of reproductive activity. It is certainly true that undoubted *K*-strategists like the wandering albatross (enormous size, high survivorship, 9–11 years of immaturity, one egg every other year) and undoubted *r*-strategists like the pest army worm, *Spodoptera exempta* (600 eggs, generation-time 3 weeks), do fit into this pattern (Southwood 1976). On the other hand, we must remember that the *r*-*K* concept is, above all, a comparative one. It would be unwarranted to conclude, for example, that the cod is an *r*-strategist *because* it has a type III survivorship curve.

However, there are undoubtedly some cases in which the evidence in favour of the accepted scheme is good. McNaughton (1975), for instance, examined the adaptive properties of *Typha* (cattail or reed mace) populations in relation to climatic gradients, and some of his results are presented in Table 6.4. This table is particularly concerned with a specialist southerly species *T. domingensis*, and a specialist northerly species *T. angustifolia,* which were taken from sites in Texas and North Dakota respectively, and grown side-by-side under the same conditions. In addition, McNaughton quantified certain aspects of the long- and short-growing-season habitats in which these species are found. (Actually, 'recolonization' and 'density variation' were measured in low- and high-altitude sites, rather than at low and high latitudes.) It is clear from these environmental measurements that the short-growing-season sites are relatively *r*-selecting and the long-growing-season sites relatively *K*-selecting. And it is equally clear from Table 6.4 that the species inhabiting

Table 6.4 Adaptive properties of *Typha* (cattail) populations, and ecosystem properties of the environments in which they grow. (After McNaughton 1975.) For further details see text.

		Growing season	
Ecosystem property	Measured by	Short	Long
Climatic variability	$s^2/\bar{x}$ frost-free days per year	3.05	1.56
Competition	biomass above ground (g m^{-2})	404	1336
Annual recolonization	winter rhizome mortality (%)	74	5
Annual density variation	$s^2/\bar{x}$ shoot numbers m^{-2}	2.75	1.51
Plant traits		*T. angustifolia*	*T. domingensis*
1 Days before flowering		44	70
2 Mean foliage height (cm)		162	186
3 Mean genet weight (g)		12.64	14.34
4 Mean number of rhizomes/genet		3.14	1.17
5 Mean weight of rhizomes (g)		4.02	12.41
6 Mean number of fruits/genet		41	8
7 Mean weight of fruits (g)		11.8	21.4
8 Mean total weight of fruits (g)		483	171

these sites conform to the accepted scheme. *T. angustifolia* (short growing season) matures earlier (trait 1), is smaller (traits 2 and 3), makes a larger reproductive effort (traits 3 and 8), and produces more, smaller 'offspring' both asexually (traits 4 and 5) and sexually (traits 6 and 7). The reverse is true of *T. domingensis*.

In a similar vein, Law *et al.* (1977) compared the life-history strategies adopted by two types of annual meadow grass (*Poa annua*) population in north-west England and North Wales. The first type of population ('opportunist') contained individuals at low density with large areas of bare ground between them, and had been maintained in that condition for some time by continuous disturbance. The second type contained individuals that were closely packed with others of their own or other species, and had also been maintained in that way for some time, often as permanent 'pasture'. The first type were taken to be experiencing predominantly density-independent limitation, and were, therefore, essentially *r*-selecting; the second were taken to be experiencing predominantly density-dependent regulation, and were, therefore, essentially *K*-selecting. Samples from each population were grown from seed under controlled, uncrowded conditions, and the life-histories of individuals monitored. Some of the results are illustrated in Fig. 6.11.

The *P. annua* populations clearly conform to the accepted scheme. In the *r*-selecting environment, individuals mature earlier (Fig. 6.11a), are smaller (Fig. 6.11b), and they make a greater reproductive effort earlier, leading to a shorter lifespan (Fig. 6.11c). The reverse is true in the *K*-selecting environment.

In certain other cases, while the evidence in favour of the accepted scheme is persuasive, it involves only a few of the relevant traits. This is true, for instance, of Forsyth & Robertson's (1975) study of the reproductive strategy of the sarcophagid fly, *Blaesoxipha fletcheri*. Many sarcophagids are, as larvae, scavengers on dead animals, parasites of vertebrates or parasitoids of invertebrates. *B. fletcheri*, by contrast, is extremely unusual in its ecology. Its larvae inhabit the tubular leaves of carnivorous pitcher plants (*Sarracenia*), living on the surface of the liquid contained within them, and

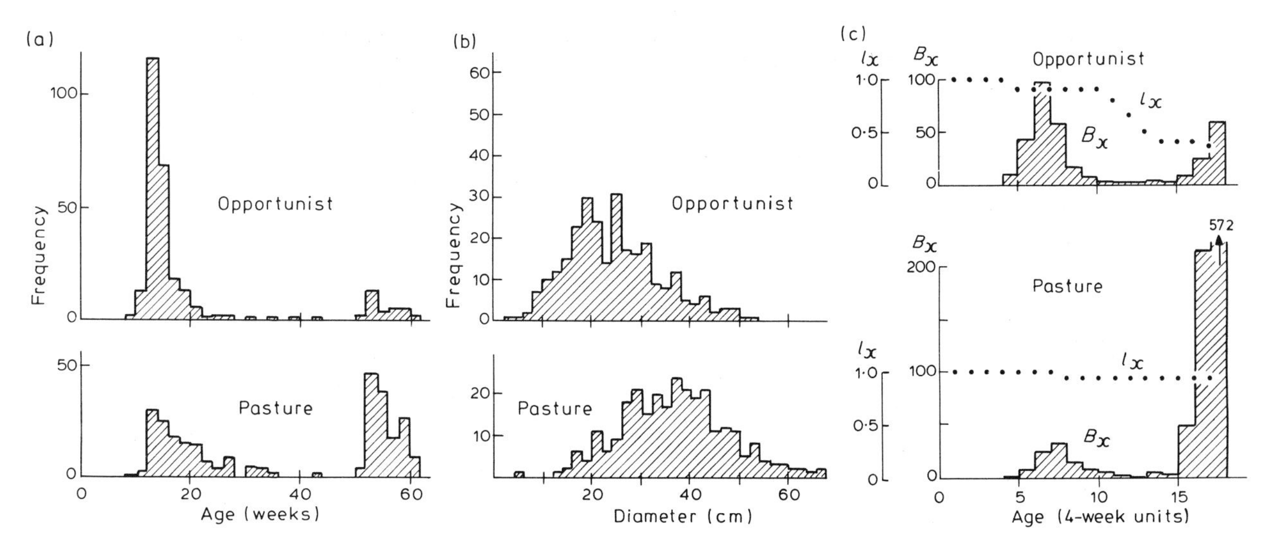

Fig. 6.11. (a) The distribution of pre-reproductive periods in *Poa annua* plants from opportunistic and pasture environments. (b) The distribution of sizes in *Poa annua* plants from opportunistic and pasture environments. (c) Rates of reproduction (B_x) and survival (l_x) in samples of *Poa annua* from an opportunistic and a pasture environment. (After Law *et al.* 1977.)

feeding voraciously on the insects that are caught and drowned. Moreover, there is intense intraspecific aggression between larvae, such that each of the limited number of pitchers can support only a single individual. In other words, there is density-dependent mortality in *B. fletcheri* which is peculiarly intense compared to sarcophagids generally. And as Table 6.5 shows, this relatively high degree of *K*-selection is reflected in *B. fletcheri*'s reproductive biology in the very way that the accepted scheme would suggest. Compared to other sarcophagids, *B. fletcheri*'s reproductive effort is devoted to a few large, rather than many small, offspring.

By now, it should be obvious that there are studies that 'fit' the accepted scheme. Yet Stearns (1977), reviewing the data on life-history traits, found that of the 35 studies that were detailed enough to allow reasonably unambiguous conclusions to be drawn, only 18 actually conformed closely to the ideal; and the 17 studies that did not do so were, if anything, more thorough than those that did. The explanatory powers of the accepted scheme based on the *r*/*K* dichotomy are obviously limited.

6.8 The limitations of the accepted scheme: a study of *Littorina*

The explanation for the *r*/*K* concept's inability to account for all life-history strategies is exceptionally simple and is best illustrated by a single example. This concerns two species of rough winkle, *Littorina nigrolineata* (Naylor & Begon 1982) and *L. rudis* (Hart & Begon 1982), which until recently were grouped with two further species under the specific name *Littorina saxatilis*. *L. nigrolineata* females are oviparous (winkles have separate sexes), eggs being deposited in jelly-like masses within which metamorphosis and development occur until miniature free-living snails emerge. *L. rudis*, on the other hand, is a viviparous species, retaining eggs within a specialized brood pouch until metamorphosis is complete and miniature snails can be released. At a single location on the exposed west coast of Holy Island, Anglesey (Wales), both species occupy two rather different habitats. The first is a gently sloping rocky shore covered with medium to large boulders. The second is a large, near-vertical, rocky stack indented with many narrow crevices, within which the winkles are

Table 6.5 The larval size (a) and fecundity (b) of *Blaesoxipha fletcheri* compared to other sarcophagid species. (After Forsyth & Robertson 1975.)

(a)

Species	Range (mm)	Mean (mm)
Blaesoxipha fletcheri	6.9–7.0	6.95
Sarcophaga agryostoma	3.1–5.3	4.63
S. bullata	1.9–3.2	2.75
S. cimbicis	3.3–3.7	3.50
S. haemorrhoidalis	2.6–4.1	3.52
S. latisetosa	1.2–3.5	2.12
S. l. herminieri	2.8–4.5	3.55
S. melanura	2.4–4.6	3.41
S. pusiola	2.2–3.3	2.61
S. rapax	1.9–2.9	2.39
S. scoparia nearctica	3.6–5.4	4.21
S. ventricosa	2.4–2.9	2.70

(b)

Species	No. larvae per female
Blaesoxipha fletcheri	11
Bellieria melanura	94
Pseudosarcophaga affinis	100
Ravinia striata	52
Sarcophaga cooleyi	50
S. haemorrhoidalis	84
S. nearctica	50
Wohlfarhtia magnifica	170

almost entirely confined. Females of both species were collected from the habitats, and the pertinent results are illustrated in Fig. 6.12 and summarized in Table 6.6.

It is immediately clear from Table 6.6 that the two species each have different life-histories in the different habitats, and that they vary 'in parallel' with one another. It is also clear, however, that the *L. rudis* strategies, far from fitting the accepted scheme, are actually compounded from *r*- and *K*-type characteristics. There is, moreover, no indication of their lying on a continuum between *r* and *K*: in some respects they are undoubtedly '*r*' and in others undoubtedly '*K*'. Yet this apparent paradox can be resolved by ridding our minds of the *r*/*K* preconceptions, and looking instead at the winkles' ecology.

Raffaelli & Hughes (1978) and Emson & Faller-Fritsch (1976) both established that the population size of winkles in crevice habitats is limited by crevice-availability. By artificially adding more and larger crevices to a habitat, they increased the size of the population, and also the size of the largest individuals. On exposed rocky faces, crevices provide protection against abiotic mortality factors (wave action, desiccation) and biotic mortality factors (bird predation)—but only for those winkles that are small enough to inhabit them. There is competition for limited space, which becomes increasingly intense as winkle size increases. With this in mind we can reconsider the type of evolutionary strategic decision that confronted *Poa annua, Balanus balanoides* and *Armadillidium vulgare* in Section 6.2.2: a choice between present reproduction with reduced growth and subsequently modest reproduction on the one hand, and increased growth with delayed but greatly increased reproduction on the other. All organisms with indeterminate growth must strike some balance between the advantages and disadvantages of these two alternatives; but for crevice winkles, the advantage of sacrificing reproduction for growth are somewhat dubious, since large winkles have relatively little chance of finding protection in a crevice. It is, therefore, to be expected that selection will favour less-delayed maturity, greater reproductive effort and smaller size in crevice populations, and this is exactly what was found. Yet the selective forces have nothing to do with those aspects of the environment conventionally associated with '*r*-ness' and '*K*-ness'.

Turning to the boulders population, the winkles are subject to two very important mortality factors, crushing and predation, against which thick shells *and* large size are undoubtedly a protection (Raffaelli & Hughes 1978; Elner & Raffaelli 1980). This explains the differences in shell thickness between crevice and boulder individuals (Fig. 6.12, Table 6.6). Yet it may also have an indirect effect on reproductive effort and the delay in reaching maturity. Energy devoted to large size and to the production of a thick shell must, to some extent, be diverted away from reproduction and the development of

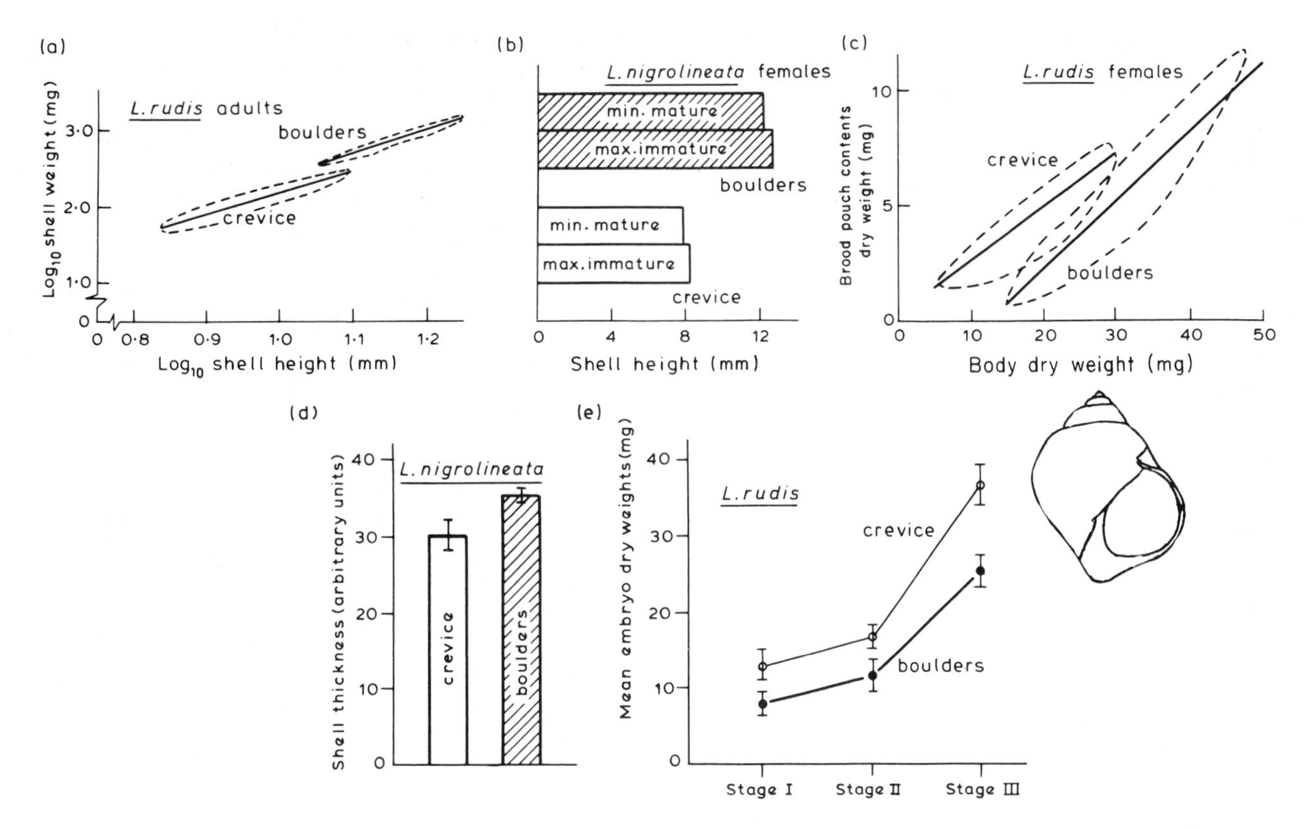

Fig. 6.12. Selected reproductive traits in *Littorina rudis* (after Hart & Begon 1982) and *L. nigrolineata* (after Naylor & Begon 1982) from two contrasting habitats: 'crevice' and 'boulders'. (a) *L. rudis* boulders adults are larger and appear to have thicker shells. (b) *L. nigrolineata* crevice females mature at a smaller size. (c) *L. rudis* crevice females make a greater reproductive effort in that a greater proportion of their total weight is represented by brood pouch contents. (d) *L. nigrolineata* boulders adults have thicker shells (individuals within the *same* size-range were compared). (e) Individual *L. rudis* crevice embryos are larger (means±2 s.e.).

Table 6.6 Summary of life-history traits displayed by *Littorina rudis* (Hart & Begon 1982) and *L. nigrolineata* (Naylor & Begson 1982) in two contrasting habitats. The expected association ('*r*' or '*K*') within the 'accepted scheme' is given in brackets.

Crevice population	Boulders population	Species
1 Thinner shells	1 Thicker shells	*rudis* and *nigrolineata*
2 Smaller individuals (*r*)	2 Larger individuals (*K*)	*rudis* and *nigrolineata*
3 Earlier maturity (*r*)	3 Relatively delayed maturity (*K*)	*rudis* and *nigrolineata*
4 Greater reproductive effort (*r*)	4 Smaller reproductive effort (*K*)	*rudis* and *nigrolineata*
5 Fewer, larger offspring (*K*)	5 More, smaller offspring (*r*)	*rudis*

gonads. It is to be expected once again, therefore, that selection will favour a relatively delayed maturity and a smaller reproductive effort in boulders populations. Yet, once again, these selective forces have nothing to do with '*r*-ness' and '*K*-ness'.

Obviously, points 1–4 in Table 6.6 are entirely compatible with the ecology of winkles, but they have little to do with *r* and *K* environments. Indeed, they actually contradict the accepted scheme, since the crevice habitat is essentially *K*-selecting (intense, density-dependent competition for 'space') while the boulders habitat is much more *r*-selective (density-independent and often catastrophic crushing and predation). However, there is no contradiction regarding point 5 in Table 6.6: the partitioning of energy within a given reproductive effort by *L. rudis* matches the accepted scheme exactly. There are more, smaller young in the *r*-selecting environment and fewer, larger young in the *K*-selecting environment.

The apparent paradox presented by these strategies has, therefore, been resolved. The winkles exhibit suites of life-history traits which are their evolved, compromise solutions to the problems posed by their environments; and while they appear, to a limited extent, to have responded to the '*r*-ness' or '*K*-ness' of these environments, they have, in other respects, responded to environmental aspects that defy any such simple classification. This conclusion has a very obvious general relevance.

The construction of the *r*/*K* concept has been a notable success in the ecologists' search for pattern; but '*r*-ness' (or '*K*-ness') is only one aspect of the environment. It may indeed be the single most useful *simple* classification (and we have seen that in some cases it is sufficiently important for strategies to reflect the *r*/*K* dichotomy and fit the accepted scheme); but it must not be imagined that life-history strategies in general can be forced into this or any other simple mould. And we must certainly not approach the study of life-history strategies with any *r*/*K* preconceptions. *r*- and *K*-selecting environments may lead to *r*- and *K*-type characteristics; but *r*- and *K*-type characteristics are by no means necessarily the result of an *r*- or *K*-selecting environment. Moreover, these characteristics are always constrained, to some extent, by the design of the organism exhibiting them. A life-history strategy, is, therefore, an organism's *unique* adaptive response to its *whole* environment; and simplistic theories must eventually give way to the detailed study of real organisms in the real world.

Chapter 7 Population Regulation

7.1 Introduction

There have, in the past, been contrasting theories to explain the abundance of animal and plant populations. Indeed, the interest has been so great, and the disagreement often so marked, that the subject has been a dominant focus of attention in population ecology throughout much of this century. The present view, however, is that the controversy has not so much been resolved, as recognized for what it really is (like so many other controversies): a product of the protagonists taking up extreme positions and arguing at cross purposes. In this chapter we shall confine ourselves to a brief summary of the main arguments in the controversy (Sections 7.2 to 7.5). This should provide the necessary background, but will avoid our making too much use of historical hindsight. It would be unwise (and unfair) to analyse the arguments in too great a detail, since most were propounded between 1933 and 1958 when understanding of population ecology was even less sophisticated than it is now. More inclusive reviews may be found in Bakker (1964), Clark *et al.* (1967), Klomp (1964) and Solomon (1964). In Section 7.6 we shall illustrate the present-day view of the problem by the extensive life-table analysis of a Colorado beetle population. Then, in Sections 7.7 to 7.10, we shall deal with a heterogeneous sequence of topics, all of which have a bearing on the general question of how population sizes are regulated and determined.

7.2 Nicholson's view

A. J. Nicholson, an Australian ecologist, is usually credited as the major proponent of the view that density-dependent, biotic interactions (which Nicholson called 'density-governing reactions') play the main role in determining population size (Nicholson 1933, 1954a,b, 1957, 1958). In his own words: 'Governing reaction induced by density change holds populations in a state of balance in their environments', and '. . . the mechanism of density governance is almost always intraspecific competition, either amongst the animals for a critically important requisite, or amongst natural enemies for which the animals concerned are requisites'. Moreover, although he recognized that '. . . factors which are uninfluenced by density may produce profound effects upon density', he considered that they only did so by '. . . modifying the properties of the animals, or those of their environments, so influencing the level at which governing reaction adjusts population densities'. Even under the extreme influence of density-independent factors '. . . density governance is merely relaxed from time to time and subsequently resumed, and it remains the influence which adjusts population densities in relation to environmental favourability' (all Nicholson 1954b). In other words, Nicholson may be taken to represent the view that density-dependent processes '. . . play a key role in the determination of population numbers by operating as stabilizing (regulating) mechanisms' (Clark *et al.* 1967).

7.3 Andrewartha and Birch's view

By contrast, the view that density-dependent processes are '. . . in general, of minor or secondary importance, and . . . play no part in determining the abundance of some species' (Clark *et al.* 1967) is most commonly attributed to two other Australian ecologists, Andrewartha and Birch. Their view is as follows (Andrewartha & Birch 1954):

> The numbers of animals in a natural population may be limited in three ways: (a) by shortage of material resources, such as food, places in which to make nests, etc.; (b) by inaccessability of these material resources relative to the animals' capacities for dispersal and searching; and (c) by shortage of time when the rate of increase r is positive. Of these three ways, the first is probably the least, and the last is probably the most important in nature. Concerning (c), the fluctuations in the value of r may be caused by weather, predators, or any other component of environment which influences the rate of increase.

Andrewartha and Birch, therefore '. . . rejected the traditional subdivision of environment into physical and biotic factors and 'density-dependent' and 'density-independent' factors on the grounds that these were neither a precise nor a useful framework within which to discuss problems of population ecology' (Andrewartha & Birch 1960). The views of Andrewartha and Birch are probably made more explicit, however, by considering one of their examples; subsequent discussion of this example will then lead on to a crystallization of the current status of the controversy.

7.4 An example: *Thrips imaginis*

Davidson & Andrewartha (1948a,b) studied population changes in a phytophagous insect, the apple-blossom thrips, *Thrips imaginis,* which lives on roses throughout southern Australia. They obtained estimates of abundance for 81 consecutive months by counting the number of thrips on a sample of 20 roses on each of approximately 20 days each month (Fig. 7.1); and then, for a further 7 years, they obtained similar estimates for spring and early summer only. In addition, they monitored local temperature and rainfall throughout the same period. By the use of a 'multiple regression' analysis (see, for instance, Poole 1978), they were able to 'account for' 78% of the variance in the yearly peak of thrips numbers by reference to four climatic factors: the suitability of temperature for development up to 31 August, the suitability of temperature for development in September and October, the suitability of temperature for development in August of the previous season, and the rainfall in September and October. In other words, knowledge of these four factors in any one year would allow the size of the peak in thrips numbers to be estimated with a good degree of statistical accuracy.

Clearly, the weather (as represented by these four factors) play a central and crucial role in the determination of *T. imaginis* numbers at their peak. Yet Andrewartha & Birch (1954) used this, and the fact that no 'density-dependent factor' had been found, to conclude that there was 'no room' for a density-dependent factor as a determinant of peak thrips numbers. The only 'balance' that Davidson and Andrewartha (1948b) recognized was '. . . a race against time with the increase in density being carried further in those years when the favourable period lasts longer, but never reaching the point where competition begins to be important'.

As Varley *et al.* (1975) point out, however, the multiple regression model is not designed to directly reveal the presence of a density-dependent factor. And

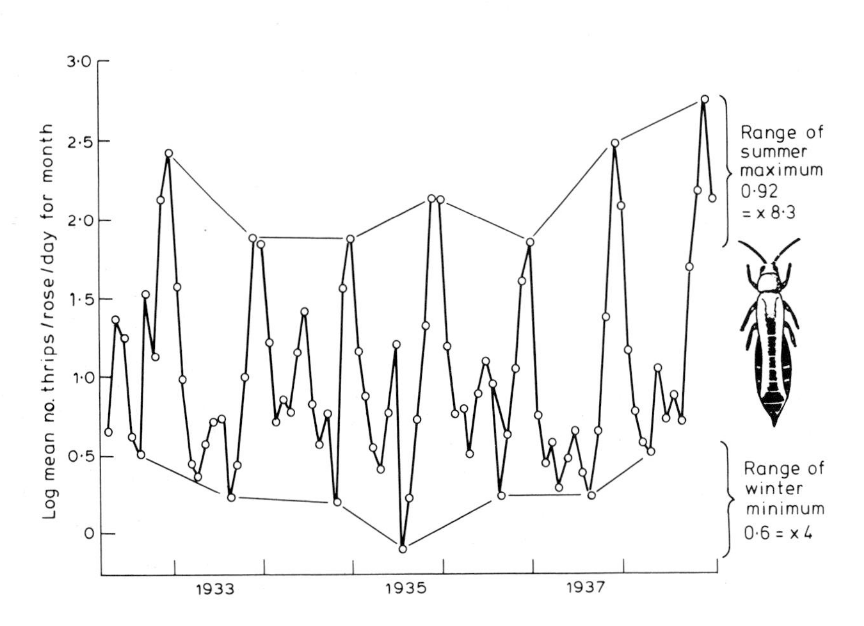

Fig. 7.1. Mean monthly population counts of adult *Thrips imaginis* in roses at Adelaide, Australia (Davidson & Andrewartha 1948a). (After Varley *et al.* 1975.)

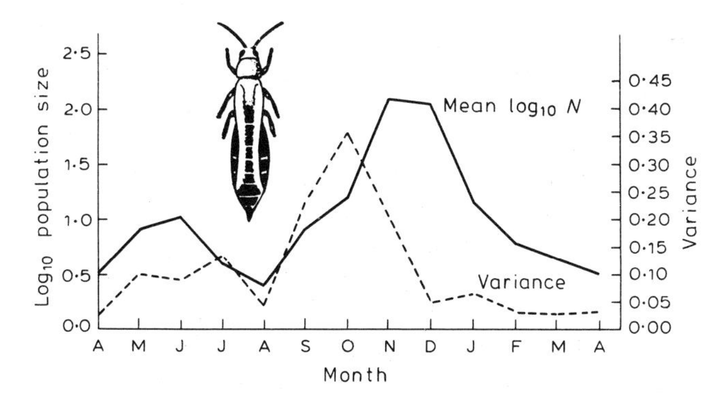

Fig. 7.2. The mean monthly logarithm of population size for *Thrips imaginis* and its variance. (After Smith 1961.)

when Smith (1961) applied methods that *were* so designed, he obtained excellent evidence of density-dependent population growth prior to the peak. He found that:

(a) there were significant negative correlations between population *change* and the population size immediately preceding the spring peak; and

(b) there was, at the same time, a rapidly decreasing variance in population size (Fig. 7.2).

Moreover, Varley *et al.* (1975) were able to suggest the existence of a strongly density-dependent *mortality* factor, *following* the peak density. They did this by the use of two hypothetical examples (Fig. 7.3). In both cases, the adult population (N_A) is given a reproductive-rate of 10, but is then subjected to a random (i.e. density-independent, perhaps climatic) factor determining the (peak) number of larvae (N_L). (Note that this random element, plotted as a *k*-factor, is the same in Fig. 7.3a and b.) In addition, however, Fig. 7.3a shows a *weakly* density-dependent mortality factor acting on the larvae to determine the next generation's adult population size, while Fig. 7.3b has a *strongly* density-dependent factor. *As a consequence of this,* the climatic factor accounts for 91% of the variaton in N_L in Fig. 7.3b, but only 32% in Fig. 7.3a. It appears, in short, as if a fairly strong density-dependent factor must also have been acting on the thrips population between the summer peak and the winter trough; and, indeed, its presence can be inferred from Fig. 7.1 itself, where the maxima are spread over an eightfold range, with the minima covering only a fourfold range. In fact, Andrewartha and Davidson themselves (1948b) felt that weather acted as a density-dependent component of the environment during the winter, by killing the proportion of the population inhabitating less favourable 'situations'. (If the number of safe sites is limited and remains roughly constant from year to year, then the number of individuals outside these sites killed by the weather will increase with density.) They also felt, however, that this did not fit the general, density-dependent theory '. . . since Nicholson (1933, pp. 135–6) clearly excludes climate from the list of possible "density-dependent factors".'

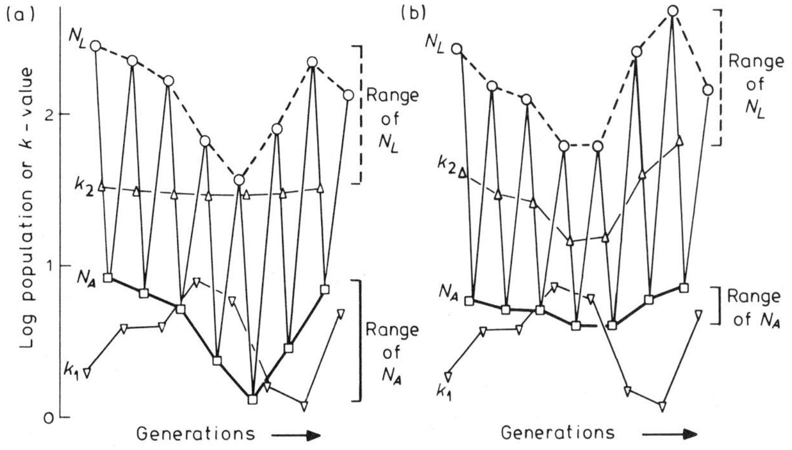

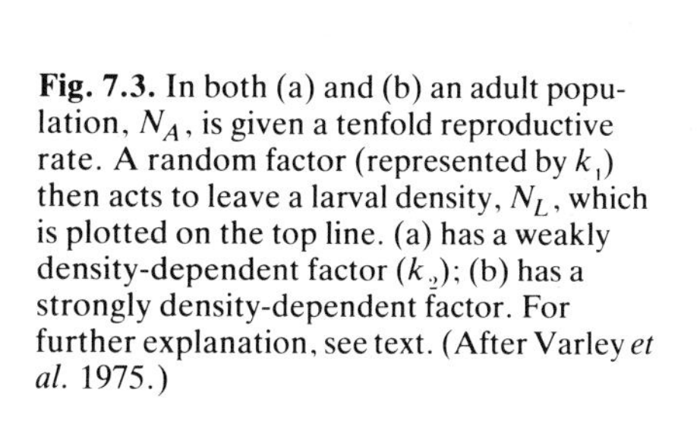
Fig. 7.3. In both (a) and (b) an adult population, N_A, is given a tenfold reproductive rate. A random factor (represented by k_1) then acts to leave a larval density, N_L, which is plotted on the top line. (a) has a weakly density-dependent factor (k_2); (b) has a strongly density-dependent factor. For further explanation, see text. (After Varley *et al.* 1975.)

7.5 Some general conclusions

There are several conclusions of general importance that we can draw from this single example:

(1) We must begin by distinguishing clearly between the *determination* of abundance and its *regulation*. Regulation has a well-defined meaning (Section 2.3), and *by definition* can only occur as a result of a density-dependent process. Abundance will be *determined,* on the other hand, by the combined effects of all the factors and all the processes that impinge on a population. Certain other terms, particularly the *control* of abundance, do not have a single, well-defined meaning. As Varley *et al.* (1975) stress, a reader should pause and consider carefully what kind of meaning is being attached to such terms.

(2) We must also recognize that any dispute as to whether climate (or anything else) *can* be a density-dependent factor is essentially beside the point. In Davidson and Andrewartha's thrips, at the appropriate time of the year, weather confined the insects to a limited amount of favourable 'space'. It is irrelevant whether we consider weather or space to be the density-dependent factor. What is important is that the two, together, drove a density-dependent *process*, and that the *effects* of weather on the thrips were, in consequence, density-dependent. Weather, like other abiotic factors, can interact with biotic components of the environment so that no single, simple 'density-dependent factor' can be isolated (cf. Section 2.3). We must concentrate on the nature of the processes and effects which emanate from such interactions.

(3) Andrewartha and Birch's view that density-dependent processes play no part in determining the abundance of some species clearly implies that the populations of such species are not regulated. Yet, as we have seen, even the data they themselves brought forward contradicts this view. Indeed it is logically unreasonable to suppose that any population is *absolutely free* from regulation. Unrestrained population growth is unknown, and unrestrained decline leading to extinction is extremely rare. Moreover, the fluctuations in almost all populations are at least limited enough for us to be able to describe them as 'common', 'rare', and so on. Thus, we can take it that all natural populations are regulated to some extent, and are therefore, to some extent, influenced by density-dependent processes.

(4) Furthermore, Andrewartha and Birch's view that density-dependent processes are *generally* of minor or secondary importance, if it is taken to imply that such processes are *un*important, is not justified either. Even in their own thrips example, density-dependent processes, apart from regulating the population, were also crucially important in *determining* abundance.

(5) On the other hand, we must remember that the weather accounted for 78% of the variation in peak thrips numbers. Thus, if we wished to predict abundance or decide why, in a particular year, one level of abundance was attained rather than another, then weather would undoubtedly be of major importance. By implication, therefore, the density-dependent processes would be of only secondary importance in this respect (see also Section 7.6).

(6) Thus, it would be unwise to go along with Nicholson wholeheartedly. Although density-dependent processes are an absolute necessity as a means of regulating populations and are generally by no means unimportant in determining abundance, they may well be of only minor importance when it comes to explaining particular observed population sizes. Moreover, because all environments are variable, the position of any 'balance-point' is continually changing. In spite of the ubiquity of density-dependent, regulating (i.e. balancing) processes, therefore, there seems little value in a view based on universal balance with rare non-equilibrium interludes. On the contrary, it is likely that *no* natural population is *ever* truly at equilibrium. This, remember, was the essential reason for our pragmatic definition of 'regulation' in Section 2.3.

7.6 A life-table analysis of a Colorado beetle population

These conclusions can be illustrated most effectively by analysing some life-table data in a way which has its roots in the work of Morris (1959) and Varley & Gradwell (1960), and is discussed more fully by Varley *et al.* (1975). As will become clear, the analysis allows us to weigh up the role and importance of each of the various mortality-factors, and also allows us to distinguish between factors that are important in determining the total mortality-rate, factors that are important in determining *fluctuations* in mortality-rate, and factors that are important in regulating a population. In the present case, the analysis is applied to a population of the Colorado potato beetle (*Leptinotarsa decemlineata*) in eastern Ontario where it has one generation per year (Harcourt 1971).

7.6.1 Life-table data

'Spring adults' emerge from hibernation around the middle of June, when the potatoes are first breaking through the ground. Within 3–4 days oviposition begins, extending for about 1 month and reaching its peak in early July. The eggs are laid in clusters (average size, 34 eggs) on the lower leaf surface, and the larvae crawl to the top of the plant where they feed throughout their development, passing through four instars. Then, when mature, they drop to the ground and form pupal cells in the soil. The 'summer adults' emerge in early August, feed, and then re-enter the soil at the beginning of September to hibernate and become the next season's 'spring adults'.

Details of the sampling methods can be found in Harcourt (1964), but it should be stressed that:

(a) on each occasion, sampling was continued until population estimates had confidence limits which were 10% or less of the mean; and

(b) the timing of samples varied from year to year, to ensure that the effects of variable climate on the insects' rate of development were allowed for.

The sampling programme provided estimates for seven age-intervals, from which a life-table could be constructed. These were: eggs, early larvae, late larvae, pupal cells, summer adults, hibernating adults and spring adults. In addition, one further category was included, 'females×2', to take account of any unequal sex ratios amongst the summer adults.

Table 7.1 lists these age-intervals and the numbers within them for a single season, and also gives the major 'mortality-factors' to which the deaths between successive intervals can be attributed. Figures obtained directly from sampling are indicated in bold type, the rest were obtained by subtraction. Amongst the eggs, predation and cannibalization were monitored directly, since both processes left behind recognizable egg remains. Reductions in hatchability, on the other hand, due either to infertility or to rainfall (mud splash), were assessed by returning samples of eggs to the laboratory and observing their progress individually. Finally, the

Table 7.1 Typical set of life-table data collected by Harcourt (1971) for the Colorado potato beetle (in this case, for Merivale 1961–62). Figures in bold type were obtained directly, the rest by subtraction.

Age interval	Numbers per 96 potato hills	Numbers 'dying'	'Mortality factor'	$\log_{10} N$	k-value	
Eggs	**11 799**	2531	Not deposited	4.072	0.105	(k_{1a})
	9268	**445**	Infertile	3.967	0.021	(k_{1b})
	8823	**408**	Rainfall	3.946	0.021	(k_{1c})
	8415	**1147**	Cannibalism	3.925	0.064	(k_{1d})
	7268	**376**	Predators	3.861	0.024	(k_{1e})
Early larvae	6892	**0**	Rainfall	3.838	0	(k_2)
Late larvae	**6892**	3722	Starvation	3.838	0.337	(k_3)
Pupal cells	**3170**	**16**	*D. doryphorae*	3.501	0.002	(k_4)
Summer adults	**3154**	**−126**	Sex (52% ♀)	3.499	−0.017	(k_5)
♀×2	3280	**3264**	Emigration	3.516	2.312	(k_6)
Hibernating adults	16	**2**	Frost	1.204	0.058	(k_7)
Spring adults	**14**			1.146	2.926	(k_{total})

figure for eggs 'not deposited' was based on the difference between the actual number of eggs and those expected on the basis of spring-adult number and mean fecundity. These five egg 'mortality-factors' have, for simplicity, been presented as acting successively in Table 7.1, although, in reality, they overlap considerably. The loss of accuracy resulting from this is generally slight (Varley *et al.* 1975).

The principal mortality factor during the first larval age-interval (from hatching to second instar) was rainfall, since, during heavy downpours, the small larvae were frequently washed from the leaves to the ground where they died in small puddles of water. This mortality was assessed by taking population counts before and after each period of rain. Amongst older larvae, on the other hand, the major mortality-factor was starvation.

Larval mortality due to parasites and predators was insignificant, and well within the sampling error. In the pupal stage, by contrast, parasitization was an important cause of mortality; the numbers of sound pupae and those containing puparia of the parasitic tachinid fly, *Doryphorophaga doryphorae* were estimated directly from the sampling data.

Amongst the summer adults, the principal cause of 'mortality' was emigration provoked by a shortage of food. This was assessed from a series of direct counts during the latter half of August. Damage by frost, the major mortality-factor acting on hibernating adults, was assessed by digging up a sample at the end of April. The number of spring adults was estimated by direct sampling in early July. There was no evidence of spring migration, nor of generation-to-generation changes in fecundity. (Where the data provided more than one estimate of the numbers in a particular stage, these were integrated into a single figure.)

As Table 7.1 shows, k-values have been computed for each source of mortality, and their mean values over ten seasons for a single population are presented in the second column of Table 7.2. These indicate the relative strengths of the various mortality factors as contributors to the total rate of mortality within a generation. Thus, the emigration of summer adults has by far the greatest proportional effect, while the starvation of older larvae, the frost-induced mortality of hibernating adults, the 'non-deposition' of eggs, the effects of rainfall on young larvae and the cannibalization of eggs all play substantial roles.

7.6.2 'Key-factor' analysis

What the second column of Table 7.2 does not tell us, however, is the relative importance of these factors as

Table 7.2 Summary of the life-table analysis for Canadian Colorado beetle populations (data from Harcourt 1971). b and a are, respectively, the slope and intercept of the regression of each k-factor on the logarithm of the numbers preceding its action; r^2 is the coefficient of determination. (See text for further explanation.)

		Mean	Coefficient of regression on k_{total}	b	a	r^2
Eggs not deposited	k_{1a}	0.095	−0.020	−0.05	0.27	0.27
Eggs infertile	k_{1b}	0.026	−0.005	−0.01	0.07	0.86
Rainfall on eggs	k_{1c}	0.006	0.000	0.00	0.00	0.00
Eggs cannibalized	k_{1d}	0.090	−0.002	−0.01	0.12	0.02
Egg predation	k_{1e}	0.036	−0.011	−0.03	0.15	0.41
Larvae 1 (rainfall)	k_2	0.091	0.010	0.03	−0.02	0.05
Larvae 2 (starvation)	k_3	0.185	0.136	0.37	−1.05	0.66
Pupae (*D. doryphorae*)	k_4	0.033	−0.029	−0.11	0.37	0.83
Unequal sex ratio	k_5	−0.012	0.004	0.01	−0.04	0.04
Emigration	k_6	1.543	0.906	2.65	−6.79	0.89
Frost	k_7	0.170	0.010	0.002	0.13	0.02

determinants of the year-to-year fluctuations in mortality (remember Sections 7.4 and 7.5). We can easily imagine, for instance, a factor that repeatedly takes a significant toll from a population, but which, by remaining constant in its effects, plays little part in determining the particular rate of mortality (and thus the particular population size) in any one year. In other words, such a factor may, in a sense, be important in determining population size, but it is certainly not important in determining changes in population size, and it cannot help us understand why the population is of a particular size in a particular year. This *can* be assessed, however, from the third column of Table 7.2, which gives the regression coefficient of each individual k-value on the total generation value, k_{total}. Podoler & Rogers (1975) have pointed out that a mortality-factor that is important in determining population change will have a regression coefficient close to unity, because its k-value will tend to fluctuate in line with k_{total} in terms of both size and direction. A mortality-factor with a k-value that varies quite randomly with respect to k_{total}, however, will have a regression coefficient close to zero. Moreover, the sum of all the regression coefficients within a generation will always be unity. Their values will, therefore, indicate their relative importance as determinants of fluctuations in mortality, and the largest regression coefficient will be associated with the *key factor causing population change* (Morris 1959; Varley & Gradwell 1960).

In the present example, it is clear that the emigration of summer adults, with a regression coefficient of 0.906, is the key factor; and other factors (with the possible exception of larval starvation) have a negligible effect on the *changes* in generation mortality, even though some have reasonably high mean k-values. (A similar conclusion could be drawn, in a more arbitrary fashion, from a simple examination of the fluctuations in k-values with time (Fig. 7.4). Note, moreover, that Podoler and Roger's method, even though it is less arbitrary than this graphical alternative, still does not allow us to assess the statistical significance of the regression coefficients, because the two variables are not independent of one another.)

Thus, while mean k-values indicate the average strengths of various factors as causes of mortality each generation, key-factor analysis indicates their relative strengths as causes of yearly changes in generation mortality, and thus measures their importance as determinants of population size.

7.6.3 Regulation of the population

We must now consider the role of these factors in the *regulation* of the Colorado beetle population. In other words, we must examine the density-dependence of each of these factors. This can be achieved most easily by using the method established in Chapter 2, of plotting k-values for each factor against the common logarithm of the numbers present before the factor acted. Thus, columns 4, 5 and 6 in Table 7.2 contain, respectively, the slopes, intercepts and coefficients of determination of the various regressions of k-values on their appropriate '$\log_{10}$ initial densities'. Three factors seem worthy of close examination.

The emigration of summer adults (the key factor) appears to act in an *over*compensating density-dependent fashion, since the slope of the regression

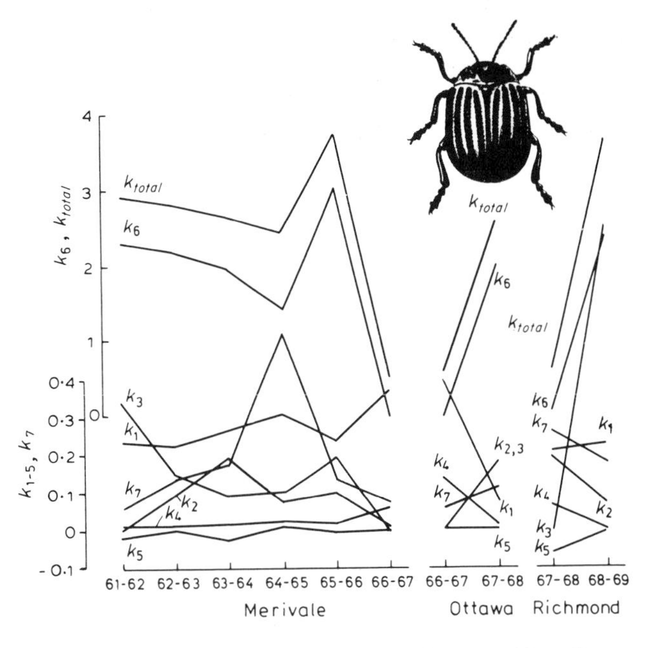

Fig. 7.4. The changes with time of the various k-values of Canadian Colorado beetle populations. Note that there are two quite separate scales on the vertical axis, and that k_6 is therefore undoubtedly the 'key-factor'. (Data from Harcourt 1971.)

(2.65) is considerably in excess of unity (Fig. 7.5a). (Once again, there are statistical difficulties in assessing the significance of this regression coefficient, but these can be overcome (Varley & Gradwell 1963; Varley *et al.* 1975), and in the present case density-dependence *can* be established statistically.) Thus, the key-factor, though density-dependent, does not so much regulate the population as lead, because of overcompensation, to violent fluctuations in abundance (actually discernible from the data: Fig. 7.7). Indeed, the potato/Colorado beetle system is only maintained in existence by humans, who prevent the extinction of the potato population by replanting (Harcourt 1971).

The rate of pupal parasitism by *D. doryphorae* (Fig. 7.5b) is apparently inversely density-dependent (though not significantly so, statistically), but because the mortality-rates are small, any destabilizing effects this may have on the population are negligible. Nevertheless, it is interesting to speculate that at low population levels, presumably prevalent before the creation of potato-monocultures by man, this parasitoid could act as an important source of beetle mortality (Harcourt 1971).

Finally, the rate of larval starvation appears to exhibit undercompensating density-dependence (though statistically this is not significant). An examination of Fig. 7.5c, however, indicates that the relationship would be far better reflected, not by a linear regression, but by a curve of the type discussed and examined in Section 3.2.2. If such a curve is fitted to the data, then the coefficient of determination rises from 0.66 to 0.97, and the slope (*b*-value) achieved at high densities is 30.95 (though it is, of course, much less than this in the density-range observed). Hence, it is quite possible that larval starvation plays an important part in regulating the population, prior to the destabilizing effects of pupal parasitism and adult emigration.

7.6.4 A population model

This type of analysis of life-table data allows us to examine the role and importance of each of the various mortality-factors acting on a population. It also illustrates the differences between factors that are important in determining year-to-year changes in mortality-rate, and factors that are important in regulating (or even destabilizing) a population. The final logical stage in such an analysis is to construct a synthetic model that will

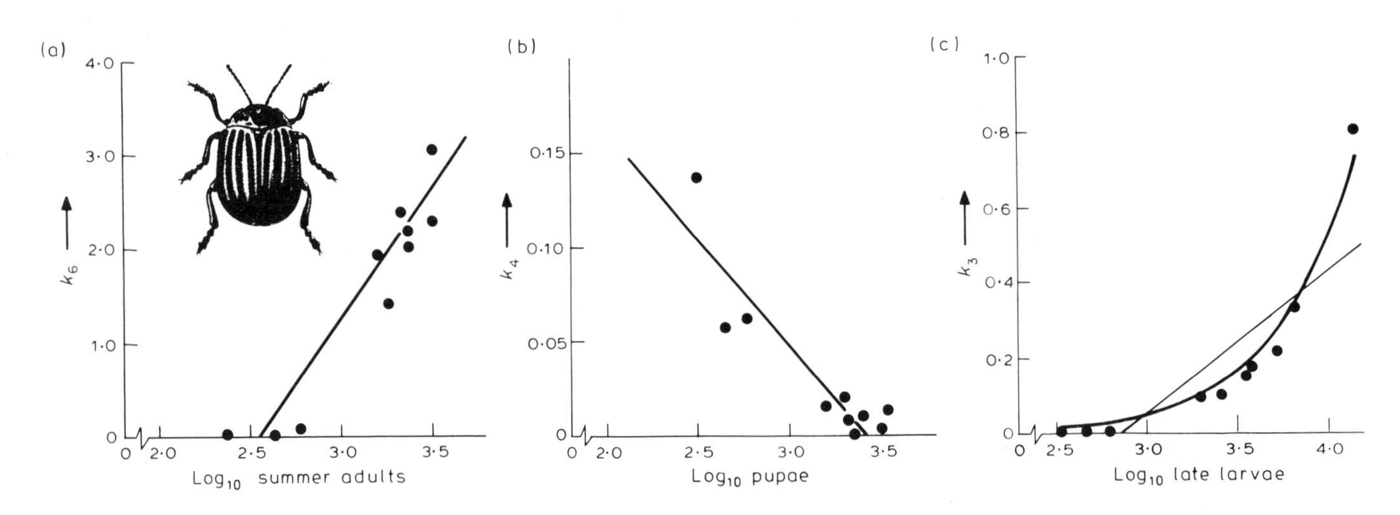

Fig. 7.5. (a) Density-dependent emigration amongst Canadian Colorado beetle summer adults which is overcompensating (slope = 2.65). (b) Inverse density-dependence in the parasitization of Colorado beetle pupae (slope = −0.11). (c) Density-dependence in the starvation of Colorado beetle larvae (straight line slope = 0.37; final slope of curve, based on equation 3.4 = 30.95). (Data from Harcourt 1971.)

allow us to predict:

(a) the probable future progress of a given population, and

(b) the consequences to the population of natural or enforced changes in any of the mortality factors.

Such a model should have as many 'steps' as there are 'stages' in the original analysis, and for each step there are two forms that the model could take. Ideally, the mortality-rate at a particular stage should be estimated directly from data on the mortality-factor itself. Thus, the mortality-rate of early larvae, for instance, should be estimated from rainfall data, and that of pupae from data on the numbers of *D. doryphorae*. Yet, the construction of these specific sub-models for each of the stages requires extensive collection of data. An imperfect but simpler alternative, and the one to which we shall be restricted in the present case, is to use the observed, empirical relationships between beetle numbers and mortality-rates, as summarized in the regression equations of columns 4 and 5 of Table 7.2. This approach literally uses what has happened in the past to predict what is likely to happen in the future, but it neglects any detailed consideration of the interactions occurring at each of the stages. (Of course, hybrid models, using different approaches at different stages, can also be constructed.)

In the present case we can argue from Table 7.2 that since, on average:

$$k_{1a} = 0.27 - 0.05 \log_{10}(\text{total potential eggs})$$

or, $\log_{10}(\text{total potential eggs}) - \log_{10}(\text{eggs deposited})$
$= 0.27 - 0.05 \log_{10}(\text{total potential eggs})$

then, $\log_{10}(\text{eggs deposited})$
$= 1.05 \log_{10}(\text{total potential eggs}) - 0.27.$

Similarly, $\log_{10}(\text{fertile eggs})$
$= 1.01 \log_{10}(\text{eggs deposited}) - 0.07,$

and so,
$\log_{10}(\text{fertile eggs})$
$= 1.01\{1.05 \log_{10}(\text{total potential eggs}) - 0.27\} - 0.07.$

If this is repeated for each stage, it allows us, eventually, to predict $\log_{10}$ (spring adults) from $\log_{10}$ (total potential eggs). In other words, we have constructed a model that will allow us to predict the level of infestation in one year given the number of eggs laid the previous year. Alternatively, since we know that each female lays, on average, 1700 eggs, we can use the number of spring adults in one year to predict the numbers in the various stages during the following year. In the present case, the model's predictions are illustrated in Fig. 7.6 (in which 'total potential eggs' is used to predict 'spring adults' in each of the ten generations studied) and Fig. 7.7 (in

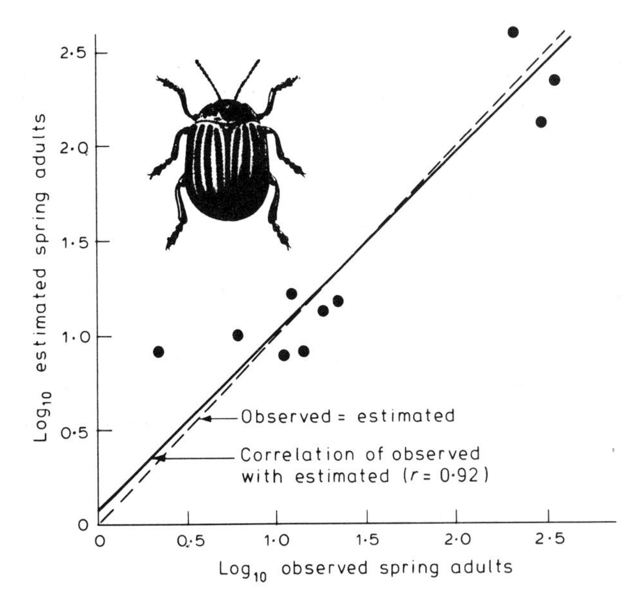

Fig. 7.6. The correlation between the observed number of Canadian Colorado beetle spring adults and the number estimated on the basis of the *k*-value model. For further explanation, see text. (Data from Harcourt 1971.)

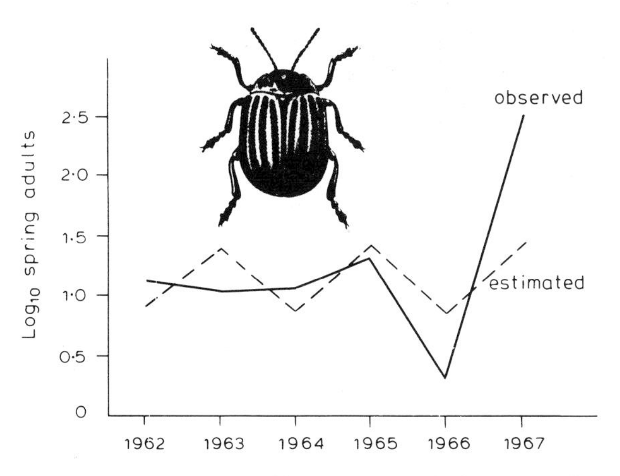

Fig. 7.7. The observed and estimated numbers of Colorado beetle spring adults at Merivale, based on the numbers observed in 1961. For further explanation, see text (Data from Harcourt 1971.)

which the number of eggs present in 1961 at a single site is used to generate an adult population curve for the following six seasons). In both cases the predictions are compared with the observed figures, and, given that in Fig. 7.7 any errors are bound to accumulate, the fit of the model is very satisfactory. This is a pleasing reflection of the amount that we can learn from such analyses of life-table data; and it serves to re-emphasize the need to consider carefully the integrated effects of all factors, both biotic and abiotic, when we seek to understand the abundance of natural populations.

7.7 Population regulation in plants

To date, natural plant populations have been subject to less intense scrutiny from population ecologists than their animal counterparts. Nevertheless, we can gain considerable insight into the nature of plant population regulation from those studies that have been made. Generally two approaches have been taken. On the one hand, populations fluctuating naturally in size have been monitored by regular census over long periods of time; and on the other hand, deliberate alterations in population size have been made to enable population responses to density to be assessed.

Annual sand dune plants provide us with good examples of species that have been rewarding to study by both techniques. Two such species are *Vulpia fasiculata* and *Androsace septentrionalis*. Both semelparous species complete their life cycle in less than a year. In *Vulpia*, seeds germinate in autumn and plants flower in late spring to shed seed in early summer. The life cycle in *Androsace* is even shorter; seedlings emerge in early March and plants have flowered and set seed by late May.

Fig. 7.8 shows the annual fluctuations observed by Symonides (1979) over an 8-year period in populations of *A. septentrionalis* in sand dunes. Annually, and with considerable consistency during the period 1968–74, over 10^5 seeds were produced per m^2. Yet at fruit dispersal in May of each year, populations fell in the restricted range of 100–300 flowering plants per m^2. Key-factor analysis (see Silvertown (1982) for details) revealed that the 'key' cause of mortality was seed loss in the 10 months intervening between seed shed in May and germination in the following March. The magnitude of this loss, however, was independent of seed density—on average only 0.4% of the seed crop was present as seedlings in early spring—and whilst the exact causes of mortality are not known, this is a clear reflection of the hostility of the sand dune environment for dormant *Androsace* seed. The only other significant period of death during the life cycle was in the establishment of

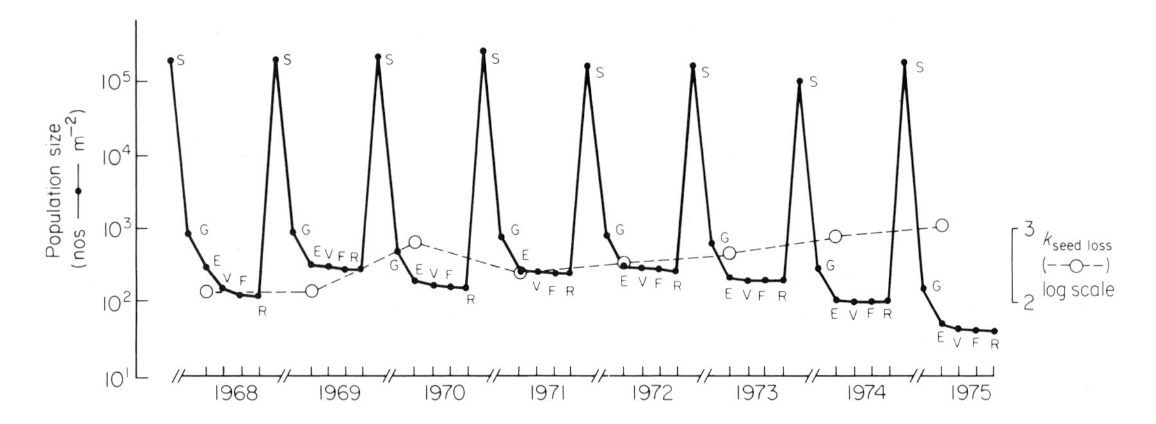

Fig. 7.8. Population fluctuations in a sand dune annual, *Androsace septentionalis*: S, seed population; G, seedling; E, established plant; V, vegetative mature plants; F, flowering plants; R, fruiting plants. Breaks in horizontal axis represent period from May to March (10 months). (Data from Symonides 1979.)

young plants from seedlings. Here density-dependence regulated the number of young plants in an undercompensatory fashion (Fig. 7.9). Such undercompensation is important to the persistence of *Androsace* in sand dunes. We can see (Fig. 7.8) that during the latter 5 years of Symonides' study, mortality of dormant seeds from primarily abiotic causes increased ($k_{seed\ loss}$ rises), and in 1975, 98% of seeds were lost after dispersal and the population had declined to 36 plants per m². Had density-dependence at the seedling stage been stronger (e.g. exact compensation) it is possible that local populations would have been very close to extinction.

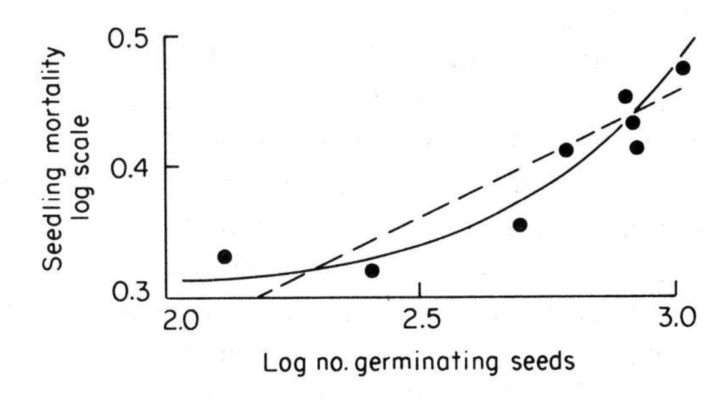

Fig. 7.9. Effect of increasing density on seedling mortality. Slope of dashed line = 0.2 (r^2 = 0.58). Fitting equation 3.4, b (final slope) = 18.5 (r^2 = 0.81). Data from Symonides 1979.)

Regulation in this species, then, is substantial in the juvenile stages (seed and seedling), although we might have expected regulation of seed number at the flowering stage as well. k-value analysis of the data revealed no evidence of this but only because the densities of flowering plants had already been restricted. This illustrates the inherent limitation of simply taking population estimates by census and seeking (albeit with a powerful tool) the causes of population regulation. If, for some reason, the range of densities of flowering plants had been greater, density-dependent regulation of seed fecundity may well have been detected.

In an endeavour to expose all possible sources of density-dependent regulation in the grass *Vulpia fasiculata* in sand dunes, Watkinson & Harper (1978) studied natural populations in which they deliberately established (by addition of seeds or removal of very young seedlings) a range of densities from 100 to 8000 plants per 0.25 m². This species contrasts with *Androsace* in having a shorter period between seed dissemination and germination; seeds 'over-summer' from July through to October, a period in which there is usually less than 1% loss of the annual seed crop. Mortality between seedling emergence and flowering varied between 7 and 41% but was not density-dependent. A range of abiotic and biotic factors were the cause of these deaths, including rabbit grazing, drought and seedling removal by wind drag. This mortality was visited on populations up to and including early flowering. *Vulpia* plants may bear up to 4 seeds, but as Fig. 7.10a shows, the actual numbers of seeds

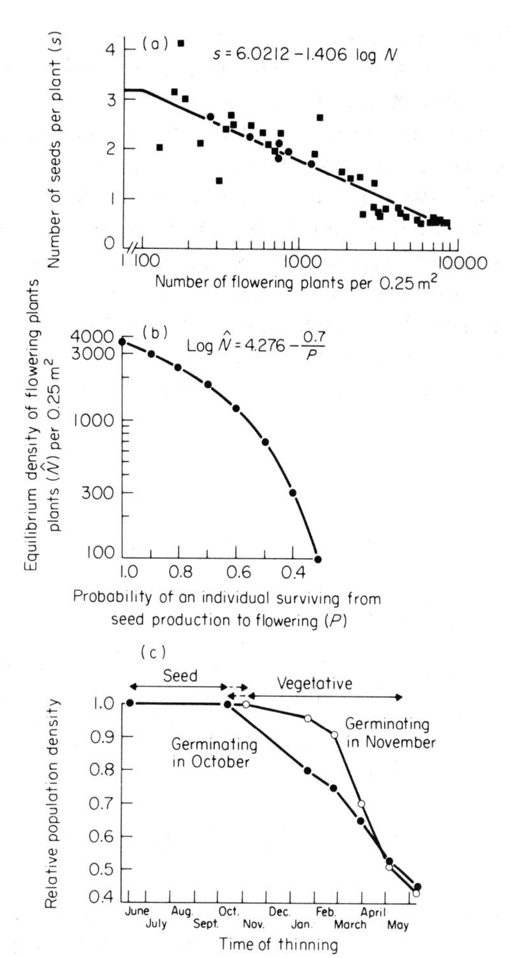

Fig. 7.10. Population regulation in *Vulpia fasiculata*. See text for details. (After Watkinson 1978, 1983.)

borne per plant was dependent on the density of plants at flowering time. (Note, though, that below 100 plants per 0.25 m^2, seed number is independent of density). To examine the stabilizing properties of this density-dependence, Watkinson and Harper proposed a model in which the density of flowering plants, N_t, was a function of the number of seeds per unit area produced in the previous generation, S_{t-1} and the probability, P, of an individual surviving from seed production (at time $t-1$) through to maturity (at time t).

Thus, (7.1)

$$N_t = P \,.\, S_{t-1}$$

But, as Fig. 7.10a illustrates, the average number of seeds borne per plant, s, was linearly related to flowering plant density N (above 100 plants per 0.25 m^2). Hence,

$$s = K - C \log N$$

(K and C are the constants describing the straight line relationship) and the seed yield per unit area of these N plants is

$$sN = (K - C \log N)N.$$

Substituting this term with the appropriate subscripts into equation 7.1 gives

$$N_t = P\,(K - C \log N_{t-1})\, N_{t-1}.$$

a model describing the changes in the number of flowering plants from generation to generation.

If density-independent mortality is constant from one generation to another we may calculate an equilibrium population density ($\hat{N} = N_t = N_{t-1}; N_t/N_{t-1} = 1$) as

$$\log \hat{N} = \frac{K}{C} - \frac{1}{CP}$$

when populations exceed 100 flowering plants per 0.25 m^2. (Below this density, there is no density-dependent regulation and hence no equilibrium population size) Fig. 7.10b shows the predicted equilibrium density for a range of P values and we can see that it becomes increasingly sensitive to lowered chances of survival. When P falls below 0.31, the population will decline; this is because the chance of an individual *Vulpia* plant replacing itself in the next generation is less than 1. Pleasingly, the values of P measured for *Vulpia* in the dunes fell in the range 0.34–0.59 supporting the prediction of the model that populations would persist in the dunes, a fact historically recorded since the beginning of the nineteenth century (Watkinson & Harper 1978).

The time at which density-independent thinning takes place in the growth of a population of plants may well be important: early thinning may offer a greater period of time than late thinning for the increased growth of survivors compensating for density reductions. The extent of compensation, however, will depend on the time when the various components of yield are formed (Section 2.5.2). Watkinson (1983) assessed the importance of this by varying the time during the life cycle when populations were thinned by a half (a figure chosen arbitrarily). Fig. 7.10c gives a summary of these effects on the calculated equilibrium population densities. Thinning during the seed phase of the life cycle has no influence on equilibrium density, since surviving plants compensate by elevated seed production (Fig. 7.10a). However, as the time of thinning was successively delayed (dotted line, Fig. 7.10c), equilibrium population densities fell as the compensatory response became increasingly muted. The individual yield components in *Vulpia*—number of fertile tillers, number of spikelets per flower and number of seeds per spikelet—were determined in accordance with the density perceived at their formation, and in consequence compensatory responses to density reductions were limited by the yield components already formed. However, the extent to which this occurred was dependent on the extent to which competition had occurred prior to the time of thinning. Plants sown at a later date (November, Fig. 7.10c) had not achieved a sufficient size to interfere with each other's growth by January and February, and thinning of populations at this time depressed population equilibria to a lesser extent than in populations sown 1 month earlier. We can appreciate, then, that the model developed earlier has the generality to encompass all situations where population regulation is achieved; yet it subsumes within it a density-dependent term which itself is determined by the level of density-independent regulation.

Both *Vulpia* and *Androsace* are plant species that lend themselves particularly well to demographic study. Generations do not overlap, either as living plants or as seeds (there is no persistent seed bank) and individuals in the population enter each stage of growth largely synchronously owing to birth (germination) at very much the same time. However, this life history is but one—and perhaps the simplest—that plant species display. Many species possess seed banks from which seedlings may be recruited over a protracted period of time. Episodic germination from a bank of seeds will result in an age-structured population which, as we have already seen (Chapter 3), requires a more sophisticated mathematical description. Yet, despite this complexity we can still unravel sources of population regulation in some instances by straightforward methods.

Fig. 7.11 illustrates the density relationships occurring within populations of the grass weed *Avena fatua* (wild oat) infesting a crop of wheat. In Britain, seed germination in this weed occurs in autumn during crop sowing as well as in spring when the crop is growing. Populations then become age-structured according to the range of seed germination times. Mature plants disseminate seed in mid-summer and die before crop harvest. Manlove (1985) sowed *A. fatua* over a wide range of seed densities on its own or in the presence of a constant density of wheat plants. He then tagged the wild oat plants as they emerged and followed their survivorship through to seed production. By plotting the recorded densities of seeds versus seedlings and seedlings versus adults we can look for regulation at these two stages in the life cycle (Fig. 7.11a,b). Density-dependent regulation is suggested at the seed/seedling stage: the scatter of points in Fig. 7.11a is below the line of unit slope. Once seedlings became established, mortality during growth to plant maturity was slight and not related to seedling density either in the presence or absence of the crop (Fig. 7.11b). Further experimental evidence showed that regulation of seedling number was primarily the result of seed loss from the soil surface in late summer and early autumn, when seed predators (birds and small mammals) foraged in the plant stubbles after harvest. The intensity of this predation increased in a density-dependent manner.

A second source of density-dependent regulation in wild oat was in seed production per plant (Fig. 7.11c). This was inversely related to mature plant density, and it was also depressed eightfold on average by wheat, across the entire density range. We can examine the stabilizing properties of these two regulatory phases by calculating the net reproductive rate, *R*, of populations at different starting seed densities (Fig. 7.12). *R* declines monotonically with density in monoculture and in wheat (square symbols) down to densities of circa 34600 and 7240 seeds per m^2 respectively. These are equilibrium

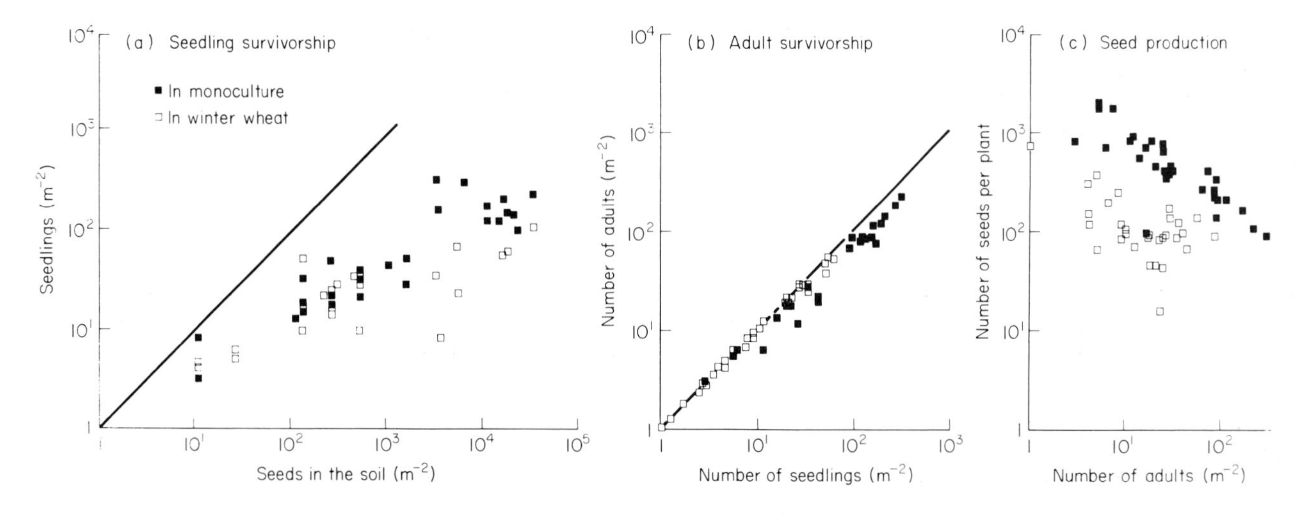

Fig. 7.11. Population regulation in *Avena fatua* in monoculture and in a crop of wheat. (a) Density-dependent regulation of seedling emergence. (b) Density-independent survivorship of adult plants. (c) Density-dependent regulation of seed production. Solid lines are of unit slope. (Data of Manlove 1985.)

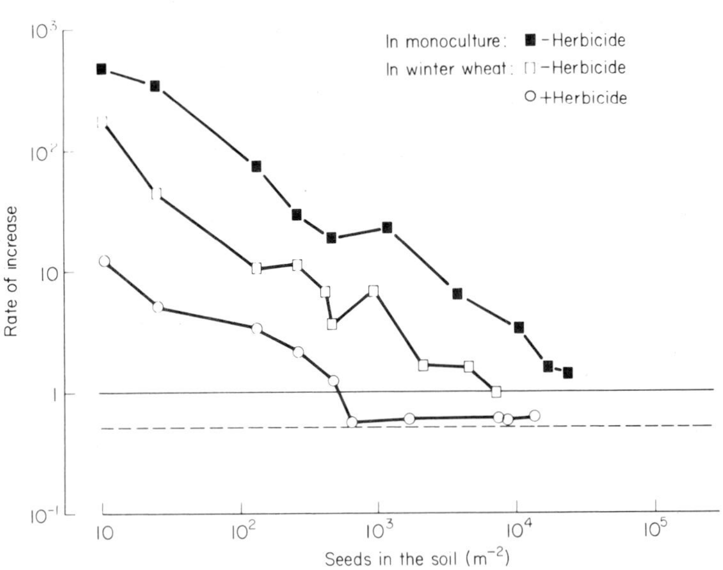

Fig. 7.12. Density-dependent and -independent regulation in *Avena fatua*. In a generation the number of seeds produced (S_t) is the arithmetic product of: the density of seeds in the soil in the autumn S_{t-1}; the probability of seedling emergence (Fig. 7.11a); the probability of seedling survival (Fig. 7.11b); and the number of seeds produced per plant (Fig. 7.11c). The rate of increase or net reproductive rate is then $S_t/S_{t-1}+0.55$. 0.55 is added as this is the fraction of dormant seeds surviving over generations. Dashed line indicates the (constant) rate of decline of the buried seed population; solid line shows rate of increase required to keep a population at equilibrium. Slopes of lines are not significantly different from one another. (Data from Manlove 1985.)

population densities–so long as all other factors remain constant from one generation to the next. Fig. 7.12 also demonstrates the effect of a density-independent control measure—a herbicide selective against wild oat which was applied in early summer before flowering. Its mode of action is to cause flower abortion in the weed (and hence seed loss) rather than plant mortality. In consequence, reproductive rates are depressed—the line moves towards the origin—and the combination of crop competition and herbicide reduced the equilibrium population density of wild oat to 470 seed per m^2.

This example, whilst supportive of what we already know regarding population regulation, illustrates particularly clearly the various roles of different components of regulation. The underlying cause of regulation is intraspecific. This arises in part from density-dependent seed predation and in part from intraspecific competition determining the seed yield of mature plants. Interspecific competition (from wheat) serves to reduce the net reproductive rate uniformly across the wild oat density range; and a density-independent control (herbicide) depresses these rates even further. We must also remember that the analysis subsumes the effect of age structure in the wild oat population. Whilst wild oats that emerge in the autumn contribute a greater number of seeds per plant to the next generation than late emergers, these differences are absorbed within the overall density response. Late emerging plants enter the hierarchy of exploitation (Section 2.5.2) later in the growing season and are proportionately disadvantaged for doing so. Nevertheless, even individuals lowest in the size hierarchy usually contributed one or two seeds to the next generation.

Part of this study also involved an examination of the dynamics of buried seed populations. Loss of seeds in the soil (through seedling germination and death) occurred at a constant rate (a type II survivorship curve) regardless of density. Slightly more than half (0.55) of the seed population in the soil survived from one generation to the next. We can easily appreciate the role of the seed bank in maintenance of plant populations by reconsidering the wild oats growing in wheat and sprayed with herbicide. Populations arising from seed densities greater than the 470 seeds per m^2 had net reproductive rates less than 1—competition and herbicide generally rendering plants at these densities barren. Yet the persistence of dormant seeds over generations enables recruitment of individuals in the succeeding generation and a rapid return to the equilibrium density.

Identifying the precise causes of population regulation in species with a *clonal* growth form is complicated by the necessity to take into account vegetative propagation by ramet fragmentation (as well as the practical problem of identifying ramets). Two species in which vegetative propagation is common are the creeping buttercup *Ranunculus repens* and the grass *Holcus lanatus*, both occurring in grasslands in Britain. In the buttercup, daughter rosettes are produced on stolons 10–12 cm apart, and in late summer (July–August) these become detached from the parent as interconnecting stolons decay. In *Holcus*, tillers are borne on shoot complexes (<5 cm apart) which continually become fragmented through natural decay and trampling of grazing animals. In grassland, establishment of new plants from seed in both species is rare. Yet as Fig. 7.13 shows, natural populations of both species undergo considerable turnover whilst maintaining relatively static population sizes. Ramet populations (tillers or rosettes) experienced a constant death risk (type II survivorship curve), the life expectancies of ramets in both species being 11–18 months (Sarukhan & Harper 1973; Weir 1985). At the same time there was recruitment of new ramets, the process occurring more or less continuously in *Holcus* but with noticeable gain and loss dominated periods in *Ranunculus*. Sarukhan and Harper were able to detect density-dependent regulation in the buttercups as populations accommodated to the pulse of recruitment (Fig. 7.14) whilst in *Holcus* this process was absent because of the fine scale periodicity in ramet replacement. This comparison leads us to a final important conclusion on population regulation in plants. In species where there are pulses of recruitment (as exemplified by unitary species and some clonal ones) there is the opportunity for density-dependent regulation to occur. On the other hand, where there is very rapid turnover at the modular level there may be little opportunity for such

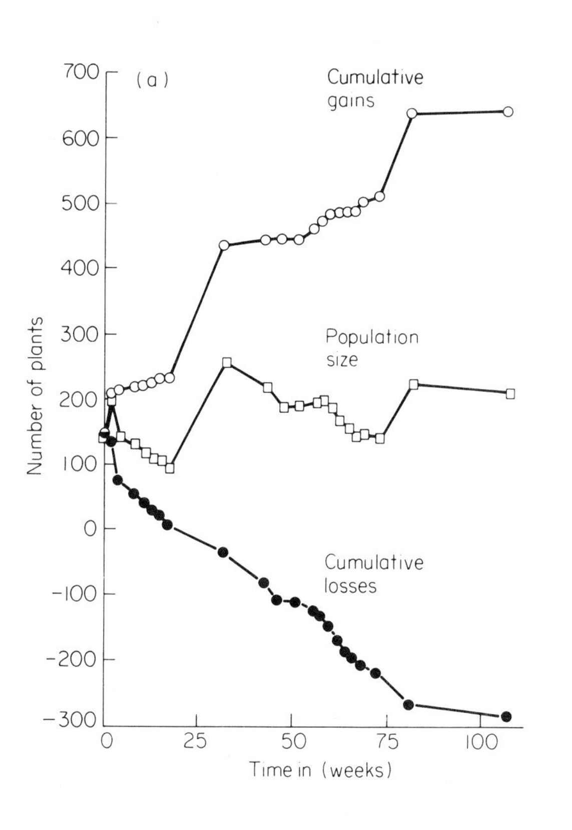

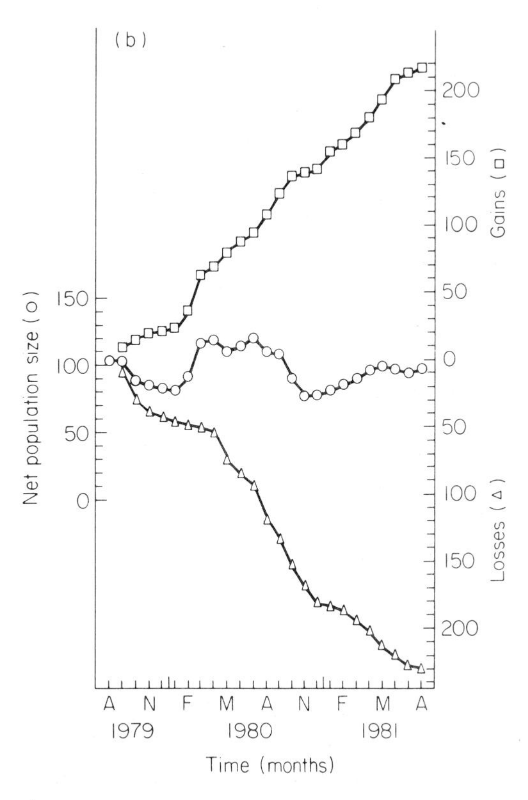

Fig. 7.13. Population turnover in (a) *Ranunculus repens* and (b) *Holcus lanatus*. (After Sarukhan & Harper 1973; Weir 1985.)

analyses for the great tit and the red grouse fail to provide clear-cut evidence of density-dependence at the appropriate stage (Podoler & Rogers 1975).

Wynne-Edwards (1962) felt that these regulatory consequences of territoriality must themselves be the root causes of territorial behaviour. He suggested that the selective advantage accrued to the population *as a whole*: that it was advantageous to the population not to over-exploit its resources. There are, however, powerful and fundamental reasons for rejecting this 'group-selectionist' explanation—essentially, it stretches evolutionary theory beyond reasonable limits (Maynard Smith 1976)—and Wynne-Edwards himself (Wynne-Edwards 1977) has subsequently recognized these reasons and accepted the rejection of his ideas. Thus, if we wish to discover the ultimate cause of territoriality, we must search, within the realms of natural selection, for some advantage accruing to the *individual.*

It must be recognized that in assessing individual advantage, we must demonstrate not merely that there are benefits, but that these exceed the costs associated with territoriality. This has been done in the case of the golden-winged sunbirds examined in Fig. 7.16. Gill & Wolf (1975) demonstrated that although the size of territories may vary enormously, the nectar-supply defended is suited to support an individual's daily energy requirements. They were able to measure the time that territory owners spent in various activities (including territory defence), and they showed that, as a result of the exclusion of other sunbirds, nectar levels per flower inside a territory were higher than in undefended flowers. Thus, Gill and Wolf found that territory owners could obtain their daily energy requirements relatively quickly, and that, overall, the energetic costs of territorial defence were easily offset by the benefits of the energy saved by a shortened daily foraging time.

A different type of individual advantage resulting from territoriality was demonstrated by Krebs (1971) for the great tit population of Wytham Woods (Fig. 7.17). There, the major predators of nestlings are weasels, *Mustela nivalis,* which may rob up to 50% of the nests in some years (Dunn 1977). But, as Fig. 7.17 shows, the closer a nest is to another nest, the greater the chance of predation. Individual selection, therefore, favours the spacing out of nests: it pays each individual to be territorial.

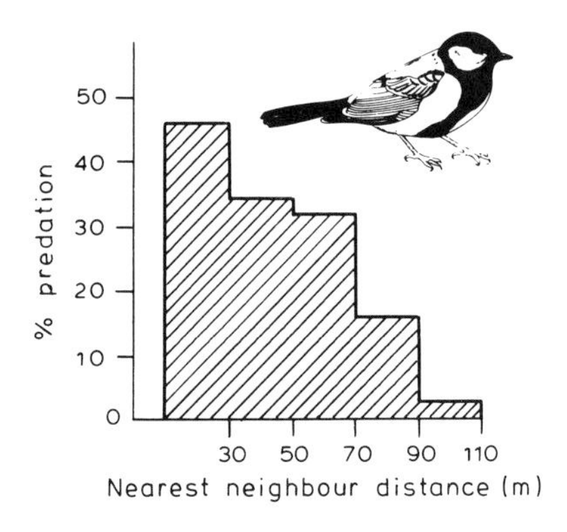

Fig. 7.17. The influence of territory size on the risk of predation in the great tit. (After Krebs 1971.)

We can suggest then, from this very limited number of examples, that territoriality has evolved as a result of the advantages accruing to territorial individuals. But as an essentially independent *consequence* of this, there is competition approaching pure contest, and therefore powerful (though not, of course, absolute) regulation of populations.

7.10 'Space capture' in plants

Although territoriality, as such, is not normally associated with plants, there is, in plants, a phenomenon which is at least analogous to territoriality. The phenomenon can be caricatured in the proverb: 'Possession is nine points of the law', and has been referred to and discussed by Harper (1977) as 'space capture'. In fact we have already discussed it briefly in Section 2.5.2 as an explanation for skewed frequency distributions of plant weights.

Fig. 7.18a shows that in experimental populations of the grass *Dactylis glomerata,* the comparatively low weights exhibited by late-emerging plants are lower than would be expected from the reduction in growing period alone (Ross & Harper 1972). The reason for this is indicated by the data in Fig. 7.18b (Ross & Harper 1972). Plants were grown from seed either under 'unrestricted' conditions: alone at the centre of a 7.4 cm diameter pot; or 'restricted' conditions: in a bare zone, 4.2 cm in diameter, surrounded by seeds sown at a seeds (there is no persistent seed bank) and individuals in the population enter each stage of growth largely synchronously owing to birth (germination) at very much the same time. However, this life history is but one—and perhaps the simplest—that plant species display. Many species possess seed banks from which seedlings may be recruited over a protracted period of time. Episodic germination from a bank of seeds will result in an age-structured population which, as we have already seen (Chapter 3), requires a more sophisticated mathematical description. Yet, despite this complexity we can still unravel sources of population regulation in some instances by straightforward methods.

Fig. 7.11 illustrates the density relationships occurring within populations of the grass weed *Avena fatua* (wild oat) infesting a crop of wheat. In Britain, seed germination in this weed occurs in autumn during crop sowing as well as in spring when the crop is growing. Populations then become age-structured according to the range of seed germination times. Mature plants disseminate seed in mid-summer and die before crop harvest. Manlove (1985) sowed *A. fatua* over a wide range of seed densities on its own or in the presence of a constant density of wheat plants. He then tagged the wild oat plants as they emerged and followed their survivorship through to seed production. By plotting the recorded densities of seeds versus seedlings and seedlings versus adults we can look for regulation at these two stages in the life cycle (Fig. 7.11a,b). Density-dependent regulation is suggested at the seed/seedling stage: the scatter of points in Fig. 7.11a is below the line of unit slope. Once seedlings became established, mortality during growth to plant maturity was slight and not related to seedling density either in the presence or absence of the crop (Fig. 7.11b). Further experimental evidence showed that regulation of seedling number was primarily the result of seed loss from the soil surface in late summer and early autumn, when seed predators (birds and small mammals) foraged in the plant stubbles after harvest. The intensity of this predation increased in a density-dependent manner.

A second source of density-dependent regulation in wild oat was in seed production per plant (Fig. 7.11c). This was inversely related to mature plant density, and it was also depressed eightfold on average by wheat, across the entire density range. We can examine the stabilizing properties of these two regulatory phases by calculating the net reproductive rate, *R*, of populations at different starting seed densities (Fig. 7.12). *R* declines monotonically with density in monoculture and in wheat (square symbols) down to densities of circa 34600 and 7240 seeds per m^2 respectively. These are equilibrium

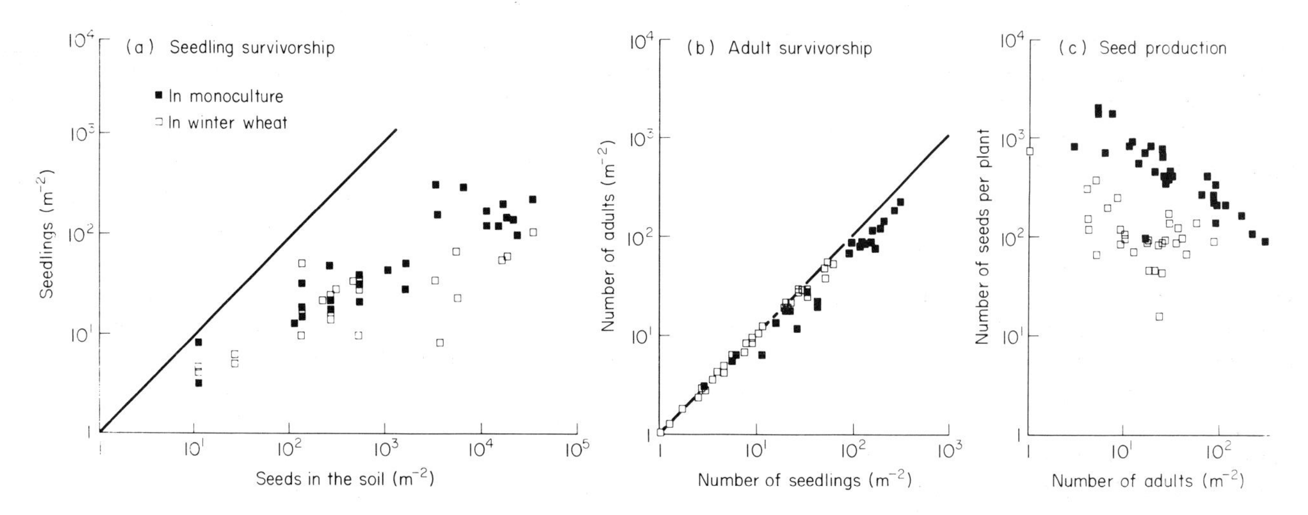

Fig. 7.11. Population regulation in *Avena fatua* in monoculture and in a crop of wheat. (a) Density-dependent regulation of seedling emergence. (b) Density-independent survivorship of adult plants. (c) Density-dependent regulation of seed production. Solid lines are of unit slope. (Data of Manlove 1985.)

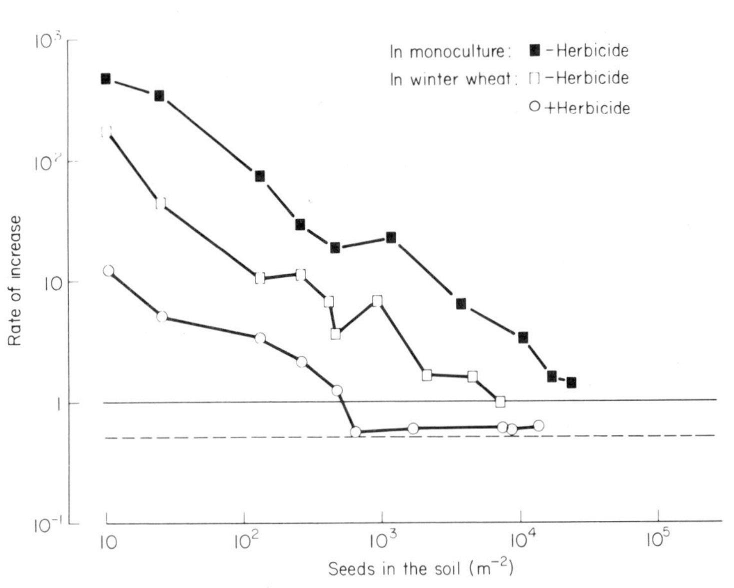

Fig. 7.12. Density-dependent and -independent regulation in *Avena fatua*. In a generation the number of seeds produced (S_t) is the arithmetic product of: the density of seeds in the soil in the autumn S_{t-1}; the probability of seedling emergence (Fig. 7.11a); the probability of seedling survival (Fig. 7.11b); and the number of seeds produced per plant (Fig. 7.11c). The rate of increase or net reproductive rate is then $S_t/S_{t-1}+0.55$. 0.55 is added as this is the fraction of dormant seeds surviving over generations. Dashed line indicates the (constant) rate of decline of the buried seed population; solid line shows rate of increase required to keep a population at equilibrium. Slopes of lines are not significantly different from one another. (Data from Manlove 1985.)

population densities–so long as all other factors remain constant from one generation to the next. Fig. 7.12 also demonstrates the effect of a density-independent control measure—a herbicide selective against wild oat which was applied in early summer before flowering. Its mode of action is to cause flower abortion in the weed (and hence seed loss) rather than plant mortality. In consequence, reproductive rates are depressed—the line moves towards the origin—and the combination of crop competition and herbicide reduced the equilibrium population density of wild oat to 470 seed per m^2.

This example, whilst supportive of what we already know regarding population regulation, illustrates particularly clearly the various roles of different components of regulation. The underlying cause of regulation is intraspecific. This arises in part from density-dependent seed predation and in part from intraspecific competition determining the seed yield of mature plants. Interspecific competition (from wheat) serves to reduce the net reproductive rate uniformly across the wild oat density range; and a density-independent control (herbicide) depresses these rates even further. We must also remember that the analysis subsumes the effect of age structure in the wild oat population. Whilst wild oats that emerge in the autumn contribute a greater number of seeds per plant to the next generation than late emergers, these differences are absorbed within the overall density response. Late emerging plants enter the hierarchy of exploitation (Section 2.5.2) later in the growing season and are proportionately disadvantaged for doing so. Nevertheless, even individuals lowest in the size hierarchy usually contributed one or two seeds to the next generation.

Part of this study also involved an examination of the dynamics of buried seed populations. Loss of seeds in the soil (through seedling germination and death) occurred at a constant rate (a type II survivorship curve) regardless of density. Slightly more than half (0.55) of the seed population in the soil survived from one generation to the next. We can easily appreciate the role of the seed bank in maintenance of plant populations by reconsidering the wild oats growing in wheat and sprayed with herbicide. Populations arising from seed densities greater than the 470 seeds per m^2 had net reproductive rates less than 1—competition and herbicide generally rendering plants at these densities barren. Yet the persistence of dormant seeds over generations enables recruitment of individuals in the succeeding generation and a rapid return to the equilibrium density.

Identifying the precise causes of population regulation in species with a *clonal* growth form is complicated by the necessity to take into account vegetative propagation by ramet fragmentation (as well as the practical problem of identifying ramets). Two species in which vegetative propagation is common are the creeping buttercup *Ranunculus repens* and the grass *Holcus lanatus*, both occurring in grasslands in Britain. In the buttercup, daughter rosettes are produced on stolons 10–12 cm apart, and in late summer (July–August) these become detached from the parent as interconnecting stolons decay. In *Holcus*, tillers are borne on shoot complexes (<5 cm apart) which continually become fragmented through natural decay and trampling of grazing animals. In grassland, establishment of new plants from seed in both species is rare. Yet as Fig. 7.13 shows, natural populations of both species undergo considerable turnover whilst maintaining relatively static population sizes. Ramet populations (tillers or rosettes) experienced a constant death risk (type II survivorship curve), the life expectancies of ramets in both species being 11–18 months (Sarukhan & Harper 1973; Weir 1985). At the same time there was recruitment of new ramets, the process occurring more or less continuously in *Holcus* but with noticeable gain and loss dominated periods in *Ranunculus*. Sarukhan and Harper were able to detect density-dependent regulation in the buttercups as populations accommodated to the pulse of recruitment (Fig. 7.14) whilst in *Holcus* this process was absent because of the fine scale periodicity in ramet replacement. This comparison leads us to a final important conclusion on population regulation in plants. In species where there are pulses of recruitment (as exemplified by unitary species and some clonal ones) there is the opportunity for density-dependent regulation to occur. On the other hand, where there is very rapid turnover at the modular level there may be little opportunity for such

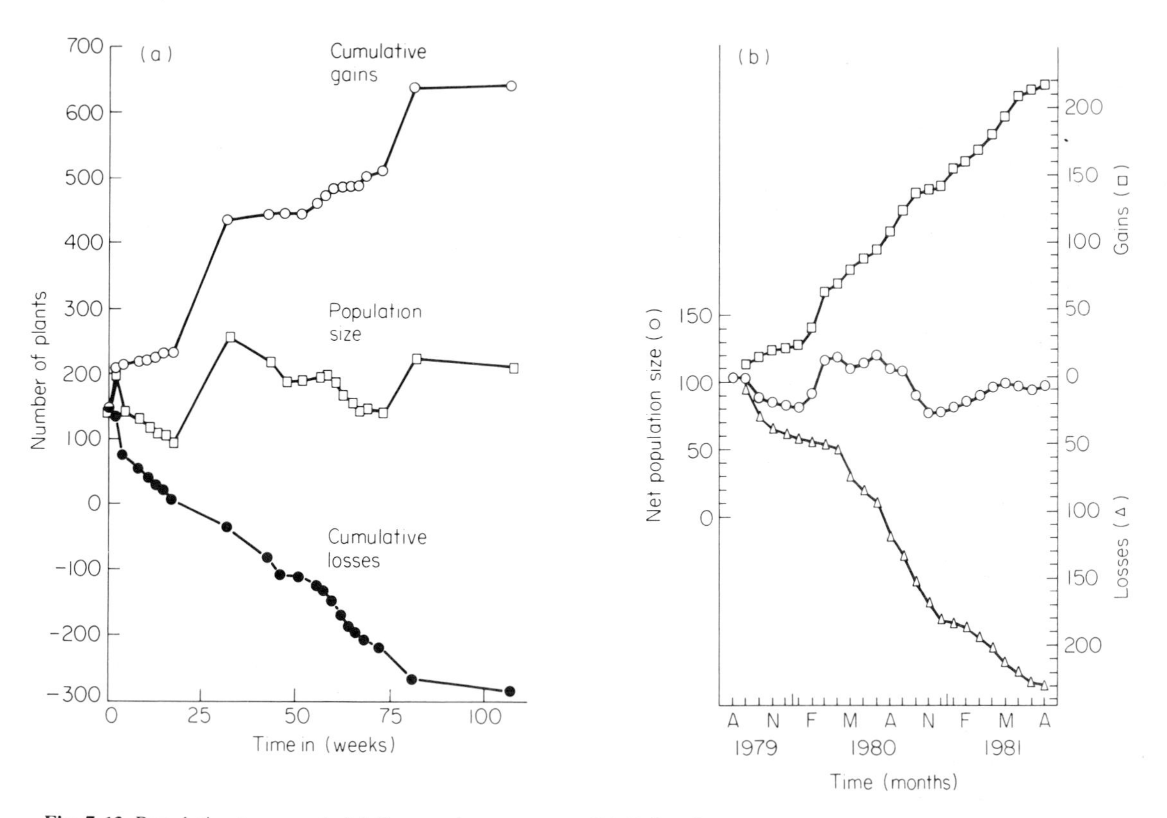

Fig. 7.13. Population turnover in (a) *Ranunculus repens* and (b) *Holcus lanatus*. (After Sarukhan & Harper 1973; Weir 1985.)

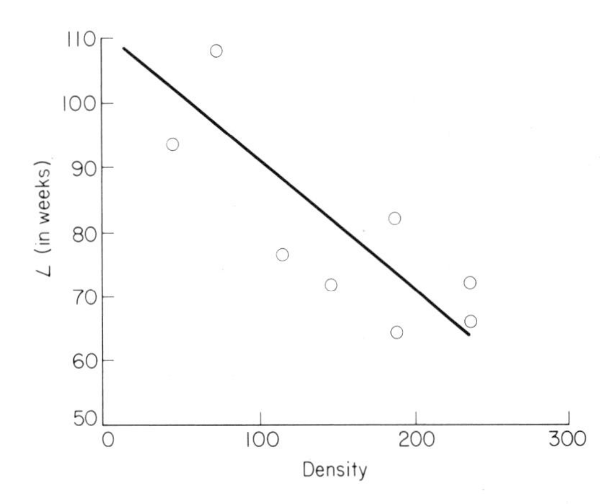

Fig. 7.14. Life expectancy (L) in weeks of vegetative propagules of *Ranunculus repens* as affected by its own species' density. Densities are average values for number of plants (m^{-2}) observed at each site in April 1969, 1970 and 1971. (After Sarukhan & Harper 1973.)

regulation because of inherent morphological and growth constraints within the growth form. In this case density-dependent regulation may be most likely at the genet level during seedling establishment and occupation of the regeneration niche (Section 4.12). It would be unwise to assume that fragmentation is a ubiquitous phenomenon amongst clonal plants. Others may simply increase in size with age and retain structural and physiological integrity. Certainly it has been shown that in some species clones are of considerable age and that they occupy large areas. For instance, Harberd (1961) found clones of a grass *Festuca ovina* in Scottish pastures up to 10 m in diameter and Oinonen (1967) has dated clones of bracken, *Pteridium aquilinum* over 450 years old extending in size over 100 m. The extent to which these clones are intact plants or populations of ramets remains an open question.

7.8 Genetic change

Discussions of population regulation are usually framed in terms of ecological time-scales; and, although it is generally accepted that the individuals concerned are the products of natural selection, genetic change, occurring, by definition, on an evolutionary time-scale, is usually neglected as a regulatory mechanism. There are some dissenting ecologists, however—and Pimentel (1961) is the most frequently quoted of these—who would suggest that such neglect is unwarranted, and that any discussion of population regulation is incomplete if it fails to take account of contemporary adaptive genetic change. There is, however, rather little concrete evidence in favour of this notion, and it is probable that the ecological and evolutionary time-scales are usually dissimilar. Nevertheless, there is some supporting evidence, and a particularly impressive example is provided by the work of Shorrocks (1970).

Shorrocks maintained populations of the fruit fly *Drosophila melanogaster* in the laboratory, and obtained regular four-generation cycles in abundance, unrelated to any regular change in an environmental variable. Pairs of flies were removed from the populations and classified as 'peak' or 'non-peak', depending on the type of abundance their parents experienced, and the numbers of offspring produced by these pairs when maintained under identical, uncrowded conditions were noted. The pairs in the 'peak' category produced significantly fewer offspring than their counterparts in the 'non-peak' category, and, more important, the difference was inherited and remained until the F_1 and F_2 generations. Clearly, demographic characters can respond to selection on an 'ecological' time-scale. In this case they did so in a way that made reproductive-rate inversely related to population size, and thus tended to regulate the population.

7.9 Territoriality

One topic which is intimately tied up with population regulation is territoriality. But as Davies (1978a) has pointed out, in a much fuller review of the subject than can be given here, there are a number of questions pertinent to territoriality which must be kept quite distinct. We can distinguish initially between 'What causes territorial behaviour?' and 'What are the consequences of territoriality?'; but even the first of these is itself the confusion of two quite separate questions, namely 'What is the ultimate cause or driving force, i.e. what is the selective advantage associated with territoriality?' and 'What is the proximate cause or mechanism through which territories are established?'.

We shall restrict ourselves here to considering the consequences of territoriality, and the selective advantages associated with it. But first we must define what is meant by a territory and by territorial behaviour. Following Davies (1978a), we will recognize a territory '. . . whenever [individuals] or groups are spaced out more than would be expected from a random occupation of suitable habitats'. Note, therefore, that territoriality can be ascribed not only to conventional cases like breeding great tits (Fig. 7.15a), but also to barnacles (Fig. 7.15b) and many plants. The rather special case of plants will be discussed in the next section.

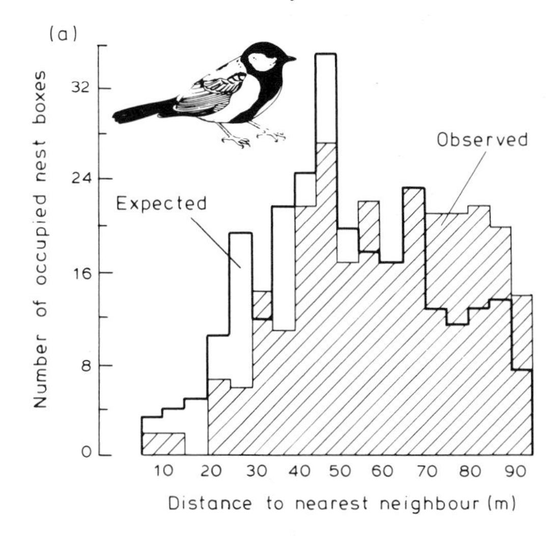

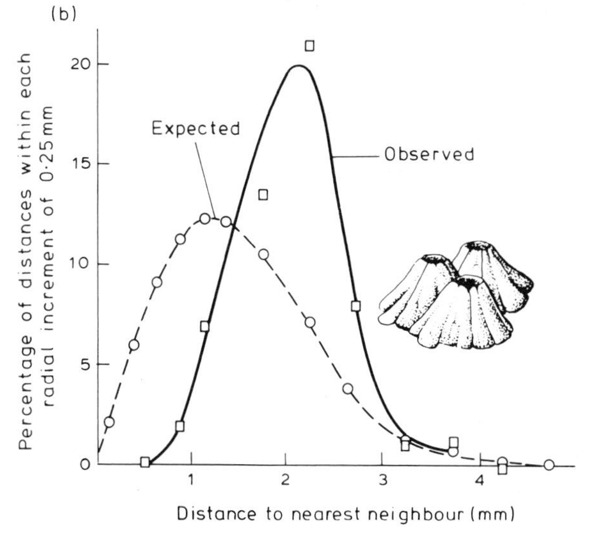

Fig. 7.15. Illustrations of territoriality. (a) Great tits (Krebs 1971) and (b) barnacles (Crisp 1961) spaced out more than would be expected from a random distribution on the available suitable habitats. (After Davies 1978a.)

The most important consequence of territoriality is population regulation. Territorial behaviour is closely allied to contest competition, and this, as we saw in Section 2.4, leads to exactly compensating density-dependence. The contest nature of territoriality is demonstrated by the fact that when territory owners die, or are experimentally removed, their places are rapidly taken by newcomers. Thus, Krebs (1971) found that in great tit populations, vacated woodland territories were reoccupied by birds coming from hedgerow territories where reproductive success was noticeably suboptimal; and Watson (1967) found that with red grouse the replacements were non-territorial individuals living in flocks, which would not have bred, and would probably have died in the absence of a territory. In both cases, therefore, overall fecundity and population size were limited by territorial behaviour: by a 'contest' for a limited number of territories. Removal experiments have demonstrated similar phenomena in mammals (Healey 1967; Carl 1971), fish (Clarke 1970), dragonflies (Moore 1964), butterflies (Davies 1978b) and limpets (Stimson 1973). It must be realized, however, that the exact number of territories is usually somewhat indeterminate in any one year, and certainly varies from year to year depending on environmental conditions (Fig. 7.16); and it is, perhaps, for this reason that life-table

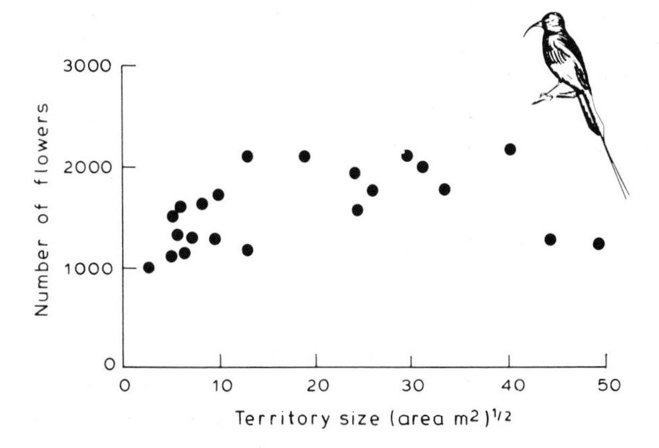

Fig. 7.16. Although the size of the territories of the golden winged sunbird, *Nectarina reichenowi*, varies enormously, each territory contains approximately the same number of *Leonotis* flowers (Gill & Wolf 1975). (After Davies 1978a.)

analyses for the great tit and the red grouse fail to provide clear-cut evidence of density-dependence at the appropriate stage (Podoler & Rogers 1975).

Wynne-Edwards (1962) felt that these regulatory consequences of territoriality must themselves be the root causes of territorial behaviour. He suggested that the selective advantage accrued to the population *as a whole*: that it was advantageous to the population not to over-exploit its resources. There are, however, powerful and fundamental reasons for rejecting this 'group-selectionist' explanation—essentially, it stretches evolutionary theory beyond reasonable limits (Maynard Smith 1976)—and Wynne-Edwards himself (Wynne-Edwards 1977) has subsequently recognized these reasons and accepted the rejection of his ideas. Thus, if we wish to discover the ultimate cause of territoriality, we must search, within the realms of natural selection, for some advantage accruing to the *individual*.

It must be recognized that in assessing individual advantage, we must demonstrate not merely that there are benefits, but that these exceed the costs associated with territoriality. This has been done in the case of the golden-winged sunbirds examined in Fig. 7.16. Gill & Wolf (1975) demonstrated that although the size of territories may vary enormously, the nectar-supply defended is suited to support an individual's daily energy requirements. They were able to measure the time that territory owners spent in various activities (including territory defence), and they showed that, as a result of the exclusion of other sunbirds, nectar levels per flower inside a territory were higher than in undefended flowers. Thus, Gill and Wolf found that territory owners could obtain their daily energy requirements relatively quickly, and that, overall, the energetic costs of territorial defence were easily offset by the benefits of the energy saved by a shortened daily foraging time.

A different type of individual advantage resulting from territoriality was demonstrated by Krebs (1971) for the great tit population of Wytham Woods (Fig. 7.17). There, the major predators of nestlings are weasels, *Mustela nivalis*, which may rob up to 50% of the nests in some years (Dunn 1977). But, as Fig. 7.17 shows, the closer a nest is to another nest, the greater the chance of predation. Individual selection, therefore, favours the spacing out of nests: it pays each individual to be territorial.

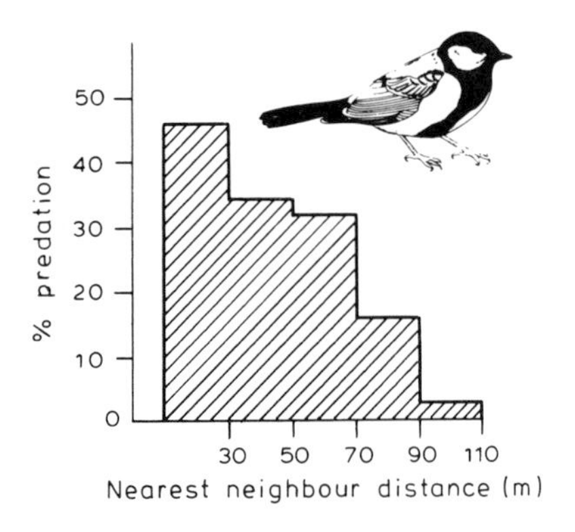

Fig. 7.17. The influence of territory size on the risk of predation in the great tit. (After Krebs 1971.)

We can suggest then, from this very limited number of examples, that territoriality has evolved as a result of the advantages accruing to territorial individuals. But as an essentially independent *consequence* of this, there is competition approaching pure contest, and therefore powerful (though not, of course, absolute) regulation of populations.

7.10 'Space capture' in plants

Although territoriality, as such, is not normally associated with plants, there is, in plants, a phenomenon which is at least analogous to territoriality. The phenomenon can be caricatured in the proverb: 'Possession is nine points of the law', and has been referred to and discussed by Harper (1977) as 'space capture'. In fact we have already discussed it briefly in Section 2.5.2 as an explanation for skewed frequency distributions of plant weights.

Fig. 7.18a shows that in experimental populations of the grass *Dactylis glomerata*, the comparatively low weights exhibited by late-emerging plants are lower than would be expected from the reduction in growing period alone (Ross & Harper 1972). The reason for this is indicated by the data in Fig. 7.18b (Ross & Harper 1972). Plants were grown from seed either under 'unrestricted' conditions: alone at the centre of a 7.4 cm diameter pot; or 'restricted' conditions: in a bare zone, 4.2 cm in diameter, surrounded by seeds sown at a

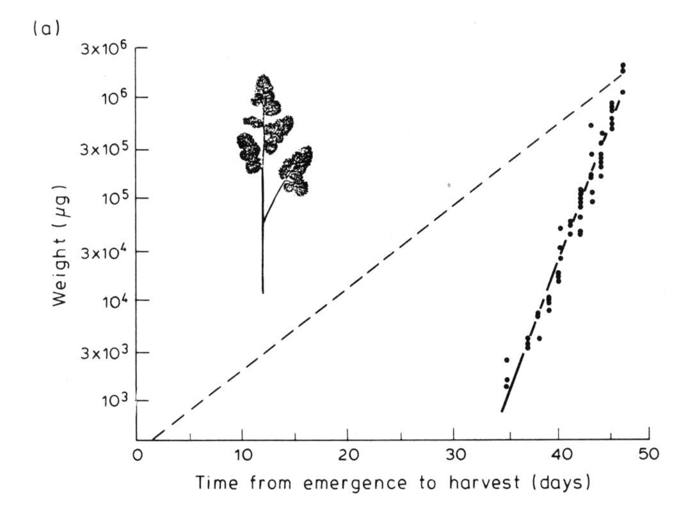

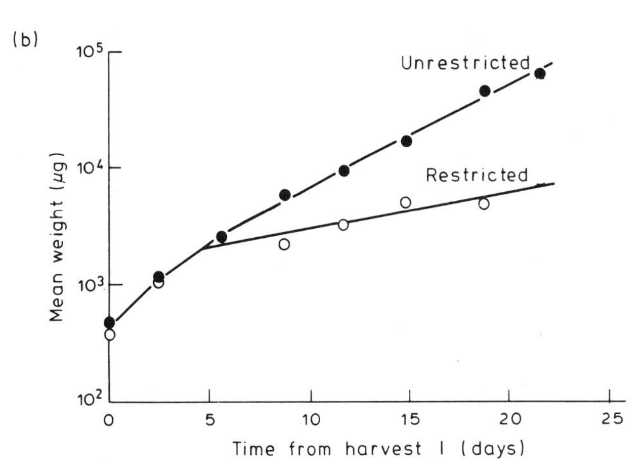

Fig. 7.18. (a) The influence of emergence time on the dry weight per plant of *Dactylis glomerata*. The dashed line shows the weights that would have been achieved had the weights of late emergers been attributable only to their reduced growth period. (b) The growth of *D. glomerata* seedlings with and without neighbours. (After Ross & Harper 1972.) For further explanation, see text.

density of 2.5 cm^{-2}. After initially growing at the same rate, the 'restricted' plants grew at a slower rate than the 'unrestricted' plants, and maintained this difference for at least 3 weeks. It appears that the growth achieved by a plant, and thus the size and *fitness* it attains, is determined early in its life by the pre-emption or 'capture' of space (or the resources implied by that space). Space is then unavailable (or, at least, relatively unavailable) to other plants, and these grow more slowly and attain a lower fitness as a consequence.

Presumably, genetic predisposition to early emergence, and a chance association with favourable microsites, both play some part in determining which plants actually capture space. In either case, however, the result is to push competition towards the contest end of the scramble-contest continuum. The plants capture

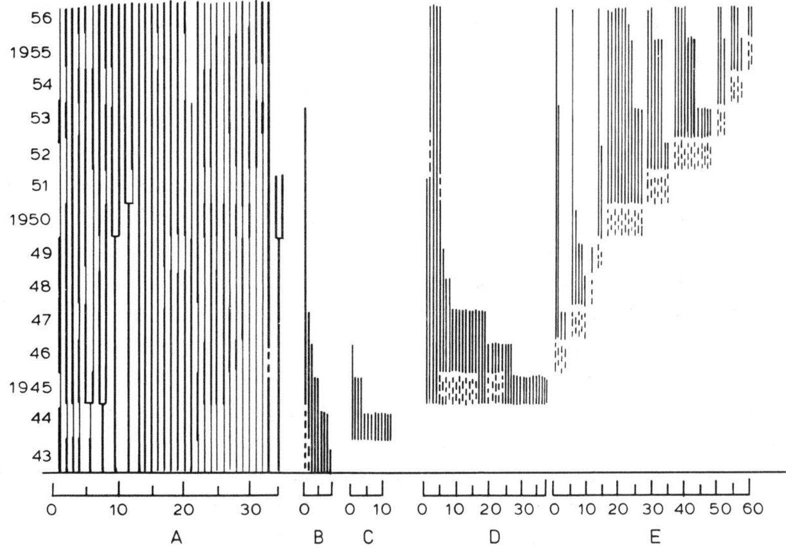

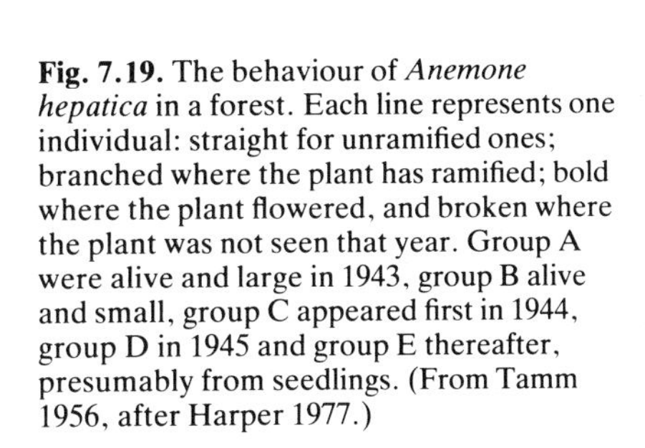
Fig. 7.19. The behaviour of *Anemone hepatica* in a forest. Each line represents one individual: straight for unramified ones; branched where the plant has ramified; bold where the plant flowered, and broken where the plant was not seen that year. Group A were alive and large in 1943, group B alive and small, group C appeared first in 1944, group D in 1945 and group E thereafter, presumably from seedlings. (From Tamm 1956, after Harper 1977.)

what is, in effect, a territory, and as a consequence there is a more exact regulation of plant numbers.

Essentially the same phenomenon is shown by the herbaceous perennial *Anemone hepatica* in Fig. 7.19 (Tamm 1956). Despite the crop of seedlings entering the population between 1943 and 1956, it is quite apparent that the most important factor determining which individuals were established in 1956 was whether or not they were established in 1943. Of the 30 specimens that had reached large or intermediate size by 1943, 28 survived until 1956, and some of these had ramified. By contrast, of the 112 plants that were either small in 1943 or appeared as seedlings subsequently, only 26 survived until 1956, and not one of these was sufficiently well established to have flowered.

Similar patterns are obvious from simple observations of tree populations. The survival-rate of the few established adults is high: that of the many seedlings and saplings is comparatively low. In all such cases it is clear that the major prerequisite for high (or indeed positive) fitness is the capture of space. We can think of this as the plant equivalent of animal territoriality, and the regulatory consequences are essentially the same.

Chapter 8 Community Structure

8.1 Introduction

The subject-matter of this book is the ecology of populations, and it would be unreasonable, in this final chapter, to attempt to cover the ground of some other book concerned with the ecology of communities. Nevertheless, it will be valuable to consider, briefly, the roles that the various processes considered in earlier chapters play in determining community structure. Specifically, we shall consider, in turn, the roles that interspecific competition, predation, disturbance, instability, and habitat size and diversity *can* play; and then draw what conclusions we can regarding their general importance in actual communities.

8.2 The role of interspecific competition

Good evidence that interspecific competition can play an important role in determining community structure is provided by an experiment carried out by Putwain & Harper (1970) on a hill grassland site in North Wales that was closely grazed by sheep. The species with which they were most concerned was sorrel, *Rumex acetosella*, which was second in abundance in the community to a grass, sheep's fescue (*Festuca ovina*). *Galium saxatile* (heath bedstraw) was also abundant, and 14 other species of grasses and herbs were present in varying numbers. Specific components of the community were experimentally removed with herbicides by setting up plots of the following types:

(a) plots sprayed with Dalapon to remove all grasses; this does little and only temporary damage to *R. acetosella* and other non-gramineous species (i.e. dicots);

(b) plots in which individual plants of all dicots except *R. acetosella* were killed by the combined application of the herbicides 2,4-D and Tordon 22K;

(c) plots sprayed with Paraquat to remove all species except *R. acetosella* which, despite having its above-ground parts scorched, regrows rapidly from buds at the base of the stem; and

(d) plots in which *R. acetosella* plants alone were killed by spot treatment with Tordon 22K.

There were also control plots that were not sprayed at all.

Spraying took place on 2 June 1965 and then, to distinguish between the effects of treatment on vegetative and seedling growth, seeds were sown in parts of the plots on 20 September 1965. Abundance of *R. acetosella* was monitored throughout the year following treatment, and its dry weight under each regime determined on 5 July 1966. The results are shown in Figs 8.1 and 8.2.

It appears that the growth of mature sorrel plants was unaffected by the removal of dicots, was increased only slightly by the removal of grasses, but was very significantly increased by the removal of both grasses and dicots. The rate of seedling establishment was increased by the removal of grasses, or of dicots, or of the mature plants of *R. acetosella* itself.

The probable explanation is illustrated in Fig. 8.3, which is a diagrammatic representation of *R. acetosella*'s niche relationships within the community. *R. acetosella*, since it exists within the sward, obviously has a realized niche; but it is competitively excluded from a substantial portion of its fundamental niche in this community by the combined action of the dicots, and from a similar but even larger portion by the combined action of the grasses. It is, therefore, only when dicots and grasses are both absent that significant competitive release occurs. It appears, moreover, that the fundamental niche of the sorrel seedlings lies largely within the combined realized niche of the grasses, though other, smaller portions lie within the realized niches of the dicots and the *R. acetosella* adults.

This experiment shows clearly that the distribution and abundance of sorrel is determined to a significant extent by the interspecific competitive interactions occurring within the grassland community. Of course,

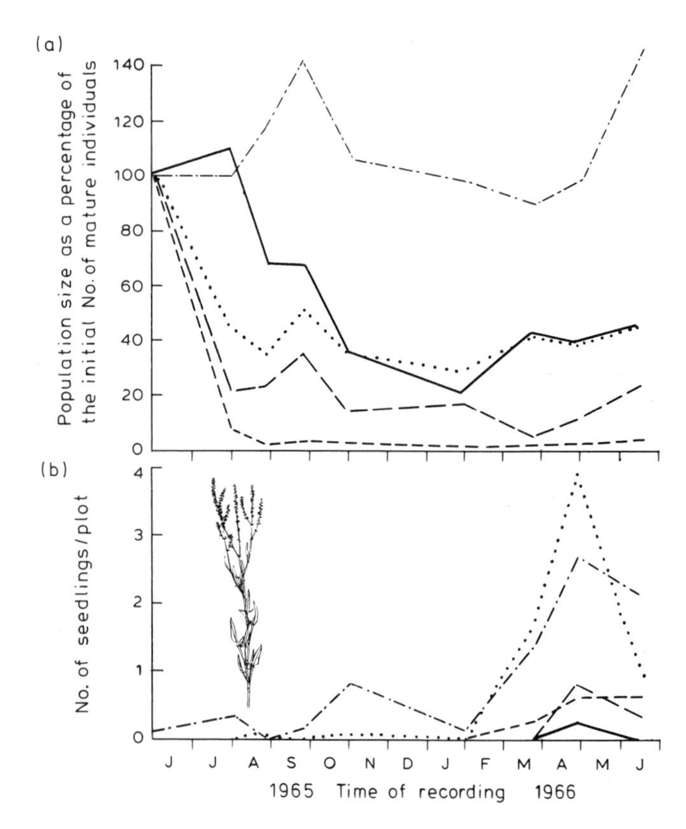

Fig. 8.1. (a) The response of a population of *Rumex acetosella* in a mixed grassland sward to the removal of certain components of the sward, expressed as a percentage of the population in June 1965. (b) Numbers of seedlings per plot. Key to treatments: ——— control; - - - - - *R. acetosella* removed; — — — all dicots except *R. acetosella* removed; ······· grasses removed; –·–·– all species except *R. acetosella* removed. (After Putwain & Harper 1970.)

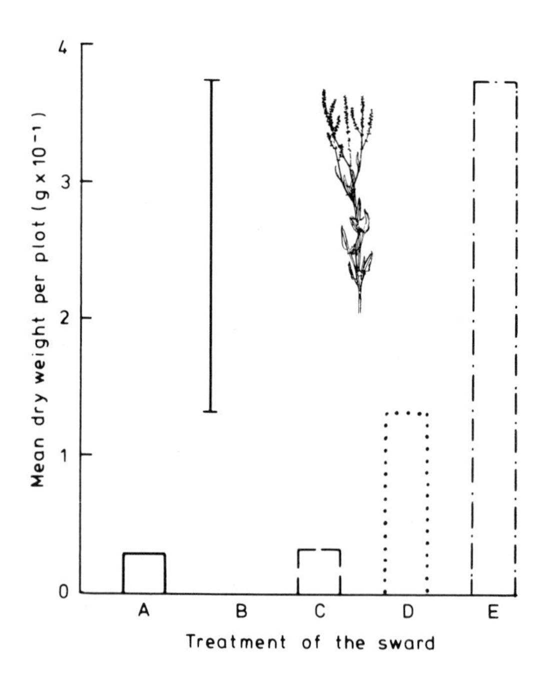

Fig. 8.2. The dry weight (g) of *Rumex acetosella* per plot at the end of the period of observation. For key to sward treatment, see legend to Fig. 8.1. L.S.D. indicated at $p = 0.05$. (After Putwain & Harper 1970.)

plants showing temporal heterogeneity in resource-utilization, and so on. In the present context, the importance of these examples is that, in all such cases, the precise design of the niches in Fig. 8.3 is quite arbitrary and their important dimensions are not even dimly understood. Nevertheless, the figure does serve to illustrate how communities must often be structured by species being competitively confined to small, realized portions of their fundamental niches.

We have, remember, seen several similar examples in Chapter 4: seed-eating desert ants showing differentiation in size and foraging strategy; *Panicum* and *Glycine* showing niche differentiation with respect to nitrogen; barnacles partitioning space; bumble-bees specializing on flowers of different corolla lengths;

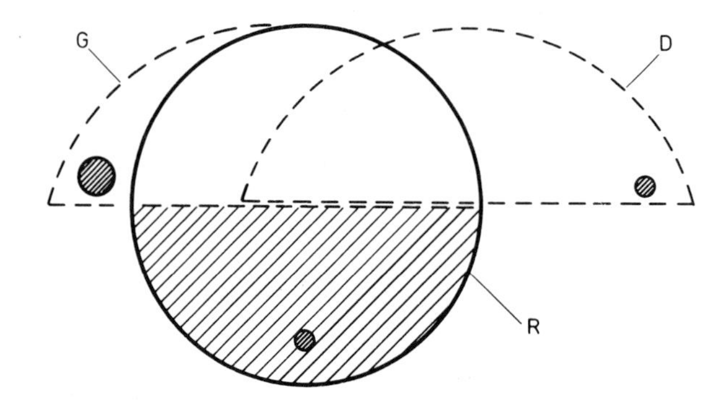

Fig. 8.3. The diagrammatic fundamental niches of *Rumex acetosella* (R), grasses (G) and dicots (D). Hatched area: the realized niche of *R. acetosella*; shaded areas: the fundamental niche of *R. acetosella* seedlings. (Modified from Putwain & Harper 1970.)

the communities are structured, and species-diversity increased, by resource-partitioning based on competitive exclusion.

8.3 The role of predation

The most famous piece of evidence supporting the importance of predators in determining community structure is provided by the work of Paine (1966), and their role is succinctly stated (as a hypothesis) by Paine himself: 'Local species diversity is directly related to the efficiency with which predators prevent the monopolization of the major environmental requisites by one species'. Paine presented some correlational support for this hypothesis, but his most persuasive evidence was experimental.

On rocky shores of the Pacific coast of North America the community is dominated by a remarkably constant association of mussels, barnacles and one starfish; and Fig. 8.4 illustrates the trophic relationships of this portion of the community as observed by Paine at Mukkaw Bay, Washington. The data are presented both as the numbers and as the total calories consumed by the two carnivorous species in the subweb: the starfish *Pisaster ochraceus* and a whelk *Thais emarginata*. Apparently this food web is based on a barnacle economy, with both major predators consuming them in quantity. Note, however, that in terms of calories the barnacles are only about one-third as important to *Pisaster* as either *Mytilus californianus*, a bivalve, or the browsing chiton *Katherina tunicata*.

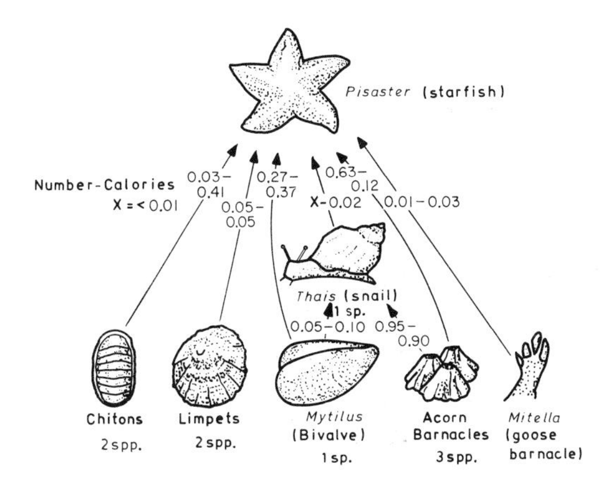

Fig. 8.4. The feeding relationships by numbers and calories of the *Pisaster* dominated subweb at Mukkaw Bay. The specific composition of each predator's diet is given as a pair of proportions: numbers on the left, calories on the right. (After Paine 1966.) 1 calorie (non-SI unit) = 4.186 joules.

For several years from June 1963, Paine excluded all *Pisaster* from a 'typical' piece of shoreline at Mukkaw Bay about 8 m long and 2 m in vertical extent. An adjacent control area was left unaltered; and line transects across both areas were taken irregularly, and the number and density of resident macroinvertebrate and benthic algal species measured. The appearance of the control area did not alter. Adult *Mytilus californianus*, *Balanus cariosus* (an acorn barnacle) and *Mitella polymerus* (a goose-necked barnacle) formed a conspicuous band in the mid-intertidal; while at lower levels the diversity increased abruptly, with immature individuals of the above species, *Balanus glandula* in scattered clumps, a few anemones of one species, two chiton species (browsers), two abundant limpets (browsers), four macroscopic benthic algae (*Porphyra, Endocladia, Rhodomela* and *Corallina*), and the sponge *Haliclona* (often browsed upon by *Anisodoris*, a nudibranch) all present.

Where the *Pisaster* were excluded, however, the situation changed markedly. *B. glandula* settled successfully throughout much of the area, and by September 1963 it had occupied 60–80% of the available space. By the following June, however, the *Balanus* themselves were being crowded out by small, rapidly growing *Mytilus* and *Mitella*; and this process of successive replacement by more efficient occupiers of space continued, leading eventually to an experimental area dominated by *Mytilus*, its epifauna, and scattered clumps of adult *Mitella*. The benthic algae, with the exception of *Porphyra*, tended to disappear due to a lack of space, while the chitons and larger limpets tended to emigrate because of an absence of space and a lack of appropriate food.

Interpretation of Paine's experiment must be tempered by the admission that the altered system may not have reached an equilibrium (Paine 1966). Nevertheless, it is clear that the removal of *Pisaster* led to a marked decrease in diversity, despite an actual increase in the size of the standing crop. There was a change from a 15-species system to a trophically simpler eight-species

Table 8.1 Effects of density on seed-mortality amongst tropical trees. All published observations and experiments known from tropical forests are included. (After Connell 1979.)

Location	Vegetation type	Plant species	% Mortality Dense	% Mortality Sparse	Authority
Queensland, Australia	Evergreen rainforest	*Cryptocarya corrugata*	100	99	Connell 1971
Costa Rica	Deciduous forest	*Acacia farnesiana*	79.7	79.6	Janzen 1975
Costa Rica	Deciduous forest	*Scheelea rostrata*	35.7	6.1	Wilson & Janzen 1972

system, and of the species that disappeared, some were and some were not in the normal diet of *Pisaster*. It seems, then, that the influence of *Pisaster* on the community is at least partly indirect; by eating masses of barnacles and the competitively dominant *Mytilus*, and thus keeping space open, *Pisaster* enhances the ability of other species to inhabit the area. When space is available, other organisms, for instance chitons, settle or move in, and form major portions of *Pisaster's* nutrition. Thus, in the absence of predation there is an increased tendency for competition at lower trophic levels to go to completion, driving species to extinction; but by its presence, *Pisaster* keeps many of these populations well below their carrying capacity. Competitive exclusion is, therefore, commonly avoided, and the diversity of the community enhanced. In short, it seems that, in the

Table 8.2 Survivorship of seeds or seedlings either near or far from adult trees of the same species. All known published field experiments or observations in tropical forests are listed, but in some cases typical, rather than total, results are presented. (After Connell 1979.)

Location	Vegetation type	Plant	Fitness parameter	Near adult	Far from adult	Authority
Queensland, Australia	Evergreen rain forest	*Crytocarya corrugata*	% seed mortality	99.8	99.5	Connell 1971
Queensland, Australia	Evergreen rain forest	*Eugenia brachyandra*	% germination in first year	14.8	14.0	Connell 1979
Queensland, Australia	Evergreen rain forest	*Planchonella* sp.	% seedling mortality in trenched plots	68	21	Connell 1971
Puerto Rico	Evergreen rain forest	*Euterpe globosa*	% non-viable seeds 1970	95–100	83–100	Janzen 1972a
			1971	0–11	0–20	
Costa Rica	Deciduous forest	*Scheelea rostrata*	% seed mortality	33.8	35.7	Wilson & Janzen 1972
Costa Rica	Deciduous forest	*Sterculia apetala*	Number of herbivorous bugs per seed	5.2	1.8	Janzen 1972b
Costa Rica	Deciduous forest	*Spondias mombin*	% seed mortality	50	45	Janzen 1975
Costa Rica	Deciduous forest	*Dioclea megacarpa*	% shoot tips eaten	86.4	16.7	Janzen 1971

present case at least, Paine's hypothesis is correct: predation prevents competitive exclusion and, therefore, increases community diversity.

In theory, predation can have an even more potent effect on species-diversity when it is frequency-dependent, i.e. when there is predator switching leading to a 'type 3' functional response (Section 5.7.4). Prey species will then be disproportionately affected when they are common, and this should lead to a large number of rare prey species. Unfortunately, in practice, there is little positive evidence that this occurs (Connell 1979). Nevertheless, by examining all the available data on seed and seedling mortality in tropical trees, Connell (1979) was able to show that, in many cases, there is decreased survivorship (a) when density is high (Table 8.1), and (b) in the immediate presence of established adults (Table 8.2). On the other hand, these tables also show that there were several other cases when this was not so. Overall, while it is clear that such frequency-dependent predation does occur in nature (leading, no doubt, to increased diversity), it is equally clear that its occurrence is by no means the general rule.

8.4 The role of disturbance

Following Connell (1979), we shall take 'disturbance' to be the indiscriminate, catastrophic removal of all individuals from an area. As such it may take a wide variety of forms: e.g. lightning, storms, land-slips or even indiscriminate predation. In general terms, its effect will be to prevent communities from reaching an equilibrium; parts of them, at least, will be repeatedly returned to early, colonizing stages of succession. Its more specific effect on species-diversity, however, will depend on the nature of the equilibrium community itself; and this, in turn, will depend on the various processes—competition, predation and so on—also discussed in this chapter. Nevertheless, a plausible, general relationship between disturbance and diversity has been proposed by Connell (1979)—the 'intermediate disturbance hypothesis'—and the role of disturbance in determining community structure can be usefully discussed in this context.

Connell recognized, essentially, three levels of disturbance (Fig. 8.5a). Where disturbances are frequent and large, the community will tend to be dominated by

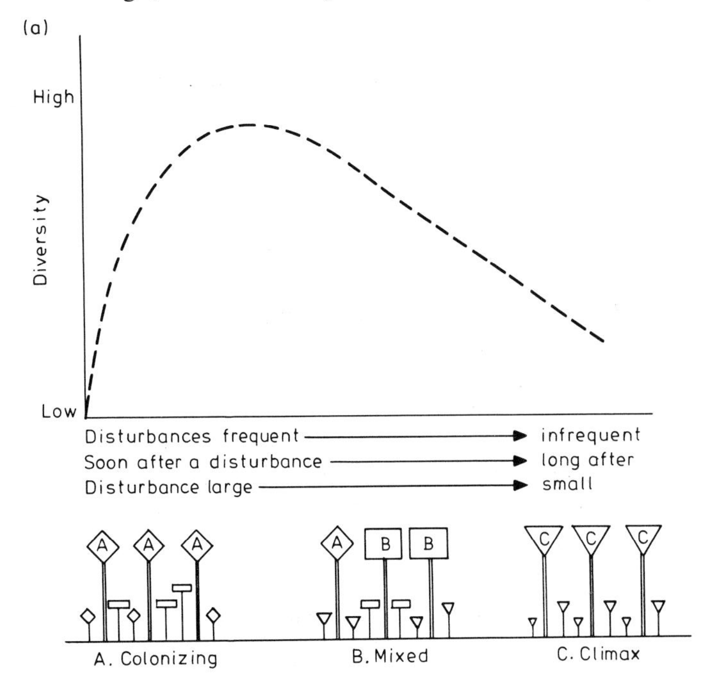

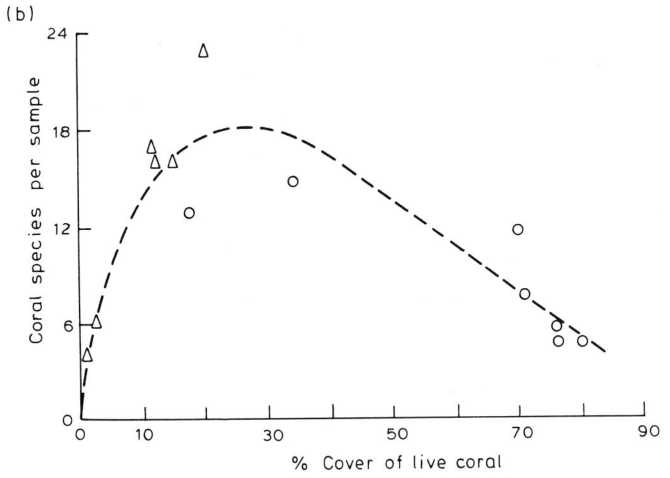

Fig. 8.5. (a) Connell's 'intermediate disturbance' hypothesis, involving opportunistic species (A), secondarily colonizing species (B) and climax species (C). (b) Data in support of the hypothesis from Heron Island, Queensland from damaged (Δ) and undamaged (O) sites. (After Connell 1979.)

opportunistic, fast-colonizing species, with, perhaps, a few individuals of intermediate, secondarily colonizing species, probably present as juveniles. Such a community will have a simple structure and a low diversity (left-hand side of Fig. 8.5a). At the other extreme, where disturbances are rare and small, the diversity will depend on the importance of what Connell calls 'compensatory mechanisms', i.e. predation, resource-partitioning and so on. As we have seen, where these are prevalent diversity will be high. In their absence, however, only highly competitive, late-succession species will be able to survive, and diversity will be low (right-hand side of Fig. 8.5a). At intermediate levels of disturbance, on the other hand, even in the absence of compensatory mechanisms, there will probably be a few adults of fast-colonizing species, many individuals of mid-succession, secondarily colonizing species, and even some individuals, possibly juveniles, of late-succession species. Overall, therefore, diversity will be high (centre of Fig. 8.5a); in the comparative absence of compensatory mechanisms, the species-diversity of communities will be highest at intermediate levels of disturbance.

Some of Connell's evidence in support of this hypothesis is shown in Fig. 8.5b, where the data come from observations on a coral reef off the coast of Queensland, Australia. Disturbance, resulting either from hurricane damage or from the effects of anchoring boats on the reef, is measured as the percentage of a site that is devoid of any live coral, and it is indeed apparent that diversity (number of species per sample) is highest at intermediate levels of disturbance. As further evidence, Connell points out that even in tropical rainforests, which we tend to think of as exhibiting very high diversity, areas that are largely undisturbed (like the Budongo forest in Uganda) come to be dominated by a single species of tree (in this case, ironwood).

Overall, therefore, we can accept that, in some cases, diversity will be highest at intermediate levels of disturbance; and that generally, large frequent disturbances will tend to decrease diversity. On the other hand, the effects of disturbance on diversity will be much less clear-cut whenever compensatory mechanisms are sufficiently potent to ensure that stable climax communities exhibit a high degree of diversity themselves.

8.5 The role of instability

All populations are, to a greater or lesser extent, liable to become extinct; and whenever this occurs, the structure of the community containing that population will obviously change. However, this liability is bound to be greater in some communities than others, and in this sense some communities must be more unstable than others. Yet the communities with structures conferring stability are the ones most likely to be observed, because they persist. Structural instability must, therefore, be an important determinant of observed community structure.

The search for what, inherently, leads to instability has been the province of theoretical ecologists, and, to paraphrase May (1979), two intertwined conclusions have emerged:

(1) In 'randomly constructed' model ecosystems, an increase in the number of species in a community is associated with an increased dynamical fragility and a diminished ability to withstand a given level of environmental disturbance. Thus, relatively stable or predictable environments may permit fragile, species-rich communities to exist; while relatively unstable or unpredictable environments will support only a dynamically robust, and therefore relatively simple, ecosystem.

(2) Real ecosystems are not assembled randomly. They are the products of long-running evolutionary processes. We are therefore bound to ask: What special structural features of real ecosystems may help to reconcile community complexity with dynamical stability? In other words, since instability will tend to simplify communities, what observable features of community-structure can be deemed to exist by virtue of the stability they confer on complex, species-rich systems? The proposed 'role of instability' will then be the 'selection' of these features.

Attempting to discover what these features might be has also been the province of theoretical ecologists; and, as yet, these attempts have been largely speculative. Nevertheless, there are several interesting possibilities (May 1979). May (1972), for instance, and more recently Goh (1978) have suggested (from the analysis of models) that ecosystems will be more robust if they consist of 'loosely coupled subsystems'. This term describes a situation in which a community consists of several parts ('subsystems'), *within* which there is considerable biological interaction, but *between* which there is very little interaction. This, according to Lawton & Pimm (1978) and Beddington & Lawton (1978), is at least consistent with the observation that most insect herbivores are monophagous or oligophagous, giving rise to relatively discrete food chains even in species-rich plant communities. However, empirical evidence generally fails to give positive support to the hypothesis (Pimm & Lawton 1980).

Another feature of natural communities possibly subject to selection by instability is the length of food chains, which rarely consist of more than four or five trophic levels. The conventional explanation for this is that length is limited by the inefficiency of energy-flow from one trophic level to the next (there is insufficient energy left to support the higher trophic levels). Yet, as Pimm & Lawton (1977) have pointed out, this cannot, by itself, explain why food chains are about as long in the tropics (where energy input is high) as they are in the barren Arctic (where energy input is low). An alternative explanation, however, was provided by Pimm and Lawton themselves. By studying the stability properties of various Lotka–Volterra models, they argued that long food chains may typically result in population fluctuations that are too severe for top predators to exist. In other words, only relatively short food chains are sufficiently stable to be observed in natural communities.

Finally, Pimm & Lawton (1978) have explored the relationship between omnivory and stability by studying model ecosystems based on Lotka–Volterra equations. Broadly speaking, they conclude that omnivory and overall dynamical stability are easier to reconcile if the omnivores and their prey are of similar size and population density, a situation that most commonly pertains to insect parasitoids. As May (1978b, 1979) suggested, this may account for the diversity of insects in general and the diversity of parasitoids in particular.

Overall, then, while its precise role remains largely the subject of theoretical speculation, it is quite clear that instability can play a crucial part in determining the structure of natural communities.

8.6 The role of habitat size and diversity

As Gorman (1979) has pointed out, Great Britain has 44 species of indigenous terrestrial mammals, extant or recently extinct, but Ireland, just 20 miles further into the Atlantic, has only 22; and while this might conceivably reflect the difficulties the mammals have in crossing water, it actually affects bats as much as any other group: only 7 of Britain's 13 species breed in Ireland. Furthermore, of Britain's 171 species of breeding birds, only 126 are recorded as breeding in Ireland, and 24 of these do so only occasionally. For example, there are no woodpeckers in Ireland (though there are plenty of trees), no little or tawny owls, and no marsh or willow tits.

The most likely explanation is that Great Britain is far larger than Ireland. But size can exert its effects in two quite separate ways. Perhaps the most obvious explanation is that differences in habitat size are important because large habitats are more diverse. But there is a second explanation that applies whenever habitats can be thought of as islands (either real islands, or 'habitat islands' of one type surrounded by a 'sea' of another habitat type). Larger islands support larger populations that have a relatively low probability of becoming extinct. In addition, larger islands represent a larger 'target' for colonization by species not already present (MacArthur & Wilson 1967). On two counts, therefore, extinction and immigration, we can expect larger islands (i.e. larger habitats) to support more species. Note, too, that this is an explanation for the fact that (small) islands generally support fewer species than a nearby (larger) mainland.

A typical relationship between the number of species living on an island and the island's area is illustrated in Fig. 8.6, for the amphibians and reptiles living on oceanic islands in the West Indies (MacArthur & Wilson 1967). The logarithm of species number rises with the logarithm of island area in a remarkably linear fashion, and the slope, 0.30, is very much in line with those obtained in other examples. For organisms ranging from birds to ants to land plants, in both real and habitat islands, the slopes of such log–log plots mostly fall within the range 0.24–0.34 (Gorman 1979). The role of island (i.e. habitat) size as a determinant of species number (and thus community structure) is, therefore, well-established.

An indication of the fact that size acts through two mechanisms, however, is provided by data in Fig. 8.7 (Williams 1964). This, too, is a plot of log species number against log habitat size, but size in this case pertains to arbitrary sampling areas within a mainland. Once again, on this log–log plot, the number of flowering plants rises linearly with the size of sampling areas in England. But the slope—around 0.1—is noticeably lower than those from the island examples, and falls near the range typical for mainland studies: 0.12–0.17 (MacArthur & Wilson 1967). The crucial point is that habitat size can only act via habitat diversity in such cases. These arbitrary areas are continually exchanging organisms with surrounding areas, and they are not, therefore, subject to the considerations of extinction and colonization that apply to isolated islands. Thus, mainland slopes from 0.12 to 0.17 reflect the effects of habitat diversity, while the increased slopes on islands reflect the additional size-effects peculiar to island biogeography.

Overall we can see that an increase in habitat size will lead to an increase in species number, and thus to an increase in the complexity of community structure. This may result from the indirect effects of habitat diversity, or from effects peculiar to the island nature of many habitats; and while it is often difficult to partition the total effect into these two components, there is no doubt that both are of very widespread importance.

8.7 Conclusions

We have seen that a variety of factors *can* influence community structure. Yet, in truth, it has to be admitted that precise statements as to their relative potencies must await further advances in our knowledge and understanding. Nevertheless, certain tentative conclusions can be drawn.

Perhaps the most significant of these is that a good case can be made for the *constancy, predictability and productivity of the abiotic environment* being of absolutely crucial importance in determining community structure. This conclusion stems from a number of considerations.

(a) In general terms, diverse, basically fragile ecosystems appear to be relatively stable in constant, predictable environments (Section 8.5).

(b) More specifically, niches can be stably packed more tightly in predictably productive environments (Section 4.15). This suggests that interspecific competition will be most potent as a mechanism promoting diversity under such circumstances (Section 8.2).

(c) Equally specifically, the stable existence of top

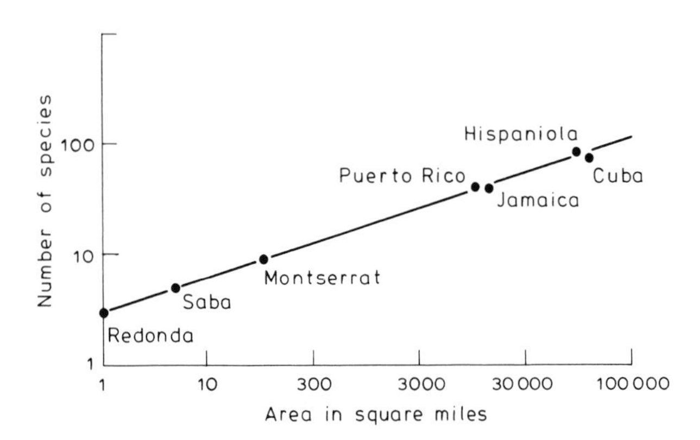

Fig. 8.6. The number of amphibian and reptile species living on oceanic West Indian islands of various sizes. (After MacArthur & Wilson 1967.)

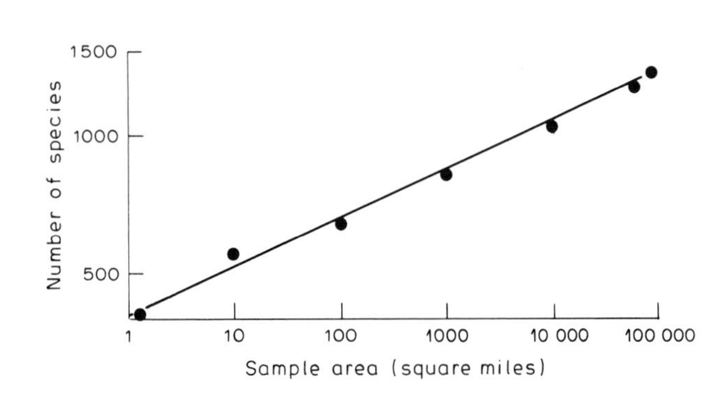

Fig. 8.7. The species–area curve for the number of flowering plants found in sample areas of England (Williams 1964). (After Gorman 1979.)

predators will be favoured in predictably productive environments (Paine 1966). It is, therefore, in such cases that they will be most potent in keeping potential competitors below their carrying capacities, and thus promoting diversity still further (Section 8.3).

(d) Finally, these other mechanisms will tend to reinforce one another. High diversity at a lower trophic level will certainly provide for niche-diversification (and thus increased diversity) at the next highest trophic level; and it is possible that this will lead to an increased intensity of predation, and thus a further increase in diversity, at the lower trophic level. Small 'inherent' differences in community structure are, therefore, likely to become exaggerated. Note, however, that this reinforcement will occur whatever the cause of the inherent differences. Note, too, as another aspect of this reinforcement, that those effects of habitat size that are attributable to habitat diversity (Section 8.6) will themselves be influenced by increases in the diversity of the biotic aspects of a habitat.

In short, there is good reason to believe that the constancies, predictabilities and productivities of abiotic environments are crucial, underlying determinants of community structure; and that competition, predation and ecosystem instability are *mechanisms* through which they exert their influence. This is almost certainly the explanation for the single most important cline of increasing diversity: from the poles to the tropics.

This view is opposed, to some extent, by Connell's 'intermediate disturbance hypothesis' (Section 8.4), since a constant, predictable environment is likely to be one with a low level of disturbance. On the other hand, Connell's 'disturbance' requires the indiscriminate removal of species from an area, and an environment can be inconstant and unpredictable without this happening. It is, therefore, possible that this hypothesized mechanism acts independently of the other factors considered, and influences diversity in a wide range of environments.

Finally, the 'island' effects of habitat size (Section 8.6) are likely to superimpose their influences on community structure wherever they occur; and, to the extent that all environments are patchy, they are likely to occur everywhere.

To summarize, then, we know a reasonable amount about the potentialities of the various factors determining community structure, but rather less about their actual potencies and the patterns of their action in nature. Discovering the rules through which communities are constructed from populations is one of the many exciting challenges that confronts population ecology today.

References

Abrams P. (1976) Limiting similarity and the form of the competition coefficient. *Theoretical Population Biology*, **8**, 356–75

Ackert J. E., Graham G. L., Nolf L. O. & Porter D. A. (1931) Quantitative studies on the administration of variable numbers of nematode eggs (*Ascaridia lineata*) to chickens. *Transactions of the American Microscopical Society*, **50**, 206–14

Agricultural Research Council (1965) *The Nutritional Requirements of Farm Livestock. 2. Ruminants.* Agricultural Research Council, London

Akinlosotu T. A. (1973) The role of *Diaeretiella rapae* (McIntosh) in the control of the cabbage aphid. Unpublished Ph.D. thesis, University of London

Allee W. C. (1931) *Animal Aggregations. A Study in General Sociology.* University of Chicago Press, Chicago

Anderson R. M. (1974) Population dynamics of the cestode *Caryophyllaeus laticeps* (Pallas, 1781) in the bream (*Abramis brama* L.). *Journal of Animal Ecology*, **43**, 305–21

Anderson R. M. (1979) The influence of parasitic infection on the dynamics of host population growth. In *Population Dynamics*, (Anderson R. M., Turner B. D. & Taylor L. R. eds), pp. 245–81. Blackwell Scientific Publications, Oxford

Anderson R. M. (1981) Population ecology of infectious disease agents. In *Theoretical Ecology: Principles and Applications*, 2nd edn (R. M. May ed.), pp. 318–355. Blackwell Scientific Publications, Oxford

Anderson R. M. & May R. M. (1978) Regulation and stability of host–parasite population interactions: I. Regulatory processes. *Journal of Animal Ecology*, **47**, 219–47

Andrewartha H. G. & Birch L. C. (1954) *The Distribution and Abundance of Animals.* University of Chicago Press, Chicago

Andrewartha H. G. & Birch L. C. (1960) Some recent contributions to the study of the distribution and abundance of insects. *Annual Review of Entomology*, **5**, 219–42

Antonovics J. & Levin D. A. (1980) The ecological and genetic consequences of density-dependent regulation in plants. *Annual reviews of ecology and systematics*, **11**, 411–452

Aspinall D. (1960) An analysis of competition between barley and white persicaria. II. Factors determining the course of competition. *Annals of Applied Biology*, **48**, 637–54

Aspinall D. & Millthorpe F. L. (1959) An analysis of the competition between barley and white persicaria. I. The effects of growth. *Annals of Applied Biology*, **47**, 156–72

Auer C. (1968) Erste Ergebnisse einfacher stochasticher Modeluntersuchungen über die Ursachen der Populations-bewegung des grauen Larchenwicklers *Zeiraphera diniana*, Gn. (= *Z. griseana* Hb). in Oberengadin 1949/66. *Zeitschrift fuer Angewandte Entomologie*, **62**, 202–35

Ayala F. J., Gilpin M. E. & Ehrenfeld J. G. (1973) Competition between species: theoretical models and experimental tests. *Theoretical Population Biology*, **4**, 331–56

Bakker K. (1961) An analysis of factors which determine success in competition for food among larvae of *Drosophila melanogaster. Archives Neerlandaises de Zoologie*, **14**, 200–81

Bakker K. (1964) Backgrounds of controversies about population theories and their terminologies. *Zeitschrift fuer Angewandte Entomologie*, **53**, 187–208

Banks C. J. (1957) The behaviour of individual coccinellid larvae on plants. *Animal Behaviour*, **5**, 12–24

Barnes H. (1962) So-called anecdysis in *Balanus balanoides* and the effect of breeding upon the growth of calcareous shell of some common barnacles. *Limnology and Oceanography*, **7**, 462–73

Beddington J. R. (1979) Harvesting and population dynamics. In *Population Dynamics* (Anderson R. M., Turner B. D. & Turner L. R. eds), pp. 307–20. Blackwell Scientific Publications, Oxford.

Beddington J. R., Free C. A. & Lawton J. H. (1975) Dynamic complexity in predator–prey models framed in difference equations. *Nature*, **225**, 58–60

Beddington J. R., Free C. A. & Lawton J. H. (1978) Modelling biological control: on the characteristics of successful natural enemies. *Nature*, **273**, 513–9

Beddington J. R. & Lawton J. H. (1978) On the structure and behaviour of ecosystems. *Journal de Physique*, **39**(c) 5–39

Begon M. (1976) Temporal variations in the reproductive condition of *Drosophila obscura* Fallen and *D. subobscura* Collin. *Oecologia*, **23**, 31–47

Begon M. (1984) Density and individual fitness: assymetric competition. In *Evolutionary Ecology* (Shorrocks B., ed), pp. 175–194. Blackwell Scientific Publications, Oxford

Bell A. D. & Tomlinson P. B. (1980) Adaptive architecture in rhizomatous plants. *Botanical Journal of the Linnean Society*. **80**, 125–160

Beverton R. J. H. & Holt S. J. (1957) On the dynamics of exploited fish populations. *Fishery Investigations, London* (Series II), **19**, 1–533

Birkhead T. R. (1977) The effect of habitat and density on breeding success in the common guillemot (*Uria aalge*). *Journal of Animal Ecology*, **46**, 751–64

Blackman G. E. (1968) The applications of the concepts of growth analysis to the assessment of productivity. In *Functioning of Terrestrial Ecosystems at the Primary Production Level* (Eckardt F. E. ed.) pp. 243–59. Unesco, Liège

Bonner J. T. (1965) *Size and Cycle: an essay on the structure of biology.* Princeton University Press, Princeton

Bradshaw A. D. (1965) Evolutionary significance of phenotypic plasticity in plants. *Advances in Genetics*, **13**, 115–55

Branch G. M. (1975) Intraspecific competition in *Patella cochlear* Born. *Journal of Animal Ecology*, **44**, 263–81

Brougham R. W. (1955) A study in rate of pasture growth. *Australian Journal of Agricultural Research*, **6**, 804–12

Brougham R. W. (1956) The rate of growth of short-rotation ryegrass pastures in the late autumn, winter and early spring. *New Zealand Journal of Science and Technology*, **A38**, 78–87

Brown J. H. & Davidson D. W. (1977) Competition between seed-eating rodents and ants in desert ecosystems. *Science*, **196**, 880–2

Burnett T. (1954) Influences of natural temperatures and controlled host densities on oviposition of an insect parasite. *Physiological Zoology*, **27**, 239–48.

Burnett T. (1956) Effects of natural temperatures on oviposition of various numbers of an insect parasite (Hymenoptera, Chalcididae,

Tenthredinidae). *Annals of the Entomological Society of America*, **49**, 55–9
Burnett T. (1958) Dispersal of an insect parasite over a small plot. *Canadian Entomologist*, **90**, 279–83

Carl E. (1971) Population control in arctic ground squirrels. *Ecology*, **52**, 395–413
Cates R. G. & Orians G. H. (1975) Successional status and the palatability of plants to generalized herbivores. *Ecology*, **56**, 410–18
Caughley G. & Lawton J. H. (1981) Plant-herbivore systems. In *Theoretical Ecology: Principles and Applications*, 2nd edn (R. M. May ed.), pp. 132–66. Blackwell Scientific Publications, Oxford
Cavers P. B. & Harper J. L. (1966) Germination polymorphism in *Rumex crispus* and *Rumex obtusifolius*. *Journal of Ecology*, **54**, 367–82
Clark L. R. (1963) The influence of predation by *Syrphus* spp. on the numbers of *Cardiaspina albitextura* (Psyillidae). *Australian Journal of Zoology*, **11**, 470–87
Clark L. R., Geier P. W., Hughes R. D. & Morris R. F. (1967) *The Ecology of Insect Populations in Theory and Practice*. Methuen, London
Clarke T. A. (1970) Territorial behaviour and population dynamics of a pomacentrid fish, the Garibaldi, *Hypsypops rubicunda*. *Ecological Monographs*, **40**, 189–212
Clements F. E., Weaver J. E. & Hanson H. C. (1929) Competition in cultivated crops. *Carnegie Institute of Washington Publications*, **398**, 202–33
Clough G. C. (1965) Viability in wild meadow voles under various conditions of population density, season and reproductive activity. *Ecology*, **46**, 119–34
Cody M. L. (1966) A general theory of clutch size. *Evolution*, **20**, 174–84
Comins H. N. & Hassell M. P. (1979) The dynamics of optimally foraging predators and parasitoids. *Journal of Animal Ecology*, **48**, 335–51
Connell J. H. (1961) The influence of interspecific competition and other factors on the distribution of the barnacle *Chthamalus stellatus*. *Ecology*, **42**, 710–23
Connell J. H. (1970) A predator–prey system in the marine intertidal region. I. *Balanus glandula* and several predatory species of *Thais*. *Ecological Monographs*, **40**, 49–78
Connell J. H. (1971) On the role of natural enemies in preventing competitive exclusion in some marine animals and in rain forest trees. In *Dynamics of Populations* (den Boer P. J. & Gradwell G. R. eds), pp. 298–312. Centre for Agricultural Publishing and Documentation, Wageningen
Connell J. H. (1973) *Biology and Geology of Coral Reefs*. Vol. 2. (Jones, O. A. & Endean, R. eds) Academic Press, New York
Connell J. H. (1979) Tropical rain forests and coral reefs as open non-equilibrium systems. In *Population Dynamics* (Anderson R. M., Turner B. D. & Taylor L. R. eds), pp. 141–63. Blackwell Scientific Publications, Oxford
Cook R. M. & Cockrell B. J. (1978) Predator ingestion rate and its bearing on feeding time and the theory of optimal diets. *Journal of Animal Ecology*, **47**, 529–48
Cremer K. W. (1965) Dissemination of seed for *Eucalyptus regnans*. *Australian Forestry*, **29**, 33–7
Crisp D. J. (1961) Territorial behaviour in barnacle settlement. *Journal of Experimental Biology*, **38**, 429–46
Crofton H. D. (1971) A model of host-parasite relationships. *Parasitology*, **63**, 343–64
Crombie A. C. (1947) Interspecific competition. *Journal of Animal Ecology*, **16**, 44–73
Curio E. (1976) *The Ethology of Predation*. Springer-Verlag, Berlin

Davidson D. W. (1977a) Species diversity and community organization in desert seed-eating ants. *Ecology*, **58**, 711–24
Davidson D. W. (1977b) Foraging ecology and community organization in desert seed-eating ants. *Ecology*, **58**, 725–37
Davidson D. W. (1978) Size variability in the worker caste of a social insect (*Veromessor pergandei* Mayr) as a function of the competitive environment. *American Naturalist*, **112**, 523–32
Davidson J. (1938) On the growth of the sheep population in Tasmania. *Transactions of the Royal Society of South Australia*, **62**, 342–6
Davidson J. & Andrewartha H. G. (1948a) Annual trends in a natural population of *Thrips imaginis* (Thysanoptera). *Journal of Animal Ecology*, **17**, 193–9
Davidson J. & Andrewartha H. G. (1948b) The influence of rainfall, evaporation and atmospheric temperature on fluctuations in the size of a natural population of *Thrips imaginis* (Thysanoptera). *Journal of Animal Ecology*, **17**, 200–22
Davidson J. L. & Donald C. M. (1958) The growth of swards of subterranean clover with particular reference to leaf area. *Australian Journal of Agricultural Research*, **9**, 53–72
Davies N. B. (1977) Prey selection and social behaviour in wagtails (Aves: Motacillidae). *Journal of Animal Ecology*, **46**, 37–57
Davies N. B. (1978a) Ecological questions about territorial behaviour. In *Behavioural Ecology: an evolutionary approach* (Krebs J. R. & Davies N. B. eds), pp. 317–50. Blackwell Scientific Publications, Oxford
Davies N. B. (1978b) Territorial defence in the speckled wood butterfly (*Pararge aegeria*), the resident always wins. *Animal Behaviour*, **36**, 138–47
DeBach P. & Smith H. S. (1941) The effect of host density on the rate of reproduction of entomophagous parasites. *Journal of Economic Entomology*, **34**, 741–5
Diamond J. M. (1975) Assembly of species communities. In *Ecology and Evolution of Communities* (Cody M. L. & Diamond J. M. eds), pp. 342–444. Harvard University Press, Cambridge
Dixon A. F. G. (1971a) The role of aphids in wood formation. I. The effect of the sycamore aphid *Drepanosiphum platanoides* (Schr.) (Aphididae) on the growth of sycamore, *Acer pseudoplatanus* (L.). *Journal of Applied Ecology*, **8**, 165–79
Dixon A. F. G. (1971b) The role of aphids in wood formation II. The effect of the lime aphid *Eucallipterus tiliae* L. (Aphididae), on the growth of the lime, *Tilia×vulgaris* Hayne. *Journal of Applied Ecology*, **8**, 393–409
Donald C. M. (1951) Competition among pasture plants. I. Intra-specific competition among annual pasture plants. *Australian Journal of Agricultural Research* **2**, 355–76
Donald C. M. (1961) Competition for light in crops and pastures. In *Mechanisms in Biological Competition* (Milthorpe F. L. ed.) Symposium of the Society of Experimental Biology, No. 15, pp. 283–313

Donald C. M. (1978) Summative Address. In *Plant Relations in pastures* (Wilson J. R. ed.) pp. 411–21. CSIRO

Dunn E. (1977) Predation by weasels (*Mustela nivalis*) on breeding tits (*Parus* spp.) in relation to the density of tits and rodents. *Journal of Animal Ecology*, **46**, 634–52

Eberhardt L. (1960) *Michigan Department of Conservation Game Division Report* No 2282

Eis S., Garman E. H. & Ebel L. F. (1965) Relation between cone production and diameter increment of Douglas fir (*Pseudotsuga menziessii* (Mirb.) Franco), grand fir (*Abies grandis* Dougl.) and Western White pine (*Pinus monticola* Dougl.) *Canadian Journal of Botany*, **43**, 1533–9

Elner R. W. & Hughes R. N. (1978) Energy maximisation in the diet of the Shore Crab, *Carcinus maenas* (L.). *Journal of Animal Ecology*, **47**, 103–16

Elner R. W. & Raffaelli D. G. (1980) Interactions between two marine snails, *Littorina rudis* Maton and *Littorina nigrolineata* Gray, a predator, *Carcinus maenas* (L.), and a parasite, *Microphallus similis* Jagerskiold. *Journal of Experimental Marine Biology and Ecology*, **43**, 151–60

Emson R. H. & Faller-Fritsch R. J. (1976) An experimental investigation into the effect of crevice availability on abundance and size structure in a population of *Littorina rudis* (Maton): Gastropoda; Prosobranchia. *Journal of Experimental Marine Biology and Ecology*, **23**, 285–97

Errington P. L. (1946) Predation and vertebrate populations. *Quarterly Review of Biology*, **21**, 144–77

Esau K. (1953) *Plant Anatomy*. Wiley, New York

Fenchel T. (1974) Intrinsic rate of natural increase: the relationship with body size. *Oecologia*, **14**, 317–26

Fenchel T. (1975) Character displacement and coexistence in mud snails (Hydrobiidae). *Oecologia*, **20**, 19–32.

Fernando M. H. J. P. (1977) Predation of the glasshouse red spider mite by *Phytoseiulus persmilis* A.-H. Unpublished Ph.D. thesis, University of London

Finlayson L. H. (1949) Mortality of *Laemophloeus* (Coleoptera, Cucujidae) infected with *Mattesia dispora* Naville (Protozoa, Schizogregarinaria). *Parasitology*, **40**, 261–4

Firbank L. G. & Watkinson A. R. (1985) On the analysis of competition within two-species mixtures of plants. *Journal of Applied Ecology*, **22**, 503–17

Ford E. D. (1975) Competition and stand structure in some even-aged plant monocultures. *Journal of Ecology*, **63**, 311–33

Forsyth A. B. & Robertson R. J. (1975) K-reproductive strategy and larval behaviour of the pitcher plant sarcophagid fly, *Blaesoxipha fletcheri*. *Canadian Journal of Zoology*, **53**, 174–9

Free C. A., Beddington J. R. & Lawton J. H. (1977) On the inadequacy of simple models of mutual interference for parasitism and predation. *Journal of Animal Ecology*, **46**, 543–54

Fujii K. (1967) Studies on interspecies competition between the azuki bean weevil *Callosobruchus chinensis* and the southern cowpea weevil, *C. maculatus*. II. Competition under different environmental conditions. *Researches in Population Ecology*, **9**, 192–200

Fujii K. (1968) Studies on interspecies competition between the azuki bean weevil and the southern cowpea weevil: III, some characteristics of strains of two species. *Researches in Population Ecology*, **10**, 87–98

Gatsuk, L. E., Smirnova, O. V., Vorontzova, L. I., Zaugolnova, L. B. & Zhukova, L. A. (1980) Age states of plants of various growth forms: a review. *Journal of Ecology*, **68**, 675–696.

Gause G. F. (1934) *The Struggle for Existence*. Williams and Wilkins, Baltimore. (Reprinted 1964, by Hafner, New York.)

Gill F. B. & Wolf L. L. (1975) Economics of feeding territoriality in the golden-winged sunbird. *Ecology*, **56**, 333–45

Goh B. S. (1978) Robust stability concepts for ecosystem models. In *Theoretical Systems Ecology* (Halfon E. ed.). Academic Press, New York

Gorman M. L. (1979) *Island Ecology*. Chapman and Hall, London

Gottleib L. D. (1984) Genetics and morphological evolution in plants. *American Naturalist*, **123**, 681–709

Griffiths K. J. (1969) Development and diapause in *Pleolophus basizonus* (Hymenoptera: Ichneumonidae.). *Canadian Entomologist*, **101**, 907–14

Gross A. D. (1947) Cyclic invasions of the Snowy Owl and the migration of 1945–1946. *Auk*, **64**, 584–601

Grubb P. J. (1977) The maintenance of species richness in plant communities: the importance of the regeneration niche. *Biological Reviews*, **52**, 107–45

Gulland J. A. (1962) The application of mathematical models to fish populations. In *The Exploitation of Natural Animal Populations* (Le Cren E. D. & Holdgate M. W. eds), Symposium of the British Ecological Society No. 2, pp. 204–17. Blackwell Scientific Publications, Oxford

Haines B. (1975) Impact of leaf cutting ants on vegetation development at Barro Colorado Island. In *Tropical Ecological Systems—Trends in Terrestrial and Aquatic Research* (Golley F. B. & Medina E. eds) pp. 99–111. Springer Verlag, New York

Haldane J. B.S. (1949) Disease and evolution. Symposium sui fattori ecologici e genetici della speciazone negli animali. *Ric. Sci.* **19** (suppl.), 3–11

Hall R. L. (1974) Analysis of the nature of interference between plants of different species. I. Concepts and extension of the de Wit analysis to examine effects. *Australian Journal of Agricultural Research*, **25**, 739–47

Hall R. L. (1978) The analysis and significance of competitive and non competitive interference between species. In *Plant Relations in Pastures* (Wilson J. R. ed.) pp. 163–74. CSIRO

Hancock D.A. (1979) *Population dynamics and management of shellfish stocks* I.C.E.S. Special Meeting on Population Assessment of Shellfish

Harberd D. J. (1961) Some observations on natural clones in *Festuca ovina L. New Phytol.* **60**, 184–206

Harcourt D. G. (1964) Population dynamics of *Leptinotarsa decemlineata* (Say) in eastern Ontario. II. Population and mortality estimation during six age intervals. *Canadian Entomologist*, **96**, 1190–8

Harcourt D. G. (1971) Population dynamics of *Leptinotarsa decemlineata* (Say) in eastern Ontario. III. Major population processes. *Canadian Entomologist*, **103**, 1049–61

Harper J. L. (1961) Approaches to the study of plant competition. In *Mechanisms in Biological Competition* (Milthorpe F. L. ed.). Symposium of the Society of Experimental Biology, No. 15, pp. 1–39

Harper J. L. (1977) *The Population Biology of Plants*. Academic Press, London and New York

Harper J. L. (1981) The population biology of modular organisms. In

Theoretical Ecology (May R. ed.). Blackwell Scientific Publications, Oxford
Harper J. L. (1984) Modules, branches and the capture of resources. In review. Yale University Press.
Harper J. L. & Bell A. D. (1979) The population dynamics of growth form in organisms with modular construction. In *Population Dynamics* (Anderson R. M., Turner B. D. & Taylor L. R. eds), pp. 29–52. Blackwell Scientific Publications, Oxford
Hart A. & Begon M. (1982) The status of general life-history strategy theories, illustrated in winkles. *Oecologia,* **52**, 37–42
Hassell M. P. (1971a) Mutual interference between searching insect parasites. *Journal of Animal Ecology,* **40**, 473–86
Hassell M. P. (1971b) Parasite behaviour as a factor contributing to the stability of insect host–parasite interactions. In *Dynamics of Populations*, (den Boer P. J. & Gradwell G. R. eds), pp. 366–79. Centre for Agricultural Publishing and Documentation, Wageningen
Hassell M. P. (1975) Density-dependence in single-species populations. *Journal of Animal Ecology,* **44**, 283–95
Hassell M. P. (1976) *The Dynamics of Competition and Predation.* Edward Arnold, London
Hassell M. P. (1978) *The Dynamics of Arthropod Predator–Prey Systems.* Princeton University Press, Princeton
Hassell, M. P. & Comins, H. N. (1976) Discrete time models for two-species competition. *Theoretical Population Biology,* **9**, 202–21
Hassell M. P. & Comins H. N. (1978) Sigmoid functional responses and population stability. *Theoretical Population Biology,* **14**, 62–7.
Hassell M. P., Lawton J. H. & Beddington J. R. (1977) Sigmoid functional responses by invertebrate predators and parasitoids. *Journal of Animal Ecology,* **46**, 249–62
Hassell M. P., Lawton J. H. & May R. M. (1976) Patterns of dynamical behaviour in single species populations. *Journal of Animal Ecology,* **45**, 471–86
Hassell M. P. & May R. M. (1973) Stability in insect host–parasite models. *Journal of Animal Ecology,* **42**, 693–736
Hassell M. P. & Rogers D. J. (1972) Insect parasite responses in the development of population models. *Journal of Animal Ecology,* **41**, 661–76
Hassell M. P. & Varley G. C. (1969) New inductive population model for insect parasites and its bearing on biological control. *Nature,* **223**, 1133–6
Hatto J. & Harper J. L. (1969) The control of slugs and snails in British cropping systems, specially grassland. *International Copper Research Association Project* 115(A), 1–25
Healey M. C. (1967) Aggression and self regulation of population size in deermice. *Ecology,* **48**, 377–92
Heed W. B., Starmer W. T., Miranda M., Miller M. W. & Phaff H. S. (1976) An analysis of the yeast flora associated with cactiphilic *Drosophila* and their host plants in the Sonoran Desert and its relation to temperate and tropical associations. *Ecology,* **57**, 151–60
Hodgson G. L. & Blackman G. E. (1956) An analysis of the influence of plant density on the growth of *Vicia faba* 1. The influence of density of the pattern of development. *Journal of experimental Botany,* **7**, 147–65
Holling C. S. (1959) Some characteristics of simple types of predation and parasitism. *Canadian Entomologist,* **91**, 385–98
Holling C. S. (1965) The functional response of predators to prey density and its role in mimicry and population regulation. *Memoirs of the Entomological Society of Canada,* **45**, 43–60
Holling C. S. (1966) The functional response of invertebrate predators to prey density. *Memoirs of the Entomological Society of Canada,* **48**, 1–86
Horn H. S. (1971) *The Adaptive Geometry of Trees.* Princeton University Press, Princeton
Horn H. S. (1978) Optimal tactics of reproduction and life-history. In *Behavioural Ecology: an evolutionary approach* (Krebs J. R. & Davies N. B. eds), pp. 411–29. Blackwell Scientific Publications, Oxford
Horton K. W. (1964) Deer prefer Jack Pine. *Journal of Forestry,* **62**, 497–9
Hubbard S. F. (1977) Studies on the natural control of *Pieris brassicae* with particular reference to parasitism by *Apanteles glomeratus.* Unpublished D.Phil. thesis, University of Oxford
Huffaker C. B. (1958) Experimental studies on predation: dispersion factors and predator–prey oscillations. *Hilgardia,* **27**, 343–83
Huffaker C. B. & Kennett C. E. (1959) A 10 year study of vegetational changes associated with biological control of Klamath weed species. *Journal of Range Management,* **12**, 69–82
Huffaker C. B., Shea K. P. & Herman S. G. (1963) Experimental studies on predation. *Hilgardia,* **34**, 305–30
Hughes T. P. (1984) Population dynamics based on individual size rather than age: a general model with a reef coral example. *American Naturalist,* **123**, 778–95
Hutchinson G. E. (1957) Concluding remarks. *Cold Spring Harbour Symposium on Quantitative Biology,* **22**, 415–27
Hutchings M. J. (1979) Weight–density relationships in ramet populations of clonal perennial herbs with special reference to the −3/2 power law. *Journal of Ecology,* **67**, 21–33

Iles T. D. (1973) Interaction of environment and parent stock size in determining recruitment in the Pacific sardine as revealed by analysis of density-dependent 0-group growth. *Rapports et Procès-verbaux, Conseil international pour l'Explortion de la Mer,* **164**, 228–40
Inglesfield C. (1979) Migration and microdistribution in *Drosophila subobscura*, Collin. Unpublished Ph.D. thesis, University of Liverpool
Inouye D. W. (1978) Resource partitioning in bumblebees: experimental studies of foraging behaviour. *Ecology,* **59**, 672–8

Janzen D. H. (1966) Coevolution of mutualism between ants and acacias in Central America. *Evolution,* **20**, 249–75
Janzen D. H. (1971) Escape of juvenile *Dioclea megacarpa* (Leguminosae) vines from predators in a deciduous tropical forest. *American Naturalist,* **105**, 97–112
Janzen D. H. (1972a) Association of a rainforest palm and seed-eating beetles in Puerto Rica. *Ecology,* **53,** 258–61
Janzen D. H. (1972b) Escape in space by *Sterculia apetala* seeds from the bug *Dysdercus fasciatus* in a Costa Rican deciduous forest. *Ecology,* **53**, 350–61
Janzen D. H. (1975) Interactions of seeds and their insect predators/parasitoids in a tropical deciduous forest. In *Evolutionary Strategies of Parasitic Insects and Mites*, (Price P. W. ed.), pp. 154–86. Plenum, New York
Johnson C. G. (1969) *Migration and Dispersal of Insects by Flight.* Methuen, London

Kays S. & Harper J. L. (1974) The regulation of plant and tiller density in a grass sward. *Journal of Ecology,* **62**, 97–105

Keith L. B. (1963) *Wildlife's Ten-year Cycle*. University of Wisconsin Press, Madison

Khan M. A. (1963) Physiological and genetic analysis of varietal differences within *Linum usitatissimum* (Flax and Linseed). Unpublished Ph. D. thesis, University of Wales

Khan M. A., Putwain P. D. & Bradshaw A. D. (1975) Population interrelationships. 2. Frequency dependent fitness in *Linum*. *Heredity*, **34**, 145–63

Kira T., Ogawa H. & Shinozaki K. (1953) Intraspecific competition among higher plants. I. Competition-density-yield interrelationships in regularly dispersed populations. *Journal of the Polytechnic Institute, Osaka City University*, **4**. (4), 1–16

Klomp H. (1964) Intraspecific competition and the regulation of insect numbers. *Annual Review of Entomology*, **9**, 17–40

Kluyver H. N. (1951) The population ecology of the Great Tit, *Parus m. major* L., *Ardea*, **38**, 1–135

Krebs J. R. (1971) Territory and breeding density in the great tit, *Parus major* L. *Ecology*, **52**, 2–22

Krebs J. R. (1978) Optimal foraging: decision rules for predators. In *Behavioural Ecology: an evolutionary approach* (Krebs J. R. & Davies N. B. eds), pp. 23–63. Blackwell Scientific Publications, Oxford

Krebs J. R., Kacelnik A. & Taylor P. J. (1978) Test of optimal sampling by foraging great tits. *Nature*, **275**, 27–31

Kuchlein J. H. (1966) Mutual interference among the predacious mite *Typhlodromus longipilus* Nesbitt (Acari, Phytoseiidae). I. Effects of predator density on oviposition rate and migration tendency. *Meded. Rijksfar. Landbwet. Gent.*, **31**, 740–6

Lack D. (1954) *The Natural Regulation of Animal Numbers*. Clarendon Press, Oxford

Lack D. (1971) *Ecological Isolation in Birds*. Blackwell Scientific Publications, Oxford

Lanciani C. A. (1975) Parasite-induced alterations in host reproduction and survival. *Ecology*, **56**, 689–95

Langer R. H. M. (1956) Growth and nutrition of timothy (*Phleum pratense*). The life history of individual tillers. *Annals of applied Biology*, **44**, 166–87

Law R. (1975) Colonisation and the evolution of life histories in *Poa annua*. Unpublished Ph.D. thesis, University of Liverpool

Law R. (1979a) Harvest optimisation in populations with age distributions. *American Naturalist*, **114**, 250–9

Law R. (1979b) The cost of reproduction in annual meadow grass. *American Naturalist*, **113**, 3–16

Law R., Bradshaw A.D. & Putwain P. D. (1977) Life history variation in *Poa annua*. *Evolution*, **31**, 233–46

Lawlor L. R. (1976) Molting, growth and reproductive strategies in the terrestrial isopod, *Armadillidium vulgare*. *Ecology*, **57**, 1179–94

Lawrence W. H. & Rediske J. H. (1962) Fate of sown douglas-fir seed. *Forest Science*, **8**, 211–18

Lawton J. H., Beddington J. R. & Bonser R. (1974) Switching in invertebrate predators. In *Ecological Stability* (Usher M. B. & Williamson M. H. eds), pp. 141–58. Chapman and Hall, London

Lawton J. H. & McNeill S. (1979) Between the devil and the deep blue sea: on the problem of being a herbivore. In *Population Dynamics* (Anderson R. M., Turner B. D. & Taylor L. R. eds), pp. 223–44. Blackwell Scientific Publications, Oxford

Lawton J. H. & Pimm S. L. (1978) Population dynamics and the length of food chains. *Nature*, **272**, 190

Leslie P. H. (1945) On the use of matrices in certain population mathematics. *Biometrika*, **33**, 183–212

Levins R. (1968) *Evolution in Changing Environments*. Princeton University Press, Princeton

Lotka A. J. (1925) *Elements of Physical Biology*. Williams and Wilkins, Baltimore

Lowe V. P. W. (1969) Population dynamics of the red deer (*Cervus elaphus* L.) on Rhum. *Journal of Animal Ecology*, **38**, 425–57

MacArthur R. H. (1965) Patterns of species diversity. *Biological Reviews*, **40**, 1510–33

MacArthur R. H. (1972) *Geographical Ecology*. Harper and Row, New York

MacArthur R. H. & Levins R. (1967) The limiting similarity, convergence and divergence of coexisting species. *American Naturalist*, **101**, 377–85

MacArthur R. H. & Pianka E. R. (1966) On optimal use of a patchy environment. *American Naturalist*, **100**, 603–9

MacArthur R. H. & Wilson E. O. (1967) *The Theory of Island Biogeography*. Princeton University Press, Princeton

MacLulick D. A. (1937) Fluctuations in the numbers of the varying hare (*Lepus americanus*). *University of Toronto Studies, Biology Series*, **43**, 1–136

McNaughton S. J. (1975) r- and k-selection in *Typha*. *American Naturalist*, **109**, 251–61

McNeill S. & Southwood T. R. E. (1978) The role of nitrogen in the development of insect/plant relationships. In *Biochemical Aspects of Plant and Animal Coevolution* (Harborne J. B. ed.), pp. 77–98. Academic Press, London and New York

Manlove, R. J. (1985) *On the population ecology of* Avena fatua *L.* Unpublished Ph. D. thesis, University of Liverpool

Marshall D. R. & Jain S. K. (1969) Interference in pure and mixed populations of *Avena fatua* and *A. barbata*. *Journal of Ecology*, **57**, 251–270

May R. M. (1972) Will a large complex system be stable? *Nature*, **238**, 13–14

May R. M. (1973) *Stability and Complexity in Model Ecosystems*. Princeton University Press, Princeton

May R. M. (1975) Biological populations obeying difference equations: Stable points, stable cycles and chaos. *Journal of Theoretical Biology*, **49**, 511–24

May R. M. (1977) Thresholds and breakpoints in ecosystems with a multiplicity of stable states. *Nature*, **269**, 471–7

May R. M. (1978a) Host–parasitoid systems in patchy environments: a phenomenological model. *Journal of Animal Ecology*, **47**, 833–43

May R. M. (1978b) The dynamics and diversity of insect faunas. In *Diversity of Insect Faunas*, (Mound L. A. & Waloff N. eds), pp. 188–204. Blackwell Scientific Publications, Oxford

May R. M. (1979) The structure and dynamics of ecological communities. In *Population Dynamics*, (Anderson R. M., Turner B. D. & Taylor L. R. eds), pp. 385–407. Blackwell Scientific Publications, Oxford

May R. M. & Anderson R. M. (1978) Regulation and stability of host–parasite population interactions. II. Destabilizing processes. *Journal of Animal Ecology*, **47**, 249–68

Maynard Smith J. (1976) Group selection. *Quarterly Review of Biology*, **51**, 277–83

Mead-Briggs A. R. & Rudge A. J. B. (1960) Breeding of the rabbit flea, *Spilopsyllus cuniculi* (Dale): requirement of a 'factor' from a pregnant rabbit for ovarian maturation. *Nature*, **187**, 1136–7

Mertz R. W. & Boyce S. G. (1956) Age of oak 'seedlings'. *Journal of Forestry*, **54**, 774–75

Michelakis S. (1973) A study of the laboratory interaction between *Coccinella septempunctata* larvae and its prey *Myzus persicae*. Unpublished M.Sc. thesis, University of London

Monro J. (1967) The exploitation and conservation of resources by populations of insects. *Journal of Animal Ecology*, **36**, 531–47

Moore N. W. (1964) Intra- and interspecific competition among dragonflies. *Journal of Animal Ecology*, **33**, 49–71

Morris R. F. (1959) Single-factor analysis in population dynamics. *Ecology*, **40**, 580–8

Mortimer A. M. (1974) Studies of germination and establishment of selected species with special reference to the fates of seeds. Unpublished Ph.D. thesis, University of Wales

Müller H. J. (1970) Formen der Dormanz bei Insekten. *Nova Acta Leopoldina*, **191**, 1–27

Murdoch W. W. (1969) Switching in general predators: experiments on predator specificity and stability of prey populations. *Ecological Monographs*, **39**, 335–54

Murdoch W. W. & Oaten A. (1975) Predation and population stability. *Advances in Ecological Research*, **9**, 2–131

Murphy G. I. (1967) Vital statistics of the Pacific sardine (*Sardinops caerulea*) and the population consequences. *Ecology*, **48**, 731–6

Murton R. K. (1971) The significance of a specific search image in the feeding behaviour of the Wood Pigeon. *Behaviour*, **40**, 10–42

Murton R. K., Westwood N. J. & Isaacson A. J. (1964) The feeding habits of the Wood Pigeon *Columba palumbus*, stock dove, *C. oenas* and the turtle dove, *Streptopelia turtur*. *Ibis*, **106**, 174–88

Naylor R. & Begon M. (1982) Variation within and between populations of *Littorina nigrolineata* Gray on Holy Island, Anglesey. *Journal of Conchology*, **31**, 17–30

Newsome A. E. (1969a) A population study of house-mice permanently inhabiting a reed-bed in South Australia. *Journal of Animal Ecology*, **38**, 361–77

Newsome A. E. (1969b) A population study of house-mice temporarily inhabiting a South Australian wheatfield. *Journal of Animal Ecology*, **38**, 341–60

Nicholson A. J. (1933) The balance of animal populations. *Journal of Animal Ecology*, **2**, 131–78

Nicholson A. J. (1954a) Compensatory reactions of populations to stress, and their evolutionary significance. *Australian Journal of Zoology*, **2**, 1–8

Nicholson A. J. (1954b) An outline of the dynamics of animal populations. *Australian Journal of Zoology*, **2**, 9–65

Nicholson A. J. (1957) *The self-adjustment of populations to change.* Cold Spring Harbour Symposium Quantitative Biology, No. 22, pp. 153–72.

Nicholson A. J. (1958) Dynamics of insect populations. *Annual Review of Entomology*, **3**, 107–36

Nicholson A. J. & Bailey V. A. (1935) The balance of animal populations. *Proceedings of the Zoological Society of London*, **3**, 551–98

Noyes J. S. (1974) The biology of the leek moth, *Acrolepia assectella* (Zeller). Unpublished Ph.D. thesis, University of London

Noy-Meir I. (1975) Stability of grazing systems: an application of predator–prey graphs. *Journal of Ecology*, **63**, 459–83

Obeid M., Machin D. & Harper J. L. (1967) Influence of density on plant to plant variations in Fiber Flax, *Linum usitatissimum*. *Crop Science*, **7**, 471–3

Oinonen E. (1967) The correlation between the size of Finnish bracken (*Pteridium aquilinum* (L.) Kuhn) clones and certain periods of site history. *Acta For. Fenn.*, **83**, 1–51

Paine R. T. (1966) Food web complexity and species diversity. *American Naturalist*, **100**, 65–75

Palmblad I. G. (1968) Competition studies on experimental populations of weeds with emphasis on the regulation of population size. *Ecology*, **49**, 26–34

Paris O. H. & Pitelka F. A. (1962) Population characteristics of the terrestrial isopod *Armadillidium vulgare* in California grassland. *Ecology*, **43**, 229–48

Park T. (1954) Experimental studies of interspecific competition. II. Temperature, humidity and competition in two species of *Tribolium*. *Physiological Zoology*, **27**, 177–238

Pearl R. (1925) *The Biology of Population Growth.* Knopf, New York

Pearl R. (1927) The growth of populations. *Quarterly Review of Biology*, **2**, 532–48

Pearl R. (1928) *The Rate of Living.* Knopf, New York

Perrins C. M. (1965) Population fluctuation and clutch-size in the great tit (*Parus major* L.). *Journal of Animal Ecology*, **34**, 601–47

Pianka E. R. (1970) On *r*- and *k*-selection. *American Naturalist*, **104**, 592–7

Pimentel D. (1961) On a genetic feed-back mechanism regulating populations of herbivores, parasites and predators. *American Naturalist*, **95**, 65–79

Pimm S. L. & Lawton J. H. (1977) Number of trophic levels in ecological communities. *Nature*, **268**, 329–31

Pimm S. L. & Lawton J. H. (1978) On feeding on more than one trophic level. *Nature*, **275**, 542–4

Pimm S. L. & Lawton J. H. (1980) Are foodwebs divided into compartments? *Journal of Animal Ecology*, **49**, 879–98

Podoler H. & Rogers D. (1975) A new method for the identification of key factors from life-table data. *Journal of Animal Ecology*, **44**, 85–114

Poole R. W. (1978) *An Introduction to Quantitative Ecology.* McGraw-Hill, New York

Porter J. R. (1983a) A modular approach to analysis of plant growth. I. Theory and principles. *New Phytologist*, **94**, 183–90

Porter J. R. (1983b) A modular approach to analysis of plant growth. II. Methods and results. *New Phytologist*, **94**, 191–200

Pratt D. M. (1943) Analysis of population development in *Daphnia* at different temperatures. *Biological Bulletin*, **85**, 116–40

Precht H., Cristophersen J., Hensel H. & Larcher W. (1973) *Temperature and Life.* Springer-Verlag, Berlin

Price P. W. (1980) *Evolutionary Biology of Parasites.* Princeton University Press, Princeton

Puckridge D. W. & Donald C. M. (1967) Competition among wheat plants sown at a wide range of densities. *Australian Journal of Agricultural Research*, **17**, 193–211

Putwain P. D. & Harper J. L. (1970) Studies on the dynamics of plant populations. III. The influence of associated species on populations of *Rumex acetosa* L. and *R. acetosella* L. in grassland. *Journal of Ecology*, **58**, 251–64

Radovich J. (1962) Effects of sardine spawning stock size and environment on year-class production. *California Department Fish and Game Bulletin*, **48**, 123–40

Raffaelli D. G. & Hughes R. N. (1978) The effects of crevice size and availability on populations of *Littorina rudis* and *Littorina neritoides*. *Journal of Animal Ecology*, **47**, 71–83

Randolph S. E. (1975) Patterns of distributions of the tick *Ixodes trianguliceps* Birula, on its host. *Journal of Animal Ecology*, **44**, 451–74

Richards O. W. & Waloff N. (1954) Studies on the biology and population dynamics of British grasshoppers. *Anti-Locust Bulletin*, **17**, 1–182

Richman S. (1958) The transformation of energy by *Daphnia pulex*. *Ecological Monographs*, **28**, 273–91

Rigler F. H. (1961) The relation between concentration of food and feeding rate of *Daphnia magna* Straus. *Canadian Journal of Zoology*, **39**, 857–68

Root R. B. (1967) The niche exploitation pattern of the blue-gray gnatcatcher. *Ecological Monographs*, **37**, 317–50

Rosen B. R. (1979) Modules, members and communes: a postcript introduction to social organisms. In *Biology and Systematics of Colonial Organisms* (Larwood G. & Rosen B. R. eds). Academic Press, London

Rosenzweig M. L. & MacArthur R. H. (1963) Graphical representation and stability conditions of predator-prey interactions. *American Naturalist*, **97**, 209–23

Ross M. A. & Harper J. L. (1972) Occupation of biological space during seedling establishment. *Journal of Ecology*, **60**, 77–88

Rotheray G. E. (1979) The biology and host searching behaviour of a Cynipoid parasite of aphidophagous syrphid larvae. *Ecological Entomology*, **4**, 175–82

Sagar G. R. & Mortimer A. M. (1976) An approach to the study of the population dynamics of plants with special reference to weeds. *Applied Biology*, **1**, 1–43

Salisbury E. J. (1942) *The Reproductive Capacity of Plants*. Bell, London

Salisbury E. J. (1961) *Weeds and Aliens*. Collins, London

Sarukhan J. & Gadgil M. (1974) Studies on plant demography: *Ranunculus repens* L., *R. bulbosus* L., and *R. acris* L. III. A mathematical model incorporating multiple modes or reproduction. *Journal of Ecology*, **62**, 921–36

Sarukhan J. & Harper J. L. (1973) Studies on plant demography: *Ranunculus repens* L., *R. bulbosus* L., and *R. acris* L. I. Population flux and survivorship. *Journal of Ecology*, **61**, 675–716

Schaefer M. B. (1957) *A study of the dynamics of the fishery for yellowfin tuna in the eastern tropical Pacific ocean*. Inter-American Tropical Tuna Commission Bulletin 2, No. 6 1957

Schaefer M. B. (1968) Methods of estimating effects of fishing on fish populations. *American Fish Society Transactions*, **97**, 231–41

Schaffer W. M. (1974) Optimal reproductive effort in fluctuating environments. *American Naturalist*, **108**, 783–90

Schoener T. W. (1974) Resource partitioning in ecological communities. *Science*, **185**, 27–39

Searle S. R. (1966) *Matrix Algebra for the Biological Sciences*. John Wiley & Sons, New York

Sharitz R. R. & McCormick J. F. (1973) Population dynamics of two competing annual plant species. *Ecology*, **54**, 723–40

Shinozaki K. & Kira T. (1956) Intraspecific competition among higher plants. VII. Logistic theory of the C–D effect. *Journal of Institute Polytechnic of Osaka City University*, **7**, 35–72

Shorrocks B. (1970) Population fluctuations in the fruit fly (*Drosophila melanogaster*) maintained in the laboratory. *Journal of Animal Ecology*, **39**, 229–53

Silliman R. P. & Gutsell J. S. (1958) Experimental exploitation of fish populations. *Fishery Bulletin, Fish Wildlife Service. U.S.*, **58**, 215–41

Silvertown J. W. (1982) *Introduction to Plant Population Ecology*. Longmans, London

Sinclair A. R. E. (1975) The resource limitation of trophic levels in tropical grassland ecosystems. *Journal of Animal Ecology*, **44**, 497–520

Skellam J. G. (1972) Some philosophical aspects of mathematical modelling in empirical science with special reference to ecology. In *Mathematical Models in Ecology*. Symposium of the British Ecological Society, No. 12, pp. 13–29. (Jeffers J. N. R. ed.) Blackwell Scientific Publications, Oxford

Slobodkin L. B. & Richman S. (1956) The effect of removal of fixed percentages of the newborn on size and variability in populations of *Daphnia pulicaria* (Forbes). *Limnology and Oceanography*, **1**, 209–37

Smith F. E. (1961) Density dependence in the Australian thrips. *Ecology*, **42**, 403–7

Snell T. W. & King C. E. (1977) Lifespan and fecundity patterns in rotifers: the cost of reproduction. *Evolution*, **31**, 882–90

Snyman A. (1949) The influence of population densities on the development and oviposition of *Plodia interpunctella* Hubn. (Lepidoptera). *Journal of the Entomological Society of South Africa*, **12**, 137–71

Solomon M. E. (1949) The natural control of animal populations. *Journal of Animal Ecology*, **18**, 1–35

Solomon M. E. (1964) Analysis of processes involved in the natural control of insects. *Advances in Ecological Research*, **2**, 1–58

Solomon M. E. (1969) *Population Dynamics*. Edward Arnold, London

Southern H. N. (1970) The natural control of a population of Tawny Owls (*Strix aluco*). *Journal of Zoology*, **162**, 197–285

Southwood T. R. E. (1976) Bionomic strategies and population parameters. In *Theoretical Ecology: principles and applications* (May R. M. ed.). pp. 26–48. Blackwell Scientific Publications, Oxford

Southwood T. R. E. (1977) Habitat, the templet for ecological strategies? *Journal of Animal Ecology*, **46**, 337–65

Spradbery J. P. (1970) Host finding by *Rhyssa persuasoria* (L.), an ichneumonid parasite of siricid woodwasps. *Animal Behaviour*, **18**, 103–14

Stearns S. C. (1977) The evolution of life history traits. *Annual Review of Ecology and Systematics*, **8**, 145–71

Stimson J. S. (1973) The role of territory in the ecology of the intertidal limpet *Lottia gigantea* (Gray). *Ecology*, **54**, 1020–30

Stubbs M. (1977) Density dependence in the life-cycles of animals and its importance in *K*- and *r*-strategies. *Journal of Animal Ecology*, **46**, 677–88

Symonides, E. (1979) The structure and population dynamics of psammophytes on inland dunes. III. Populations of compact psammophyte communities. *Ekologia Polska*, **27**, 235–57

Takahashi F. (1968) Functional response to host density in a parasitic wasp, with reference to population regulation. *Researches in Population Ecology*, **10**, 54–68

Tamm C. O. (1956) Further observations on the survival and flowering of some perennial herbs. *Oikos*, **7**, 274–92

Thompson D. J. (1975) Towards a predator-prey model incorporating age-structure: the effects of predator and prey size on the predation of *Daphnia magna* by *Ishnura elegans*. *Journal of Animal Ecology*, **44**, 907–16

Tilman D. (1976) Ecological competition between algae. Experimental confirmation of resource-based competition theory. *Science*, **192**, 463–5

Tilman D., Kilham S. S. & Kilham P. (1982) Phytoplankton community ecology: the role of limiting nutrients. *Annual Reviews of Ecology and Systematics*, **13**, 349–72

Tinbergen L. (1960) The natural control of insects in pinewoods. 1: Factors influencing the intensity of predation by songbirds. *Archives Neerlandaises de Zoologie*, **13**, 266–336

Trenbath B. R. (1978) Models and the interpretation of mixture experiments. In *Plant Relations in Pastures*, (J. R. Wilson ed.) pp. 145–63, CSIRO

Trenbath B. R. & Harper J. L. (1973) Neighbour effects in the genus *Avena*. I. Comparison of crop species. *Journal of Applied Ecology*, **10**, 379–400

Turnbull A. L. (1962) Quantitative studies of the food of *Linyphia triangularis* Clerck (Aranaea: Linyphiidae). *Canadian Entomologist*, **94**, 1233–49

Turnbull A. L. (1964) The searching for prey by a web-building spider *Archaeranea tepidariorum* (C. L. Kock). *Canadian Entomologist*, **96**, 568–79

Ullyett G. C. (1949a) Distribution of progeny by *Chelonus texanus* Cress. (Hymenoptera: Braconidae). *Canadian Entomologist*, **81**, 25–44

Ullyett G. C. (1949b) Distribution of progeny by *Cryptus inornatus* Pratt (Hymenoptera: Ichneumonidae). *Canadian Entomologist*, **81**, 285–99

Uranov A. A. (1975) Age spectrum of the phytocoenopopulation as a function of time and energetic wave processes. *Biologicheskie Nauki*, **2**, 7–34

Usher M. B. (1972) Developments in the Leslie matrix model. In: *Mathematical Models in Ecology* (Jeffers J. N. R. ed.). Symposium of the British Ecological Society, No. 12, pp. 29–60. Blackwell Scientific Publications, Oxford

Usher M. B. (1974) *Biological Management and Conservation*. Chapman and Hall, London

Usher M. B., Longstaff B. C. & Southall D. R. (1971) Studies on populations of *Folsomia candida* (Insecta: Collembola): the productivity of population in relation to food and exploitation. *Oecologia*, **7**, 68–79

Utida S. (1957) Cyclic fluctuations of population density intrinsic to the host parasite system. *Ecology*, **38**, 442–9

Utida S. (1967) Damped oscillation of population density at equilibrium. *Researches in Population Ecology*, **9**, 1–9

Vallis I., Haydock K. P., Ross P. J. & Henzel E. F. (1967) Isotopic studies on the uptake of nitrogen by pasture plants. III. The uptake of small additions of ^{15}N-labelled fertilizer by Rhodes grass and Townsville Lucerne. *Australian Journal of Agricultural Research*, **18**, 865–77

Vance R. R. (1972a) Competition and mechanism of coexistence in three sympatric species of intertidal hermit crabs. *Ecology*, **53**, 1062–74

Vance R. R. (1972b) The role of shell adequacy in behavioural interactions involving hermit crabs. *Ecology*, **53**, 1075–83

Vandermeer J. H. (1972) Niche theory. *Annual Review of Ecology and Systematics*, **3**, 107–32

van der Meijden E. (1971) *Senecio* and *Tyria* (*Callimorpha*) in a Dutch dune area. A study on an interaction between a monophagous consumer and its host plant. *Proceedings of the Advanced Study Institute on 'Dynamics of Numbers in Populations'* (den Boer P. J. & Gradwell G. R. eds). Pudoc, Wageningen

Varley G. C. (1947) The natural control of population balance in the knapweed gall-fly (*Urophora jaceana*). *Journal of Animal Ecology*, **16**, 139–87

Varley G. C. & Gradwell G. R. (1960) Key factors in population studies. *Journal of Animal Ecology*, **29**, 399–401

Varley G. C. & Gradwell G. R. (1963) The interpretation of insect population changes. *Proceedings of the Ceylon Association for the Advancement of Science*, **18**, 142–56

Varley G. C. & Gradwell G. R. (1968) *Population models for the winter moth*. Symposium of the Royal Entomological Society of London, No. 4, pp. 132–42

Varley G. C. & Gradwell G. R. (1970) Recent advances in insect population dynamics. *Annual Review of Entomology*, **15**, 1–24

Varley G. C., Gradwell G. R. & Hassell M. P. (1975) *Insect Population Ecology*. Blackwell Scientific Publications, Oxford

Volterra V. (1926) Variations and fluctuations of the numbers of individuals in animal species living together. (Reprinted, 1931, in R. N. Chapman, *Animal Ecology*, McGraw-Hill, New York.)

Watkinson A. R. (1980) Density-dependence in single species populations of plants. *Journal of Theoretical Biology*, **83**, 345–57

Watkinson A. R. (1981) Interference in pure and mixed populations of *Agrostemma githago*. *Journal of Applied Ecology*, **18**, 967–76

Watkinson A. R. (1983) Factors affecting the density response of *Vulpia fasiculata*. *Journal of Ecology*, **70**, 149–61

Watkinson A. R. & Harper J. L. (1978) The demography of a sand dune annual *Vulpia fasiculata*; I. The natural regulation of populations. *Journal of Ecology*, **66**, 15–33

Watson A. (1967) Territory and population regulation in the red grouse. *Nature*, **215**, 1274–5

Watt K. E. F. (1968) *Ecology and Resource Management*. McGraw-Hill, New York

Way M. J. & Cammell M. (1970) Aggregation behaviour in relation to food utilization by aphids. In *Animal Populations in Relation to their Food Resources* (Watson A. ed.), pp. 229–47. Blackwell Scientific Publications, Oxford

Weir D. A. (1985) *The population ecology and clonal structure of two grasses*. Unpublished Ph.D. thesis, University of Liverpool.

Werner P. A. (1975) A seed trap for determining patterns of seed deposition in terrestrial plants. *Canadian Journal of Botany*, **53**, 810–13

Werner P. A. (1979) Competition and coexistence of similar species. In *Topics in Plant Population Biology*, (Solbrig O. T., Jain S., Johnson G. B. & Raven P. H. eds), pp. 287–310. MacMillan, London

Werner P. A. & Platt W. W. (1976) Ecological relationships of co-occurring golden rods (*Solidago*: Compositae). *American Naturalist*, **110**, 959–71

Wesson G. & Wareing P. F. (1969) The induction of light sensitivity in weed seeds by burial. *Journal of Experimental Botany*, **20**, 413–25

White J. (1980) Demographic factors in populations of plants. In *Demography and Evolution in Plant Populations*, (Solbrig O. T. ed.), Blackwell Scientific Publications, Oxford

White J. (1981) The allometric interpretation of the self-thinning rule. *Journal of theoretical Biology*, **89**, 475–500

White T. C. R. (1978) The importance of a relative shortage of food in animal ecology. *Oecologia*, **33**, 71–86

Whittaker J. B. (1979) Invertebrate grazing, competition and plant dynamics. In *Population Dynamics*. (Anderson R. M., Turner B. D. & Taylor L. R. eds), pp. 207–22. Blackwell Scientific Publications, Oxford

Whittington R. (1984) Laying down the −3/2 power law. *Nature*, **311**, 217

Willey R. W. & Heath S. B. (1969) The quantitative relationships between plant population and crop yield. *Advances in Agronomy*, **21**, 281–321

Williams C. B. (1964) *Patterns in the Balance of Nature and Related Problems in Quantitative Ecology*. Academic Press, New York

Wilson D. E. & Janzen D. H. (1972) Predation on *Scheelea* palm seeds by bruchid beetles: seed density and distance from the parent palm. *Ecology*, **53**, 954–9

deWit C. T. (1960) On competition. *Verslagen van landbouwkundige onderzoekingen*, **66**, 1–82

deWit C. T. & van den Bergh J. P. (1965) Competition between herbage plants. *Netherlands Journal of Agricultural Science*, **13**, 212–21

deWit C. T., Tow P. G. & Ennik G. C. (1966) Competition between legumes and grasses. *Verslagen van landbouwkundige onderzoekingen*, **687**, 3–30

Wolf L. L. (1969) Female territoriality in a tropical hummingbird. *Auk*, **86**, 490–504

Wynne-Edwards V. C. (1962) *Animal Dispersion in Relation to Social Behaviour*. Oliver and Boyd, Edinburgh

Wynne-Edwards V. C. (1977) Intrinsic population control and introduction. In *Population Control by Social Behaviour*, (Ebling F. J. & Stoddart D. M. eds), pp. 1–22. Institute of Biology, London

Yoda K., Kira T., Ogawa H. & Hozumi K. (1963) Self thinning in overcrowded pure stands under cultivated and natural conditions. *Journal of Biology, Osaka City University*, **14**, 107–29

Authors Index

Organism Index

Subject Index